# Sustainable Development Goals Series

The **Sustainable Development Goals Series** is Springer Nature's inaugural cross-imprint book series that addresses and supports the United Nations' seventeen Sustainable Development Goals. The series fosters comprehensive research focused on these global targets and endeavours to address some of society's greatest grand challenges. The SDGs are inherently multidisciplinary, and they bring people working across different fields together and working towards a common goal. In this spirit, the Sustainable Development Goals series is the first at Springer Nature to publish books under both the Springer and Palgrave Macmillan imprints, bringing the strengths of our imprints together.

The Sustainable Development Goals Series is organized into eighteen subseries: one subseries based around each of the seventeen respective Sustainable Development Goals, and an eighteenth subseries, "Connecting the Goals", which serves as a home for volumes addressing multiple goals or studying the SDGs as a whole. Each subseries is guided by an expert Subseries Advisor with years or decades of experience studying and addressing core components of their respective Goal.

The SDG Series has a remit as broad as the SDGs themselves, and contributions are welcome from scientists, academics, policymakers, and researchers working in fields related to any of the seventeen goals. If you are interested in contributing a monograph or curated volume to the series, please contact the Publishers: Zachary Romano [Springer; zachary.romano@springer.com] and Rachael Ballard [Palgrave Macmillan; rachael.ballard@palgrave.com].

Simon Stewart

# Heart Disease
# and Climate Change

 Springer

Simon Stewart
University of Notre Dame Australia
Fremantle, WA, Australia

ISSN 2523-3084          ISSN 2523-3092   (electronic)
Sustainable Development Goals Series
ISBN 978-3-031-73105-1        ISBN 978-3-031-73106-8   (eBook)
https://doi.org/10.1007/978-3-031-73106-8

*I couldn't have completed this book without the vision, forbearance, encouragement and wonderful support of people like Grant Weston (Springer Press), Professors Jim Codde, Max Bulsara and Aron Murphy (University of Notre Dame Australia), my colleagues (with special mention to a Ph.D. candidate who helped me in my hours of need—the soon to be Dr Alex Chen) and, of course my beautiful family, to whom I very much dedicate this book.*

# Foreword

My professional connection with Professor Simon Stewart comes from our common interest in linking diseases to its social determinants. Our research focus has been driven by the desire to understand the epidemiology of neglected cardiopulmonary diseases in low- and middle-income countries, and the factors that define the health systems' response to them in the most underserved areas of the globe. We both see climate change as a potentially major driver to exacerbate the current disparities in access to health care between high- and low-income countries. I am therefore thrilled to have been invited to write this foreword.

I am a cardiologist born in Mozambique, working both as a clinician and a researcher. This allows me to possess a perspective of the individual but also the societal aspects of climate change that affect the world today. I have worked with Professor Stewart in local and international research projects to describe and understand the role of climate changes in perpetuating the gaps in health care and research in the less developed regions of the globe. Together, we believe that it is crucial to unveil the effects of climate change in the profile and trends of heart and lung disease in communities suffering concomitantly from poverty, rapid unplanned urbanisation, and forced displacements due to economic reasons or conflicts. By doing this we hope to be prepared to prevent multimorbidity, disability and premature deaths worldwide.

Professor Stewart's long career studying issues related to determinants of health and outcomes of heart disease gives him credibility to write this book—Heart Disease & Climate Change. It reveals his personal views on how climate change might (and does) influence the evolving burden of heart disease worldwide and across the lifespan. The reading of its ten chapters will surely foster rethinking of the current models of cardiovascular risk assessment and care delivery. In particular, the concept of *climatic vulnerability* versus *resilience* is described from an individual perspective to a broader societal picture, aiming at challenging the risks of keeping the *status quo* regarding the priorities to achieve the *United Nation's Sustainable Development Goals*. The book clearly shows why climate change needs to be part of our common effort to reduce the burden of cardiovascular disease globally.

The risks of no action are elegantly discussed throughout Chaps. 3 to 5, which deal with the biological imperative to preserve the function of cardiovascular system, due to its critical role in maintaining homeostasis. The need to survive and adapt the human body to the rapid climate change and higher exposure to weather extremes, while keeping the ability to maintain homeostasis in cold and hot extremes is highlighted. Concomitantly, the need for the health systems to improve their resilience and innovate surveillance to capture key information regarding these extreme events is clearly stated. Finally, the book discusses health services adaptations and research innovations needed to embrace the new paradigm of emergency and cardiovascular care determined by the predictable climatic transitions (seasons) and unpredictable weather conditions (heatwaves and cold spells).

Chapters 6 and 7 make the intersection between climate change and pollution. In doing so, it argues for the need to create a deeper understanding of the human interactions with weather as well as the synergistic threat of pollution as both a consequence and cause of climate change given that air pollution, usually seen as consequence of human activity- and technology-driven climate change, is by itself a cardio-toxic. Indeed, both outdoor and indoor air pollution adversely impact the cardiovascular system, which are modulated by climatic conditions, topography and human structures/activity. Accordingly, pollution has been elevated to the status of 'non-traditional major risk factor' for cardiovascular disease. Interestingly, there may be some unexpected benefits from climate change, particularly concerning infectious disease reduction, but also spread of some infectious agents to new habitats.

Finally, the last three chapters of the book summarise the pillars of our response to climate changes, assessed from a perspective of promoting change to prevent heart disease and effectively manage existing heart disease. While science can inform some policies, major gaps in knowledge persist and multidisciplinary dialogues are needed to identify the right research questions and the best buys to protect humankind from climate change. Lessons learned from the COVID-19 pandemic are analysed and possible solutions to protect us from catastrophic events due to climate change are proposed.

In this context of uncertainty regarding the full picture of climate-provoked cardiac events, this book provides structured information that should be known by the community, patients, caregivers, health professionals, civil society and policymakers from all sectors of the society. By reading the book you learn and reflect on your role in addressing this major threat, at an individual and societal level. The book is an important tool to increase contemporary knowledge and equip society as a whole to act urgently to stop and react to climate change.

The challenge of preventing heart disease related to climate change cannot be the responsibility of health professionals and/or researchers such as myself and Professor Stewart. I strongly believe that after reading this excellent piece of science, written in captivating and entertaining style, you will use it as a tool for action and advocacy. I expect many readers to join us as partners in disseminating not only the current knowledge and concerns, but also the good practices that reduce climate change and promote well-being. Considering that heart disease is the number one killer worldwide,

our individual action is important to avoid overwhelming of the already constrained health systems and to address the threat that climate change represents to both developed and low-income settings.

June 2024                                              Ana Olga Mocumbi MD, PhD, FESC
                                                       Head of the Division of Non-Communicable
                                                       Diseases at the National Public Health Institute
                                                       Professor of Cardiology
                                                       Eduardo Mondlane University
                                                       Maputo, Mozambique
                                                       ana.mocumbi@uem.ac.mz
                                                       ana.mocumbi@ins.gov.mz

# Preface

It is becoming increasingly obvious that within the ageing populations of high-income countries that cardiac events do not occur randomly. This is also the likely case among younger, more vulnerable populations living in low-to-middle-income countries. Critically, such 'random' events are not benign given they include hospital admissions and deaths that could have been prevented. Moreover, they represent a complex set of bio-behavioural interactions between an individual at risk of experiencing an acute cardiac event and the environment in which they live and have the (potential) capacity to control. Beyond households with high levels of indoor air pollution (e.g. due to the need to use poorly ventilated fossil fuels for heating and/or cooking) or exposed to high levels of external pollution due to adjacent industries, the significant and indeed ubiquitous component of that environment is the local climate and the weather conditions that typically occur in that location.

As reflected in my specific notes on the relevance of this book to the **United Nation's Sustainable Development Goals** (UNSDGs—see Chap. 1) this fundamental relationship between us and the environment/local climate in which we live and breathe, gains increasing importance when seeking to improve heart health outcomes. At the heart of this relationship is our willingness and capacity to modulate the nature of those interactions to avoid our cardiovascular system being overwhelmed.

Unfortunately, most expert guidelines and epidemiological reports ignore this fundamental fact—that cardiac events in nearly every country rise and fall with predictable climatic transitions (seasons). They also rise and fall in response to more unpredictable, acute climatic provocations to our health such as unseasonal cold spells, heatwaves and even 'asthma storms'. Complicating these challenges, is the fact we are living in an increasingly polluted world filled with the poor air quality, more noise, heavy metals and microplastics that will steadily degrade our 'resilience' to climatic provocations to our health.

As we deal with this degradation in environmental conditions, a much large issue looms—that of *climate change*. In very simple terms, pollution is merely a symptom of our excessive (and often misguided) attempts to achieve mastery of our physical environment. This mastery is driven by a biological imperative (whether it be actions to heat or cool the immediate environment) to experience a high level of comfort and, more importantly, maintain physiological homeostasis. Ironically, in achieving local control of the mini climates encapsulated by millions of habitats in every part of the world, we've lost control of the global environment. This has largely resulted from our burning of otherwise geologically trapped, carbon-rich fossil fuels that release *greenhouse gases* that end up trapping excess heat within our atmosphere and oceans.

While the narrative of climate change is around 'heat', it's important to circle back to two important fundamentals—1) As humans, we've taken thousands of years to adjust to the local climates we live in and, indeed the seasonal changes that typically occur on a regular basis and, 2) Despite those adaptations, a good majority of the population (particularly people with pre-existing cardio-metabolic disease) are already vulnerable to pro-vocative climatic conditions (i.e. even without the emergence of climate change), with cold/wintry conditions a major 'killer' in this regard.

As will be described in this book, climate change is far too complex a phenomenon to simply describe it as *'Global Warming'*—hence the aban-donment of that term! Instead, climate change will not only produce a warming planet but provoke more unpredictable weather that will exacer-bate a pre-existing phenomenon that we are yet to properly recognise and address—*'climatic vulnerability'*. Typically, and unjustly, it affects those whose past generations have contributed least to climate change—the poor-est people in the world in whom many UNSDGs are becoming increasingly unachievable.

In this context, this book, for the first time, will not only articulate why climate change will harm our collective heart health but how it has height-ened a perennial threat to the most vulnerable people in society—*climatic provocations to the heart* and by extension the whole cardiovascular system. There is still so much we don't understand about our interactions with the climate. This is especially true in an era of rapid climatic changes. New dire modelling predictions are being made almost every week. However, as will be further articulated in this book, all health professionals, researchers and policymakers with an interest in reducing the global burden of disease asso-ciated with the world's biggest killers—acute and chronic forms of heart and cerebrovascular disease—need to consider the need for a new paradigm in how we detect, characterise and then manage people who are more likely or not suffer an event or die prematurely, when challenged by the weather and other related environmental extremes.

Having identified the problem and the need for a new perspective, the book identifies the strategies that show promise in establishing *'climatic resilience'* and therefore better health outcomes. It also critically highlights what is missing from a clinical to research perspective.

I hope you (the reader) enjoy the challenges I pose in this book. If rel-evant, I also hope you will consider adopting a broader and more holistic perspective on health and cardiac conditions typically managed within the four walls of a clinic or a hospital. For me, this represents an extension of the foundational research that demonstrated the value of comprehensive ger-iatric assessments, home visits and multidisciplinary care of heart failure—all of which challenged a purely medical response to complex health issues, while leading to a more person-centred approach that improved health outcomes in a cost-effective manner.

I don't claim to know all the answers, but hopefully I'm asking the right questions and prompting you to do the same as we deal with one of humanity's greatest threats—anthropometric climate change.

Fremantle, Australia                                                                               Simon Stewart
June 2024

# Contents

**Abstract**

Before introducing the specific topic of climate change and how it might (and does) influence heart disease-related events on a global scale, this brief chapter frames this book within the bigger picture of the *United Nation's Sustainable Development Goals* (United Nations Department of Economic and Social Affairs in Sustainable development (the 17 goals); 2024. https://sdgs.un.org/goals. Accessed June 2024). It should become clear to the reader that the intersection between heart health and climatic conditions spans from the individual to the population level. Throughout, the concept of *climatic vulnerability* versus *resilience* will be mainly discussed from an individual perspective. However, the broader picture (in the face of a global threat that extends to every horizon) demands we first consider vulnerability and priorities within the global population. In specific terms, who stands to lose most from climate change and, more pertinently, who is able to respond to the health issues that arise?

**Keywords**

Climate change · United Nations sustainable development goals · Heart disease · Climatic vulnerability · Climatic resilience · Climate action · Burden of disease · Health inequality

## 1.1 Are There Any Winners When It Comes to Climate Change?

In framing the importance of this book beyond those living in high-income countries, it is important to consider the substantive and indeed evolving burden imposed by cardiovascular disease (CVD) in every part of the world (Roth et al. 2020; Vogel et al. 2021; Vaduganathan et al. 2022; Global Burden of Disease Collaborators 2023; Xu et al. 2024). On this basis, it is estimated that (80%) of CVD cases, of which heart disease (both in its communicable and non-communicable) forms is the main component, reside in low-to-middle income countries (World Heart Federation 2023). As demonstrated in Fig. 1.1, while there have been impressive declines in (reported) age-standardised death rates due to CVD on a global basis over the past 30 years (Roth et al. 2020), as identified by Pineiro colleagues (Pineiro et al. 2023), this positive trend has slowed—with a 'moribund' trend most evident in low-income countries compared to more economically 'upwardly mobile' middle-income countries (World Heart Federation 2023). Complicating this picture, of course, is the paucity of accurate data from highly populous, but extremely poor countries like Mozambique in sub-Saharan Africa (Damasceno et al. 2010; Mocumbi et al. 2019).

S. Stewart, *Heart Disease and Climate Change*, Sustainable Development Goals Series, https://doi.org/10.1007/978-3-031-73106-8_1

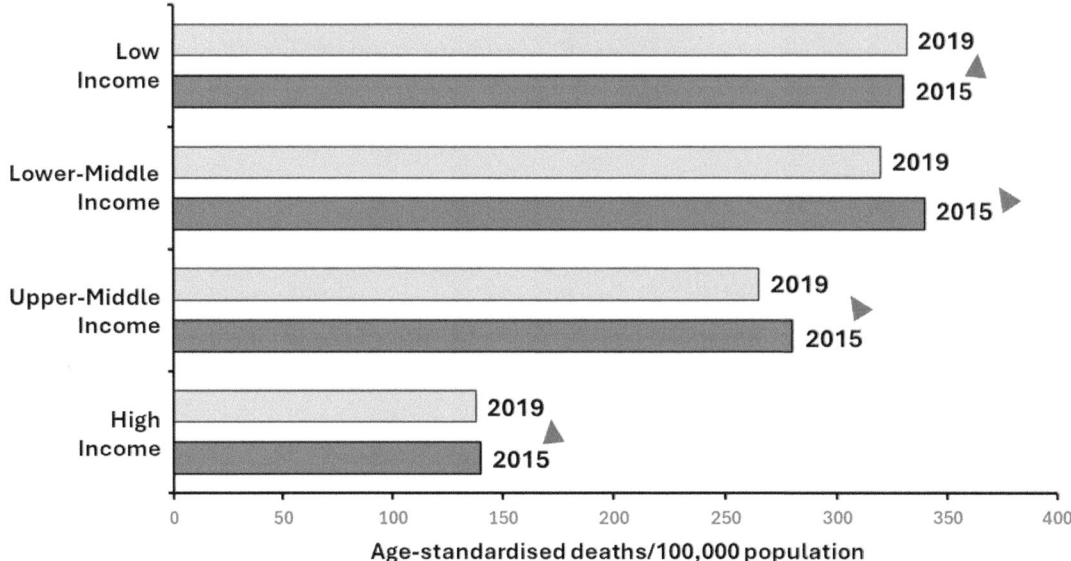

**Fig. 1.1** Burden and direction of age-standardised mortality rates due to CVD: the 'haves' versus 'have-nots'. Based on data published by Pineiro and colleagues and the 2023 World Heart Federation Report (World Heart Federation 2023; Pineiro et al. 2023) on global heart disease, this graph shows the differential impact of CVD on age-standardised deaths according to the relative wealth of regions, while indicating (red triangles) in which direction death rates are heading (2019 versus 2015)

When further considering the traditional risk factors that have been identified as the major contributor to cardiovascular-related mortality worldwide [many of which are derived from the original findings of the seminal *Framingham Heart Study* (Franklin and Wong 2013; Kannel et al. 1971) and then confirmed from other large population-based studies of cardiovascular risk and outcomes (Ikram et al. 2017; Murphy et al. 2006; Stewart et al. 2001, 2002; Tate et al. 2015)], one might argue that the climate/weather conditions were overlooked—see Fig. 1.2. However, as will be explained in the next two chapters, the biological influence of both predictable and unpredictable variations in climatic conditions (i.e. the weather!) on these risk factors cannot be denied, when there are such large variations in cardiovascular and heart health outcomes according to the time of year. This especially applies to 'dynamic' risk modulators/factors such as blood levels and air quality.

As will be specially described in Chap. 3, the influence of *climate change* on current ebbs and flows in cardiac events linked to climatic conditions/acute weather events is unlikely to be a positive one (Masselot et al. 2023; Peters and Schneider 2021). Indeed, the decreased ability to control many of the extreme cardiac provocations linked to climate change (from heatwaves to a higher burden of parasitic diseases) will mean that those living in the poorer countries and regions of the world will bear the brunt of a problem that is not of their our own making—the irony/conundrum being that the intoxicating mix of industrial and economic development at a population level, and wealth at the individual level, delivers more incentives to control the local environment and protect comfort levels while damaging the global environment for those who can least combat the adverse effects those increased comfort levels bring.

## 1.2  What Can We Expect from Climate Change?

So, how will progressive climate change influence our ability to meet the ***United Nation's Sustainable Development Goals*** [**UNSDGs** (United Nations Department of Economic and

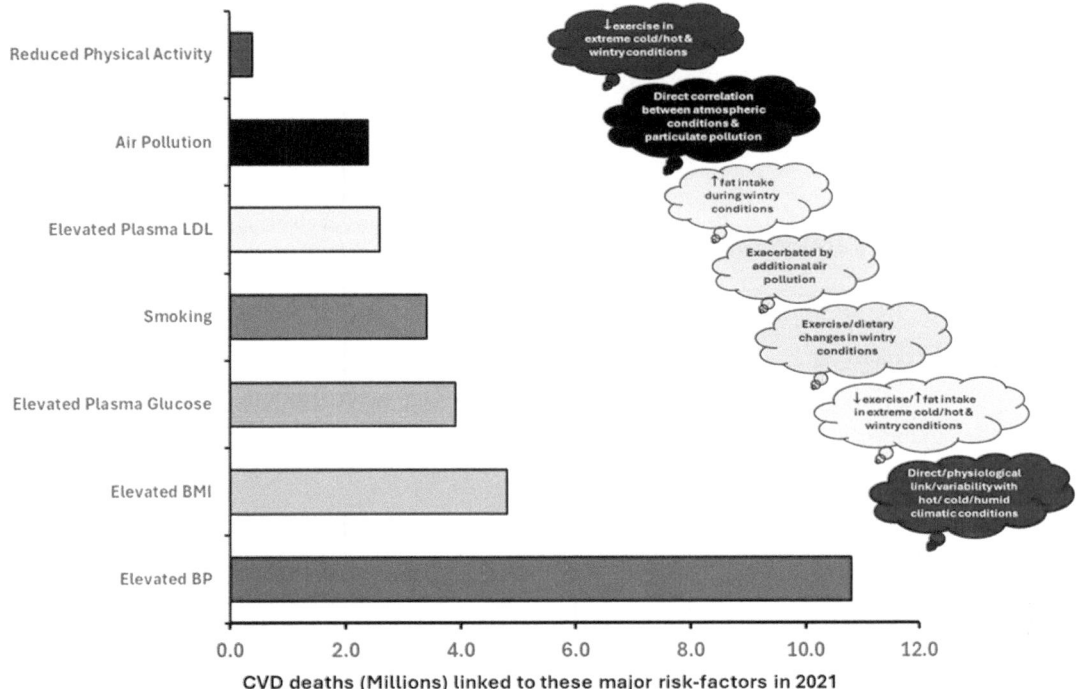

**Fig. 1.2** Major risk factor contributions to CVD-related deaths and their links to climate change. These data show the major contributors to CVD-related deaths [in estimated millions of deaths in 2021 reported by the World Heart Federation (2023)] and the extent to which climatic conditions (and therefore climate change) is likely to influence them from a bio-behavioural perspective (insert clouds)

Social Affairs 2024)]? Clearly, on many fronts! This is because there are 17 aspirational goals and 169 interconnected targets that span the spectrum of "eliminating poverty and hunger" to establishing "peace, justice and partnerships" to address inequalities and achieve these goals. As described in a series of reports produced by UN panels (United Nations 2023), there are many intersection points between these goals. They reflect the complexities of how we live and work together as a global society. In the context of this book, one may well argue that our approach to health (particularly when considering tertiary health care) largely ignores the intersection between the many facets of our individual lives (from our culture to socio-economic status and life-stage) and distils it down to a more singular biological condition that is largely divorced from this reality. In a later chapter an even more expansive perspective on our place in the biosphere and how that intersects with the rest of the planet is described.

Nevertheless, two of the 17 stated goals encompassed by the UNSDGs are—**Goal 3—Ensure healthy lives and promote well-being for all at all ages** and **Goal 13—Take urgent action to combat climate change and its impacts**. Unfortunately, there is increasing evidence that if we don't combat the health impact of climate change, we won't achieve healthier lives for all given the following:

A. Cardiovascular conditions, including common forms of heart disease, have an enormous impact on the health and longevity of the global population and;
B. People living with early-to-late forms of CVD, including increasingly prevalent forms of heart disease, are highly sensitive to a broad range of predictable to unpredictable changes in the environment/climatic conditions.

Specifically, as the leading cause of premature morality worldwide [contributing to 38% of premature deaths due to non-communicable disease [NCDs] in 2019 (World Heart Federation 2023)], effectively preventing and treating CVD forms a key aspiration of **SDG3.4** (with the specific goal of reducing NCDs by one-third by 2030) (United Nations Department of Economic and Social Affairs 2024). In absolute terms, the latest report from the *World Heart Federation* describes how:

> more than half a billion people around the world continue to be affected by cardiovascular diseases, which accounted for 20.5 million deaths in 2021—close to a third of all deaths globally (World Heart Federation 2023).

As further articulated by the *World Health Federation*, addressing the key determinants of cardiovascular health requires a more holistic perspective that encompasses the importance of broader factors such as financing, environment and education (Pineiro et al. 2023). As will be discussed in more detail in this book, when recognising the influence of climatic conditions and climate change on "who" and "when" someone will experience an otherwise preventable cardiac/cardiovascular event, and then seeking to prevent that from occurring, it will become obvious that:

1. Those living in poverty and in resource-poor countries are most vulnerable to this phenomenon and
2. We can only help establish "***climatic resilience***" if we apply the same (holistic) perspective as to why "***climatic vulnerability***" exists.

Unfortunately, as will be explored in later chapters, some of the factors that drive/contribute an individual's vulnerability to experiencing a potentially fatal cardiovascular-related or cardiac-specific event in response to provocative climatic conditions is (currently) unavoidable. This includes a person's biological sex and (advancing) age. Nevertheless, a broad range of factors are modifiable and many of these are closely linked to UNDSGs (United Nations Department of Economic and Social Affairs 2024). It is these factors we need to consider (within the entirety of our aspirations to achieve

equity in human health, wealth and potential) in responding to the threat of climate change. Specifically, as recently stated in a recent UN report on *Tackling Climate and SDG Action Together* (United Nations 2023):

> We must solve the climate emergency and sustainable development challenges together, or we will not solve them at all.

In considering that position, one must also consider the progress made thus far in meeting the SDGs—with 50% lacking sufficient progress and 30% either stalling or even going backwards (United Nations 2023).

In the background, global temperatures have risen (on average) by 1.1 °C, with a critical threshold of 1.5 °C likely to be reached in the next decade. As will be described in the next few chapters, it would be a fatal mistake to assume that climate change is all about these temperature increases. The health implications are far more complex—from an individual to societal perspective. Indicative of this truism, between 2010 and 2020, vulnerable regions of the world experienced a 15-fold increase in mortality linked to a broad range of "weather/climatic" events, including floods, droughts and storms, rather than just "heatwaves" (United Nations 2023). In attempting to quantify the inter-linking benefits of effectively tackling climate change the UN estimates, for example, that the implementation of low-emission development pathways (to curb greenhouse gas emissions) will not only help to achieve targets and goals around economic growth (91%) and reduced inequality (75%) but improve health (72%) in parallel/conjunction with these (United Nations 2023).

## 1.3 Too Little, Too Late, or Time to Intervene?

Taken to its logical conclusion, if we don't better understand or respond to the growing provocations posed by climate change on the heart health of those living in the poorest countries in the world, the aspirational goals of the UNSDGs are likely to recede further into this distance.

Moreover, the health systems of even the most financially well-off and resourced health care systems in wealthy countries already struggling with a 'tsunami' of older individuals with historically high levels of obesity, diabetes and hypertension who develop heart disease and other common forms of CVD, will crumble in the face of the additional pressures of climatic provocations to our collective heart health. In this respect, we need to understand that the poorest people in the world are an early warning system of what is to come (effectively the '*canary in the coal mine*') for the rest of the world. As such, they've already signalled their distress as climate change takes hold and erodes their health and stretches limited resources.

Throughout this book, therefore, I will be attempting to highlight what factors are most relevant to the poorest peoples in the world AND what interventions/strategies we need to apply to further protect their heart health. In fundamental terms, if we protect the heart health of the most vulnerable people in the world, they will live longer, be more productive and in turn,

be able to protect others from the more extreme climatic provocations that will inevitably arise from climate change. As shown in Fig. 1.3, the aspiration to achieve better health for all (UNSDG Goal 3) by taking climate action (UNSDG Goal 13) to reduce the unhealthy side-effects of climate change, can only be mediated if we reduce inequalities (UNSDG Goal 10) in the capacity for individuals to societies to protect themselves from climate change. This includes providing them (the poorest nations and their people) with the resources to readily access health care services that meet their needs.

If there was one challenge those seeking to improve the health of disadvantaged people around the world might adopt, it would be to develop world-first health care and support sensitive to climate provocations to heart health—thereby placing them in a unique position to teach wealthier countries how to best combat climate change.

Consider the case of Mozambique located in South-East Africa, one of the poorest countries in the world. With a rapidly growing and

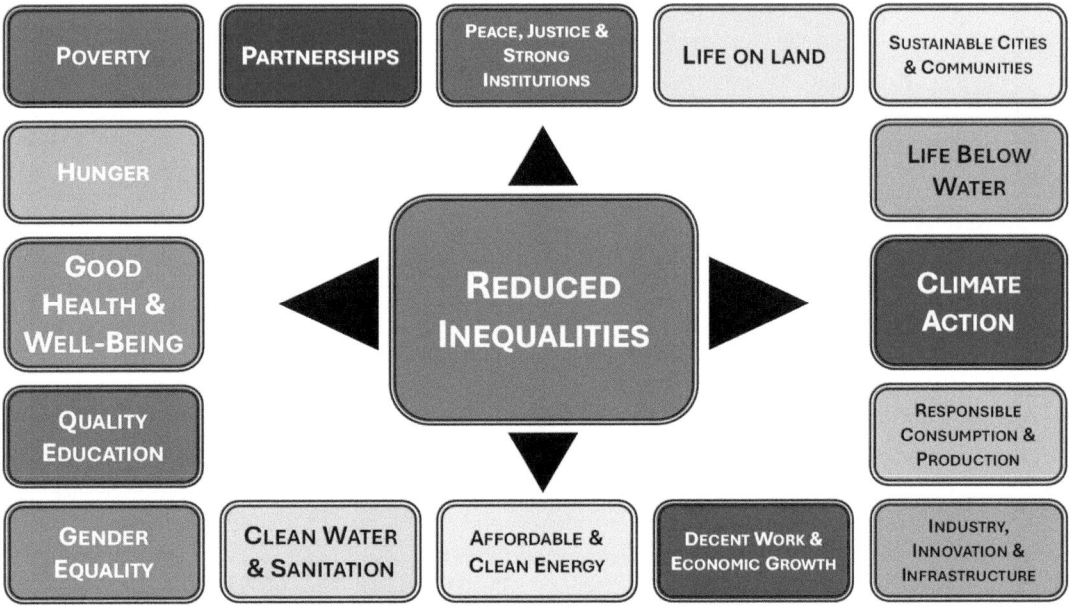

**Fig. 1.3** Health, climate action and reducing inequalities. As depicted in this 'reordering' of the 17 UNSDG goals (United Nations Department of Economic and Social Affairs 2024), there is an ever-growing link between health and well-being (or lack thereof) and climatic change requiring climate action. However, these two issues cannot be truly addressed unless inequalities across all facets of human activity (including access to life-saving health care) is addressed

young (median age ~17 years) population of ~29 million, two-thirds of whom live in rural/regional areas, Mozambique is microcosm of poverty-stricken communities worldwide, as it struggles to overcome crippling health issues. This includes a median life-expectancy hovering around 60 years. Despite burden of disease reports, that confidently provide key health metrics for Mozambique and surrounding countries, like many other low-resource countries, there are many uncertainties around the pattern of disease and how to manage health priorities (Prabhakaran et al. 2020; Manafe et al. 2019; Dzudie et al. 2019; Stewart and Sliwa 2009). It was in this context that the MOZambique snApshot of emeRging Trends (MOZART) Disease Surveillance Study (Mocumbi et al. 2019) was led by Professor Ana Mocumbi from the Nacional de Saúde and Universidade Eduardo Mondlane. As shown in Fig. 1.4, this unique study focussed on three diverse communities in Southern, Central and Northern and the profile/clinical characteristics seeking acute hospital care in two very different seasons. In this respect, Mozambique's historical climate is

classified as "Tropical Savanna" with mean temperatures maintained ~23 °C year-round but with distinctive wet and warm versus dry "cooler months" (The Koppen Climate Classification 2024). Without repeating the specific findings, as suspected the MOZART Study, while revealing some expected patterns of communicable versus non-communicable disease presentations on age, sex and even seasonal basis, exposed the fallacy of maintaining 'disease silos'. Specifically, these data revealed an increasingly complex pattern of multimorbidity that varied according to the location (and climatic conditions) (Mocumbi et al. 2019). When ignoring the artificial boundaries between countries, it soon became clear that people living in adjacent countries (living under similar conditions) share the pattern of disease and health challenges. It was on this basis that a major 'North–South' initiative was born. Consistent with the content of the final chapter of this book (Chap. 10) that is devoted to reviewing and addressing the key issues raised in the preceding chapters, the specifics of this initiative and why it is important will be presented and discussed there.

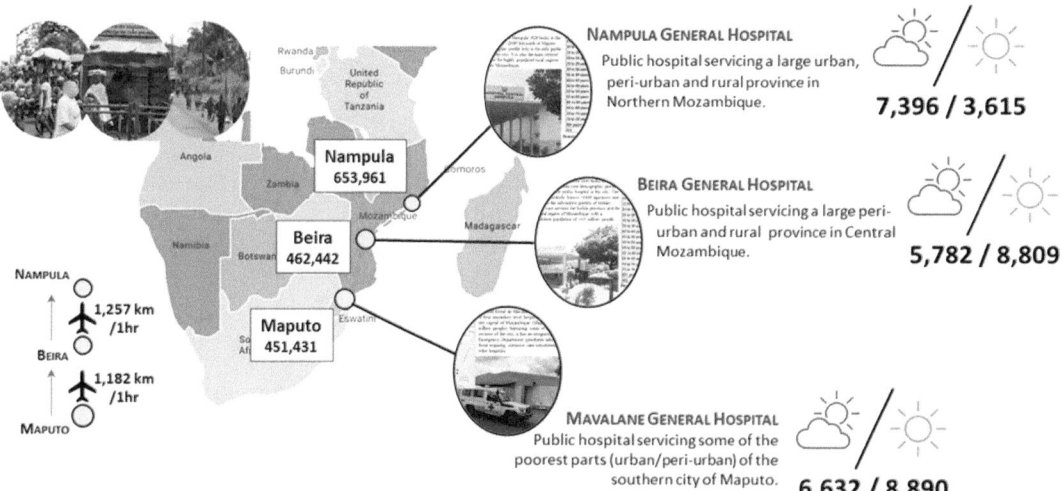

**Fig. 1.4** Insights from the MOZART study in Mozambique. This figure shows the location and pattern of (total) presentations in April (hot and humid conditions—purple symbol) versus October (generally cooler and drier conditions—orange symbol) in Maputo, Beira and Nampula. Overall, the cohort comprised 4021 males and 3788 (48.5%) females of all ages—of whom 630

(8.1%), 2070 (26.5%), 1009 (12.9%) and, 4100 (52.5%) were infants, children, adolescents and adults. Overall, a communicable disease (3914 cases/50.1%) was the most common reason for seeking acute hospital care, followed by a non-communicable disease NCD (1963/25.1%) and any type of injury (1932/24.7%) (Mocumbi et al. 2019)

Over the next few chapters, the dynamics of climate change, relative to human colonisation of nearly every corner of the world and the close links between human adaptation and cardiovascular health will be presented and discussed. Appreciating these links and the way climate change poses a growing threat to heart health, is critical to further understanding how this phenomenon is manifested from an individual to health system level.

# References

Damasceno A, Gomes J, Azevedo A, Carrilho C, Lobo V, Lopes H, Madede T, Pravinrai P, Silva-Matos C, Jalla S, Stewart S, Lunet N. An epidemiological study of stroke hospitalizations in Maputo, Mozambique: a high burden of disease in a resource-poor country. Stroke. 2010;41:2463–9.

Dzudie A, Kengne AP, Lamont K, Dzekem BS, Aminde LN, Abanda MH, Thienemann F, Sliwa K. A diagnostic algorithm for pulmonary hypertension due to left heart disease in resource-limited settings: can busy clinicians adopt a simple, practical approach? Cardiovasc J Afr. 2019;30:61–7.

Franklin SS, Wong ND. Hypertension and cardiovascular disease: contributions of the framingham heart study. Glob Heart. 2013;8:49–57.

Global Burden of Disease Collaborators. Global, regional, and national burden of diabetes from 1990 to 2021, with projections of prevalence to 2050: a systematic analysis for the Global Burden of Disease Study 2021. Lancet. 2023;402:203–34.

Ikram MA, Brusselle GGO, Murad SD, van Duijn CM, Franco OH, Goedegebure A, Klaver CCW, Nijsten TEC, Peeters RP, Stricker BH, Tiemeier H, Uitterlinden AG, Vernooij MW, Hofman A. The Rotterdam study: 2018 update on objectives, design and main results. Eur J Epidemiol. 2017;32:807–50.

Kannel WB, Gordon T, Schwartz MJ. Systolic versus diastolic blood pressure and risk of coronary heart disease: the Framingham study. Am J Cardiol. 1971;27:335–46.

Manafe N, Matimbe RN, Daniel J, Lecour S, Sliwa K, Mocumbi AO. Hypertension in a resource-limited setting: poor outcomes on short-term follow-up in an urban hospital in Maputo, Mozambique. J Clin Hypertens. 2019;21:1831–40.

Masselot P, Mistry M, Vanoli J, Schneider R, Iungman T, Garcia-Leon D, Ciscar JC, Feyen L, Orru H, Urban A, Breitner S, Huber V, Schneider A, Samoli E, et al. Excess mortality attributed to heat and cold: a health impact assessment study in 854 cities in Europe. Lancet Planet Health. 2023;7:e271–81.

Mocumbi AO, Cebola B, Muloliwa A, Sebastiao F, Sitefane SJ, Manafe N, Dobe I, Lumbandali N, Keates A, Stickland N, Chan YK, Stewart S. Differential patterns of disease and injury in Mozambique: new perspectives from a pragmatic, multicenter, surveillance study of 7809 emergency presentations. PLoS ONE. 2019;14:e0219273.

Murphy NF, MacIntyre K, Stewart S, Hart CL, Hole D, McMurray JJ. Long-term cardiovascular consequences of obesity: 20-year follow-up of more than 15,000 middle-aged men and women (the Renfrew-Paisley study). Eur Heart J. 2006;27:96–106.

Peters A, Schneider A. Cardiovascular risks of climate change. Nat Rev Cardiol. 2021;18:1–2.

Pineiro DJ, Codato E, Mwangi J, Eisele JL, Narula J. Accelerated reduction in global cardiovascular disease is essential to achieve the sustainable development goals. Nat Rev Cardiol. 2023;20:577–8.

Prabhakaran D, Perel P, Roy A, Singh K, Raspail L, Faria-Neto JR, Gidding SS, Ojji D, Hakim F, Newby LK, Stepinska J, Lam CSP, Jobe M, Kraus S, Chuquiure-Valenzuela E, et al. Management of cardiovascular disease patients with confirmed or suspected COVID-19 in limited resource settings. Glob Heart. 2020;15:44.

Roth GA, Mensah GA, Johnson CO, Addolorato G, Ammirati E, Baddour LM, Barengo NC, Beaton AZ, Benjamin EJ, Benziger CP, Bonny A, Brauer M, Brodmann M, Cahill TJ, Carapetis J, Catapano AL, Chugh SS, Cooper LT, Coresh J, Criqui M, DeCleene N, Eagle KA, Emmons-Bell S, Feigin VL, Fernandez-Sola J, Fowkes G, Gakidou E, Grundy SM, He FJ, Howard G, et al. Global burden of cardiovascular diseases and risk factors, 1990–2019: update from the GBD 2019 study. J Am Coll Cardiol. 2020;76:2982–3021.

Stewart S, Sliwa K. Preventing CVD in resource-poor areas: perspectives from the 'real-world.' Nat Rev Cardiol. 2009;6:489–92.

Stewart S, Hart CL, Hole DJ, McMurray JJ. Population prevalence, incidence, and predictors of atrial fibrillation in the Renfrew/Paisley study. Heart. 2001;86:516–21.

Stewart S, Hart CL, Hole DJ, McMurray JJ. A population-based study of the long-term risks associated with atrial fibrillation: 20-year follow-up of the Renfrew/Paisley study. Am J Med. 2002;113:359–64.

Tate RB, Cuddy TE, Mathewson FA. Cohort profile: the manitoba follow-up study (MFUS). Int J Epidemiol. 2015;44:1528–36.

The Köppen Climate Classification. https://www.mindat.org/climate.php. Accessed April 2024.

United Nations. Synergy solutions for a world in crisis: tackling climate and SDG action together report on strengthening the evidence base. 2023. https://sdgs.un.org/sites/default/files/2023-09/UN%20Climate%20SDG%20Synergies%20Report-091223B_1.pdf. Accessed May 2024.

United Nations Department of Economic and Social Affairs. Sustainable development (the 17 goals); 2024. https://sdgs.un.org/goals. Accessed June 2024.

Vaduganathan M, Mensah GA, Turco JV, Fuster V, Roth GA. The global burden of cardiovascular diseases and risk: a compass for future health. J Am Coll Cardiol. 2022;80:2361–71.

Vogel B, Acevedo M, Appelman Y, Bairey Merz CN, Chieffo A, Figtree GA, Guerrero M, Kunadian V, Lam CSP, Maas A, Mihailidou AS, Olszanecka A, Poole JE, Saldarriaga C, Saw J, Zuhlke L, Mehran R. The Lancet women and cardiovascular disease commission: reducing the global burden by 2030. Lancet. 2021;397:2385–438.

World Heart Federation. World heart report 2023: full report. 2023. https://world-heart-federation.org/resource/world-heart-report-2023/. Accessed June 2024.

Xu X, Islam SMS, Schlaich M, Jennings G, Schutte AE. The contribution of raised blood pressure to all-cause and cardiovascular deaths and disability-adjusted life-years (DALYs) in Australia: analysis of global burden of disease study from 1990 to 2019. PLoS ONE. 2024;19:e0297229.

## Abstract

Having briefly introduced this topic from the perspective of the *United Nation Development Programme Goals,* this *introductory chapter* provides a personal to scientific perspective on climate change in the modern era. As such, it will specifically describe how it (climate change) has the potential to adversely influence the heart health of the global population. In making this case, a cruel irony will be highlighted—that is, while most of the world's poorest people barely contribute(d) to the reasons why climate change occurred, unlike high-polluting countries, they still suffer the same and even worse consequences. Moreover, they have limited resources and capacity to address the difficult challenges arising from climate change. In this context, while clinicians are mostly focussed on the individuals they care for, this chapter further explores why having a "*climatic conscience*" or at least awareness of climatic conditions on health, is the pathway to better health outcomes. Specifically, it provides a rationale why health services and clinicians alike, need to acknowledge and understand the link between external conditions and the physiological status of any individual—thereby thinking beyond the four walls of an environmentally controlled hospital or GP clinic.

## Keywords

United Nations development programme goals · Global warming · Climate clock · Heart disease · King Canute · Human resilience · Industrial revolution · Greenhouse gases

## 2.1 A Rising Tide of Climate Change

Like the *Fable of King Canute*, as individuals, we are (and will continue to be) powerless to stem a '*rising tide*' of climate change. As will be explained as key focus of this book, climate change (and the other environmental dangers that essentially stem from the same causality—human industrial activity) will progressively damage our hearts (Peters and Schneider 2021). It will, perhaps, take decades and even centuries to reverse the damage to the planet we've already inflicted. This will involve the development of new 'clean' technologies and public health policies (United Nations 2023)—the depth and breadth of which is beyond the scope of this focussed book, but is well covered in other books focussed on climate change (Filho 2011; Palinkas 2020). However, with a change in mind-set in how we assess and manage those who develop and live with heart disease,

S. Stewart, *Heart Disease and Climate Change*, Sustainable Development Goals Series,
https://doi.org/10.1007/978-3-031-73106-8_2

we can start to protect them from climatic and other environmental provocations to their health. Unfortunately, at the time of writing of this book I'm unaware of any clinical guidelines (including those produced by the *European Society of Cardiology* or *American Heart Association/ American College of Cardiology*), outlining the expert prevention and/or management of heart disease that explicitly acknowledge that cardiac events do *not* occur (completely) on a random basis. Thus, whether you consider expert recommendations from a Northern or Southern Hemisphere perspective over the past decade or so (Krum et al. 2011; Kirchhof et al. 2016, 2017; Atherton et al. 2018; McDonagh et al. 2021; Humbert et al. 2022, 2023), or from contemporary and highly influential North American perspective (Ommen et al. 2024; Ford et al. 2023; Gragnano et al. 2023; Abovich et al. 2023; Virani et al. 2023; Konemann et al. 2023), you would struggle to reconcile that of the perspective of the World Heart Federation (in recognising the critical importance of environmental factors in shaping cardiovascular health) (World Heart Federation 2023; Sliwa et al. 2024). Critically, this oversight extends to burden of disease reports that routinely provide absolute numbers and rates of cardiovascular-related outcomes as if they were a 'static' statistic (Roth et al. 2020). Consider for instance, the foreword to the latest burden of disease report published by the American Heart Association (the claims of importance being essentially true)—"*the AHA's annual Heart Disease and Stroke Statistical Update continues to provide the most up-to-date statistics on cardiovascular disease (CVD). As it has evolved over the years, the report has become a preeminent resource in identifying the overall impact of all types of CVDs, including who is most affected and where it is most prevalent.*" (Martin et al. 2024) What this statement is missing, in the context of this book (and the evidence it presents), is *when* some cardiovascular events occur and *why* some people/communities are more affected. Such wider and more holistic considerations are important when one considers that, as has been shown time and time again in relation to

heart failure (Shah et al. 2018; Inglis et al. 2008; Martinez-Selles et al. 2002; Matsuda et al. 2024; Nganou-Gnindjio et al. 2021; Stewart et al. 2002; Boulay et al. 1999), and other common forms of heart disease [from acute coronary syndromes (Furukawa 2019; Keller et al. 2019; Akioka et al. 2019; Yang et al. 2017; Swampillai et al. 2012; Kriszbacher et al. 2008; Ahlbom 1979) to paroxysmal and more permanent forms of atrial fibrillation (Deshmukh et al. 2013; Sheehy et al. 2022; Ahn et al. 2018; Loomba 2015; Murphy et al. 2004; Spengos et al. 2003a; Frost et al. 2002; Kupari and Koskinen 1990)] as well as cerebrovascular events [more so ischaemic than haemorrhagic strokes (Oida et al. 2020; Xue et al. 2023; Tian et al. 2022; Fujii et al. 2022; Li et al. 2019; Han et al. 2015; Tsuji et al. 1975; Jakovljevic et al. 1996; Spengos et al. 2003b; Ansa et al. 2008; Karagiannis et al. 2010)], at the very least, they ebb and flow with seasonally driven, transitions in climatic and acute weather conditions (see Chap. 5) (Stewart et al. 2017).

Whether climate change will dramatically change existing patterns of heart failure/illness is still yet to be determined. For example, consider the conclusions of a systematic review (15 studies analysed) of seasonal/climatic influences on paroxysmal atrial fibrillation—"*The rate of occurrence of paroxysmal AF varies by seasons and is greatest during winter and least in summer.*" (Loomba 2015) One might wonder, therefore, if 'global warming' will mean less people experience potentially harmful episodes of paroxysmal atrial fibrillation? However, the story is probably too complex to make such simplistic conclusions.

Nevertheless, to date, '*the sound of tumbleweeds rolling across a barren desert*', best describes our collective response to pre-existing climate provocations to the heart health of millions of people around the world. Unfortunately, this passive response (some may it call wilful ignorance) becomes more alarming when one considers the likely impact of climate change in provoking more, otherwise preventable cardiac (and cerebrovascular) events, in those most sensitive or vulnerable to challenges to their physical and mental health. It would be dishonest for anyone to claim they know exactly how climate change will influence the pattern of cardiac events worldwide, or indeed quantify that change. However, we can be certain that climate change will *not* be a benign influence on the heart and cardiopulmonary health of millions of people worldwide (Peters and Schneider 2021; Xie et al. 2021). Fortunately, as will be identified in Chap. 4, there are many elements to what is likely to be a 'malignant' phenomenon, that are modifiable and therefore can be prevented. Moreover, some unexpected benefits from climate change may still occur.

---

**Textbox 2.1: United Nations Framework Convention on Climate Change (United Nations 2024)**

'*Climate change*' means a change of climate which is attributed directly or indirectly to human activity that alters the composition of the global atmosphere and which is in addition to natural climate variability observed over comparable time periods.

*Adverse effects of climate change* means changes in the physical environment or biota resulting from climate change which have significant deleterious effects on the composition, resilience or productivity of natural and managed ecosystems or on the operation of socioeconomic systems or on human health and welfare.

---

It would be a grave mistake if anyone applying clinical management or public health policies/strategies only considered the impact of 'global warming' on human health—noting the retirement of that unidirectional term! As articulated within a more nuanced framework, the United Nation (UN) focusses on change that generates variability over and above natural climatic variability. Thus, the impact of *climate change on human health* (i.e., its *adverse effects*), as articulated by the UN,

will not be exclusively mediated via progressively increasing temperatures and more severe heatwaves. Unfortunately, this more 'enlightened' perspective (based on hard science rather than popular perception) is yet to filter-down to the public. Much of this can be attributed to the gravitation (by media and some academics alike) towards the adverse effects of heat rather than dynamic weather events.

So, what needs to change? For all clinicians (and not just a minority of enlightened few), this requires embracing a ***paradigm change*** in how they perceive the interaction between climatic and environmental conditions and the health/risk status of those they are treating/caring for. This includes understanding the when and why someone is more vulnerable to experiencing a clinical crisis—resulting in hospitalisation and/or death. In this author's humble opinion, this means making climate change personal and understanding the broader issues that are driving individual to regional health inequities and outcomes—a global perspective if you will. Concurrently, it requires greater introspection and personal experiments/awareness of what happens to the body during weather extremes. The sections below provide such a journey from providing a personal to global perspective on dynamic climatic conditions and the role they play in determining individual to global patterns of heart health and related health outcomes.

## 2.2   A Personal Perspective on Climate Change

In my life journey as a (now) cardiovascular-focussed health-services researcher, who migrated from the UK to Australia as a teenager, and then had the privilege of travelling the world (visiting and working in many different countries along the way), it's only recently that I've really focussed on what climate change will mean for our collective heart health. The irony is that I've been studying the impact of predictable (e.g. the onset of winter) and predictably, unpredictable climate variations and weather events (e.g. heatwaves and cold snaps) since the turn of the millennium (Inglis et al. 2008; Stewart et al. 2017, 2002; Murphy et al. 2004; Mocumbi et al. 2019).

It is only many years later that I can reflect on my own childhood playing in the predictable winter snow that would fall in the beautiful countryside of Herefordshire and then the Wirral in the UK. Both locations are far removed from the mountainous regions of the Pennine, Lake District and Scottish Highlands. And, yet every garden shed had a snow-sled hanging on a hook (next to a home-made go-kart) ready to be wheeled-out each winter (often before Christmas arrived). Imagine my surprise, therefore, when returning to Glasgow, Scotland (hundreds of miles north in colder climes) in 1999 and learning that it hadn't snowed there at Christmas for more than 20 years! Similarly, at the other end of the globe, I distinctly remember a predictable pattern of unrelentingly dry, searing heat, during the summer months spent in the Mediterranean climes here in Adelaide, South Australia. While we still seem to experience hot summer months and relatively cool wet winters here in Adelaide, there is a definite sense that the seasons are no longer that predictable. In recent years, reflecting large-scale fluctuations in sea levels and air pressure [the Southern Oscillation Index (National Centers for Environmental Information 2024)] between the western and eastern borders of the Pacific Ocean (spanning the Eastern to Western coastlines of Australia to South America/California with the Polynesian Islands in the middle), Australia and South America swap "climatic polarities" from stability/predictability to weather extremes. Accordingly, in recent years both regions have felt the differential effects of La Niña (Australian Bureau of Meterology 2024) versus El Niño (Srinivas et al. 2024; Deng and Dai 2024; Jiang et al. 2024; Cordero et al. 2024) events that lead to less rain, warmer temperatures (with drought conditions extreme bush/wildfire conditions far more likely on one side of the ocean) versus more rain, colder temperatures and more dynamic temperature/weather conditions affecting those living on the other side of the ocean. Thus, weather patterns are far less predictable than ever before as the 'oscillation' between air and ocean currents vary with devastating effect (Deng and Dai 2024). One only has to consider the ring of extreme 'bushfires' surrounding the outer suburbs of Sydney (population 5.3 million) on the Eastern seaboard of Australia as part of a devastating fire season affecting the country (Mellish et al. 2024; Hasnain et al. 2024; Ong et al. 2023). This was mirrored (but not in the same year) by the devastating 'wildfires' raging through California, North America (including Beverly Hills and surrounding suburbs in Los Angeles—population 3.8 million) (Do et al. 2024; Masri et al. 2022; Heaney et al. 2022; Naqvi et al. 2023) and Chile in Southern America (Cordero et al. 2024). In between these challenging times, those recovering from the scorchingly hot winds and firestorms have had to deal with rain and floods of biblical proportions (Crompton et al. 2023).

Having lived through one of the most dynamic eras in human history, I'm particularly conscious of the dramatic changes in technology and communications that have turned us from a series of local communities with well-defined boundaries, to a highly connected global village. In just over half-a-century, pictures have turned from black and white to vivid LCED colour and communications have evolved from handwriting in blotchy ink and delivered by hand, to thoughts and emotions instantaneously delivered to any corner of the globe via smart phone technology! If we reflect on these dramatic changes through the prism of major global events such as climate change, we rapidly realise that it would have been almost impossible to understand the scope of the problem even two or three decades ago. In specific terms, it would have been impossible for anyone to collate and synthesise data from all parts of the globe to generate a coherent picture—even with the help of satellite imagery (a technology that, initially at least, could only reflect dramatic changes, such as mass-deforestation, to the health of our planet) (Mokhtar et al. 2023; Zhao et al. 2023; Smith et al. 2020; Zou et al. 2018; Norris et al. 2016; Maynard and Conway 2007; Renner 1999; The ATOC Consortium 1998). Now, meteorologists and climate researchers can access imagery and granular data from across the globe in real-time. They can also feed complex data into super-computers to produce increasingly accurate models around the 'what next', from regional weather forecasts to longer-term prognostications on major climatic variations (Tollefson 2023) and the timing/frequency of potentially life-threatening weather events such as the 'Atlantic Hurricane Season' officially running from 1st June to 30th November (National Hurricane Center 2024). They also generate increasingly more complex models seeking to explain and describe global climate change—the accuracy and veracity of which remains open to debate.

## 2.3   The King Canute Conundrum—Sit Tight or Move the Chair?

We, the lay public and amateur meteorologists/climatologists alike are similarly blessed (or cursed) with an overload of information as the notion of climate change and its consequences bleeds into our consciousness, popular culture, social media and newsfeeds. The result is that we live in a world where an event of enormous significance to the climate, but occurring in the '*middle of nowhere*', is rapidly reported in vivid sound and colour from many objective perspectives. This contrasts with the historical recounting of eyewitness accounts that are difficult to imagine and personalise, especially when the details are sketchy at best—the famous explosion of *Krakatoa* on the Indonesian archipelago (now home to a large portion of the world's population) in 1883, being a classic example (Smithsonian Institute NMoNH 2024). This volcanic eruption produced what is reportedly to be the loudest sound in human recorded history. It then produced spectacular sunsets world-wide from the volcanic material (21 cubic km in volume) thrust into the air and then circulated around the atmosphere/stratosphere for months thereafter. And, yet beyond a much-diminished crater, it is difficult to appreciate the magnitude

of this event (Smithsonian Institute NMoNH 2024). Juxtaposed against the historical mystery of *Krakatoa*, is the eruption of Hunga Tonga-Hunga Ha'apai—a hitherto unremarkable and hidden, under-sea volcano located in the middle of the Pacific Ocean in January 2022. In human history, this is the largest single (recorded) detonation of water vapour entering the atmosphere, resulting in immediate degradation (by 30% or more) of the Ozone Layer (Asher et al. 2023). However, the scientists from the US National Oceanic and Atmospheric Administration's (NOAA) Global Monitoring Laboratory, could never have established this fact, if they were not alerted immediately to that event by the seismic monitoring data and astonishing satellite pictures they were able to download in real-time (National Oceanic and Atmospheric Administration 2023).

Thus, in evaluating the science of climate change, it is important to consider the immediacy of such events given that they are rapidly and even pre-emptively reported—often with sensational pictures and videos shared on every mainstream media and social media platform around the world. This was the case for the *Hunga Tonga-Hunga Ha'apai* eruption. As just a small example of this phenomenon, in the space of just 12 h following me writing/drafting the previous paragraph, three prominent news stories appeared on the news/buzz feeds I regularly browse:

- **China issues "once in a century" flood warning for Guangdong's Bei River zone**—the accompanying article explaining that flood waters within the Pear River tributary were expected peak early in the morning, with the cities of Guangzhou (15 million people), Qingyuan (4 Million people), Shaoguan (0.8 Million people) and Huizhou (2.7 million people) would bear the brunt of the flooding. https://www.scmp.com/news/china/politics/article/3259806/china-issues-once-century-flood-warning-guangdongs-bei-river-zone
- **Europe endured record number of "extreme heat stress" days in 2023**—the report quoting a new report from the

*European Union's Copernicus climate monitoring service* and the *World Meteorological Organization* that uses the *Universal Thermal Climate Index* (Grigorieva et al. 2023; Brode et al. 2012) to measure the effect of environmental/climatic conditions on humans. https://www.aljazeera.com/news/2024/4/22/europe-endured-record-number-of-extreme-heat-stress-days-in-2023

- **2023 Was event Hotter Than Predicted, Raising Fears We're in Unchartered Territory**—the accompanying report describing how 2023 was 0.2 °C warmer than climate models predicted. The accompanying pictorial showed an alarming 'sea of red' across the globe—Fig. 2.1 (https://www.epa.gov/climate-indicators/climate-change-indicators-sea-surface-temperature). https://www.sciencealert.com/2023-was-even-hotter-than-predicted-raising-fears-were-in-uncharted-territory

On other platforms more likely to appeal to younger people, such as Tik-Tok, there is a growing sentiment (and content) highlighting the need to address a growing threat that will influence their future health (https://www.wired.com/story/climate-change-tiktok-science-communication/2021). Why is such content important to healthcare professionals? Simply put, the OECD reports a rapidly ageing healthcare workforce among its member states (e.g. by 2019 it was reported that more than one-third of physicians are aged 55 years or older) (OECD 2024). So, is the average physician likely to be aware of, and monitor, the progress of the '*Climate Clock*' unveiled in 2020 in Union Square in New York to count down how much time we have left before climate change becomes irreversible—"*the most important number in the world*" (The Climate Clock 2024). Or are they more likely to be cognisant of the more esoteric '*Doomsday Clock*' (currently sitting at 90s to midnight as the horrors of the Ukraine War and Gaza Conflict evolve) first published in 1947 by the Bulletin of the Atomic Scientists in response to the growing threat of a devastating nuclear war (The Doomsday Clock 2024). These examples illustrate why the notion of a generational disconnect in recognising and responding to

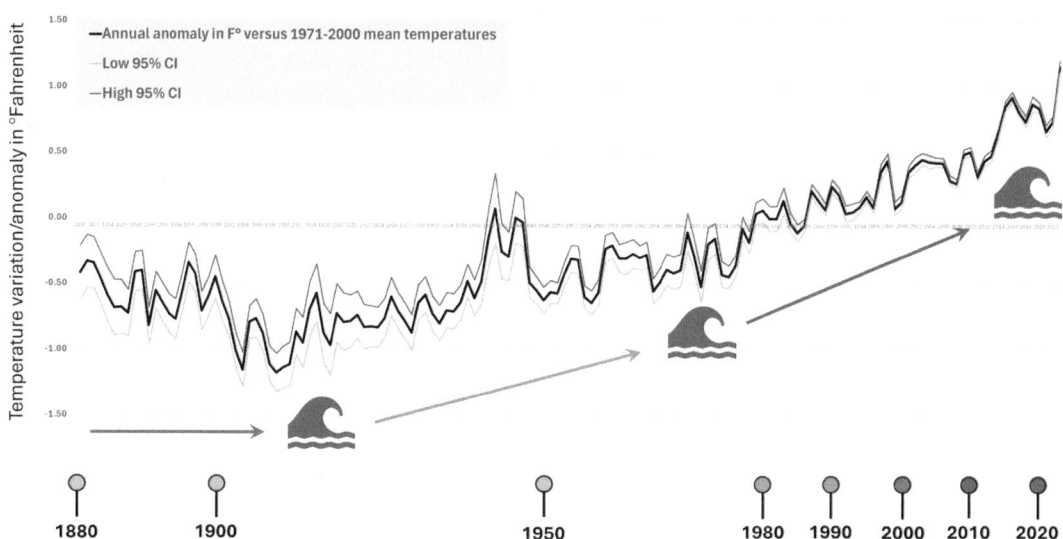

**Fig. 2.1** Based on a "Sea of Red"—NASA's thermal satellite imaging of ocean temperatures. As described on NASA's routinely updated website—sea surface temperature (SST) anomaly data from the jet propulsion laboratory (JPL) multi-scale ultra-high resolution (MUR) sea surface temperature analysis shows rising sea temperatures of historical precedence. This figure is derived from published data in July 2024 (https://www.epa.gov/climate-indicators/climate-change-indicators-sea-surface-temperature)

climate change as a matter of urgency, is something all health professionals need to reflect on. For example, consider the increasing number of Climate Change Activists on Tik-Tok who typically thrive on "*weird food that celebrities like to eat*" or "*annoying things people do at the gym*" but now post content urging more action on climate change—a phenomenon being played out on the streets of every corner of the world.

It is important to note, in this context, that this book does not purport to provide a crystal ball on the exact nature and impact of future climate change. Nor does it propose to provide compelling reasons why older health professionals should worry about the future health of younger generations and the planet—lifelong belief systems are difficult enough to change without talking about the '*what if?*'. Rather, it seeks to present robust scientific arguments why the environment and climatic conditions have *always* influenced health and longevity (particularly that relating to heart and cardiovascular disease), but this has not been a focus of clinical assessment and management.

A legitimate question at this point, is whether we (including younger climate change activists) are *catastrophising* what has been a constant (i.e. dramatic climatic fluctuations) since the earth formed oceans and an atmosphere predominantly filled with oxygen and nitrogen? Our childhood perceptions and long-term memories are notoriously unreliable; especially when it comes to retrofitting neat narratives formulated later in life. So, it would be a fallacy for me, or anyone to rely on anything I remember, without deriving hard evidence that we have entered a period of climate change—and not just simple variations in climatic patterns stimulated by geographic events beyond human control. Perhaps the most famous of these is the "Little Ice Age"—the coldest period on record during the past 10,000 years (Mann et al. 2009). This was a distinctive period of climatic cooling affecting those living in the northern hemisphere. Lasting anywhere from the fourteenth century to nineteenth century, but most evident during the sixteenth and seventeenth centuries (i.e. long before the industrial revolution), its

origins remain open to conjecture—although a combination of reduced solar activity coupled with increased volcanism affecting oceanic and atmospheric conditions/heat exchange have been identified as the possible cause (Arellano-Nava et al. 2022; Hu et al. 2022; Yang et al. 2022; Primeau et al. 2019). In contrast to this recent (in geological terms) 'ice age', the science now clearly indicates that we are indeed experiencing an accelerated period of human-induced (*anthropogenic*) climate change (United Nations 2023; United Nations Environment Programme 2024; Beggs et al. 2024a), the key characteristics and metrics of which are described below.

As always, our natural introspection and instincts for self-preservation leads us to important questions about our own health and circumstances. Of course, the interpretation of any answers one might generate is broadly influenced by our individual experiences and outlook on life. For example, studies of optimism among health workers attest to the intrinsic benefits of this attribute in dealing with the stressors of both daily medical duties and when medical emergencies arise (Boldor et al. 2012). As Boldor and colleagues explained, when reviewing the published literature—"*Optimism was also revealed one of the key components of resilience and self-efficacy*". As best explained in the next few chapters, the notion of promoting "resilience" (as opposed to vulnerability) is likely to play an important role in any therapeutic response to weather and/or broader environmental-mediated provocations to health in the future. Not surprisingly, perhaps, there is a whole literature devoted to the mind-set or 'mindfulness' of different health professions and, the broader public. Rather depressingly (pun intended), scientific reports suggest that pessimism is more powerful than optimism in determining health outcomes. For example, as part of the large, longitudinal Mayo Clinic cohort study of personality and ageing (Grossardt et al. 2009), Grossardt and colleagues concluded that "*pessimistic, anxious, and depressive personality traits were associated with increased all-cause mortality in both men and women*" among those attending the Mayo Clinic in the US. Similarly, a study of mortality over a 20-year period conducted by Whitfield and colleagues found that while age-adjusted scores on a pessimism scale were associated with all-cause and cardiovascular mortality (but not cancer-related deaths), optimism was not protective the other way around (Whitfield et al. 2020).

***Whether we are optimistic or pessimistic about the likely impact of climate change, really depends on the depth of our understanding of what is changing (within the climate) and what it means to human health—particularly the cardiovascular system***. A more simplistic perspective will focus on global warming and the impact of heat. For example, one might ask—"*Is it so bad that the world will be warmer and perhaps, more people in places like Europe and North America will require less heating?*". Extending that logic—"*Surely, exposure to less cold is a good thing?*". The problem is of course, as succinctly observed by (re-elected US President) Donald Trump who, perhaps, mimics many other sceptics about climate change when confronted by anything other than a heatwave, is that we appear to be experiencing *different weather*, not just more heat. Indeed, in the winter of 2024, Northern Europe experienced icy-cold weather not seen in years—as evidenced by Salt Bay (an extension of the Baltic Sea) on the shores of Stockholm becoming frozen in the process and the sight of people ice skating on adjacent Lake Mälaren. Meanwhile, in the Southern Hemisphere, spring temperatures in Southern Australia are the lowest recorded in the last century.

If, as suggested earlier, we are *King Canute* (i.e. sitting on a beach ordering the rising tide to recede) when it comes to rapidly addressing climate change, what is our individual potential to resist the inevitable? Should we sit and accept that we will be swamped by a rising tide of climate change, or can we at least move the chair rather than the sea?

As we are fed by a constant diet of calamitous news about the forces of nature beyond our control, it is worth considering that in many parts of the world, our latent capacity to maintain *individual* homeostasis via bio-behavioural

and environmental control mechanisms should not be under-estimated. As will be argued throughout this book, understanding these mechanisms and how they can be addressed to (re-) establish '*climatic resilience*' at the individual level should form part of the *core curriculum of all health practitioners*. Moreover, as the evidence evolves, they should be rapidly reflected in expert clinical guidelines. At minimum, much like economic reports that routinely refer to 'seasonally adjusted' figures, burden of disease reports focusing on heart disease and other common forms of cardiovascular disease should at least acknowledge that annual event rates reflect marked 'highs' and 'lows' in event rates. Once this occurs, we can appreciate how aspiring to remove the highs and ensuring the lows become the norm will have a dramatic impact on the burden of disease imposed by common forms of heart disease across the globe. This potential becomes more obvious when one considers the top 5 causes of death worldwide (as identified by the *Global Burden of Disease Study*) in the year 2021:

1. Ischaemic heart disease: estimated deaths—9.44 million/DALYs—185.0 million
2. Ischaemic stroke: estimated deaths—3.87 million/DALYs—70.2 million
3. Intracerebral haemorrhage: estimated deaths 3.46 million/DALYs—78.6 million
4. Hypertensive heart disease: estimated deaths 1.41 million/DALYs—24.9 million
5. Rheumatic heart disease: estimated deaths: 0.39 million/DALYs—13.4 Million (Vaduganathan et al. 2022)

Although there are no reliable figures for how many hospitalisations these conditions provoke, as will be described in Chap. 5, the impact of climatic provocations to an individual's heart health is not confined to the risk of death. As such, future health economic analyses of the impact of normal to anthropometric-driven variations in the climate and their provocation of heart disease-related hospitalisations, will undoubtedly identify an enormous cost-burden that will rise if we allow climate change to accelerate AND fail to address its adverse impact on vulnerable people (at least at the clinical level).

## 2.4   Taking the Plunge to Explore Human Resilience

As described above, for those of us fortunate enough to have smart devices and are wired into the world-wide web (with all its flaws), we are fed a never-ending stream of images and reports from around world that helps us imagine what it's really like to be there. However, imagining is not the same as a visceral experience. Some images may leave a lasting impression, but they can't replace lived experiences that leave a lasting impression and can prompt critical reflection. Based on my own interest in researching the limits of human resilience and adaptability from a theoretical 'distance' (see Chap. 4) and specifically preparing for this book, I decided to take the plunge as it were into cold extremes. This is something that my long-term collaborator, Professor Karen Sliwa (Past-President of the World Heart Federation and Director of the Cape Heart Institute in South Africa) would approve of. She now has a daily sojourn into the icy sea pool in nearby Camps Bay that, in her own accounts, has resulted in increasing tolerance to cold, improved mental acuity and physical vitality.

It was thus, that I undertook the increasingly popular (for tourists) activity of booking a sauna on the shores of the Oslofjorden in Southeastern Norway in the middle of a winter snowstorm (the outside temperature and wind-chill factor being well below zero and with a layer of broken ice floating calmly below the flurries of snow above). The idea of course, is to undertake what many Scandinavians do as a way of life—bake in the heat and humidity of a traditional sauna, before taking the plunge firstly into the icy atmosphere outside and the then icy-cold waters thereafter—before running back into the sauna to recover! So, what does that feel like as an outsider who has just been basking in a South Australian heatwave (with a top temperature of

43 °C on the day of departure)? Before answering, some general observations are worth considering. Firstly, the bravery and physical capacity to tolerate temperature extremes is most certainly a function of age. As articulated later in the book, age-related degradations in our cardiovascular system turn climatic resilience into vulnerability. Consequently, it was inevitable that the younger members of our party were able to spend more time in the icy-cold waters of the fjord and then repeat the cycle of sauna to fjord to sauna more easily. And yet, for me (a middle-to-older age man) this was still an amazingly invigorating activity that ultimately left me feeling more alive and, indeed more tolerant of the cold extremes that came along during the remainder of the trip. One of the key degradations of old age, is the blunting of our sensory perceptions (thereby leaving us vulnerable to not responding to potentially life-threatening conditions). It was for this reason, I was delighted (in a highly ironical manner) to feel burning pain and shock in all my limbs as I plunged into the fjord. My one and only thought was one of self-preservation as I could feel my body shut-down, and yet I persisted as long as I could—the relief of entering the hot sauna and escaping even the burning passage of feet skipping across the icy boardwalk being immense! However, here is the thing—at the start of the process, my body was breaking-down (in the form of extreme vasodilation resulting in profound sweating and a pounding headache) under the extreme heat of the sauna. This radically changed as soon as I had found myself in a more extreme environment, as my body plunged (including a brief immersion of the head) into the icy fjord. As soon as I returned to the sauna, it was no longer 'hot' and my body (now vasoconstricted and literally blue) refused to sweat. My perceptions of temperature at this point, completely turned upside down. I remained cold and clammy for quite a few minutes, before attaining a kind of physical 'euphoria' that lasted long afterwards.

This self-imposed confrontation with the elements was followed by exposure to a 'once in a generation' Arctic storm that hit the northern Norwegian city Tromsø (one of the northern most settlements within the Arctic Circle) a few days later. Rest assured, I'm unlikely to experience or voluntarily witness such ferocious winds that turned snow into bullets ever again. Nevertheless, I was able to observe the pattern of behaviour of locals versus tourists as this storm approached (one that ultimately devastated many parts of Norway). The locals had been preparing for days, warning everyone that would listen that they should be bunkered down for at least a day and night—the post-pandemic trend of working from home being invoked for anyone not physically required to service the tourists. And, what of the tourists? Well, seemingly oblivious of the dangers from wind gusts more than >100 km/h (with the wind-chill factor falling well below -10 °C) as they 'dance' on icy paths, they played the perfect 'fish out of water' in terms of adapting to the prevailing climatic conditions. So, while (most) of the tourists had equipped themselves with appropriate clothing, their adaptive behaviours clearly didn't match those with far more experience of the prevailing climate—failing badly in terms of avoiding potentially harmful provocations to their health.

In further reflecting on my experience with the combination of sauna/ice-cold plunge, I was able to then ponder what it can that tell us about our future in an increasingly more dynamic world of weather extremes? It also made me wonder if we can indeed build/grow/develop resilience to provocative climatic conditions when technology fails us, or when we are forced out into the open? As I reflected further, I wondered at what point would my body fail to make the corrections it did, indeed, at what age would an activity cause cardiogenic shock and/or a hypothermic coma because my sensory and homeostatic mechanisms were overwhelmed? In broader terms therefore, no matter how resilient we are/become, when does the human body (and particularly the cardiovascular system) fail to cope with the provocation of too much cold and the corollary of too much heat? What happens to the point of resilience/failure with normal ageing, physical fitness and cardiovascular disease? Furthermore, what does the role of lifelong/deeply ingrained cultural and behavioural adaptation play in determining when such age-related degradation will happen and at what age will any climatic/environmental extremes overwhelm us, if we leave ourselves open to the elements?

These are all poignant questions that have informed the content of this book—especially as I considered my symptoms as I took an early morning run along the cold and chilly banks of the River Mersey in Liverpool (adjacent to the Irish Sea in North-West England) soon after leaving the Artic Circle and strange days filled with just a few hours of filtered sunlight. For a start, it appeared that I had misjudged the temperature and what I needed to wear (given I had shorts and a few layers). Although it was closer to 5 °C than freezing and the sun was starting to appear, the wind running off the Mersey sucked the temperature out of my core very quickly. I could literally feel the cold creep into my bones and, as I persisted with my run. Very quickly, I could feel the blood supply escaping the exposed skin of my legs and arms and then rushing to the central parts of my body. Not unsurprisingly, after 30 min of hard exercise in the cold air, I had developed central chest pain (non-cardiac!) and I returned to my hotel a panting patchwork of marble cold and hot perfusion. It felt colder and more profoundly challenging than anything I had experienced in Norway.

My own physical (and perhaps mental) frailties were placed into stark contrast when I had the opportunity to learn first-hand how the Nomadic peoples of the Arctic Circle (the *Sami*) survive in often punishing conditions. In the absence of timber, their traditional 'Lavvu' provide a portable oasis of warmth from the historically freezing temperatures encountered in the Arctic. When combined with insulating leathers and furs (derived from the reindeer herds they tend to and eat) and their seeming impunity to the cold, this places them a world apart in terms of climatic vulnerability. They would undoubtedly struggle though if they were exposed to the opposite end of temperature/climactic extremes—and yet, in 2021, the World Meteorological Organisation reported a temperature of 38 °C in a Siberian town within the Arctic Circle (World Meteorological Organization's 2021).

As many of us with life-experience soon learn there are 'controllables' and then there are the 'uncontrollable' (much like the fable of King Canute). Clearly, the *Sami* have adapted and thrived in a harsh environment, but what happens when that traditional lifestyle and resources are denied to them? This was a line of thought that I contemplated when visiting Glasgow in Scotland. Working with academics like Professors John McMurray and Simon Capewell, I had co-authored a series of reports confirming that people in Western Scotland had some of the worst rates of heart disease in Europe (Stewart et al. 2001a, b, c, d, 2002a). Among many factors, including socio-economic status (McAlister et al. 2004), gender biases (MacIntyre et al. 2001) and even quality/capacity variations within the Scottish health system (Stewart et al. 2002b), it was increasingly clear that the (relatively) harsh climate of Western Scotland [type Cfb—moderate sea climate according to the Köppen climate classification system (https://en.climate-data.org/europe/united-kingdom/scotland-257/) was a major influence on the pattern of cardiovascular-related deaths and hospitalisations (Stewart et al. 2002; Murphy et al. 2004). I wondered what really happens when deprived populations are exposed to a climate provocations? Can they truly escape the consequences or do they 'mal-adapt'. These were all front-of-mind when I consciously paid attention to the quality of clothing (both in Liverpool and Glasgow in the UK) with those I'd observed in Sweden and Norway. Concurrently, I counter-pointed the fat-rich diets of those living in the West of Scotland (from fried Mars Bars to fish and chips!) to the seafood-rich food culture of Scandinavia. Extending this line of thought, I was struck by the pervasive stories of 'fuel poverty' in the UK (even among wealthy individuals) compared to those selling them much of the fuel they use to heat their homes—Norwegians.

The above example of the '*have and have nots*' and how it influences our choices and ability to maintain homeostasis and health at the individual level, has parallels to how climate change has developed into a planetary crisis with clear *winners and losers*—this was obviously a major theme in Chap. 1. Under the banner of "*Heatwaves are our deadliest environmental disaster*" the Australian-based 'Sweltering Cities' initiative advocates for a more pro-active response to heatwaves, while highlighting the differential socio-economic impact they will have in the future. As succinctly articulated by one of their members, Dr Rutherford—"*The rich will find their world to be more expensive, inconvenient, uncomfortable, disruptive, colourless and in general more unpleasant, but the poor will die*" (Sweltering Cities Report. https://swelteringcities.org/2024/04/04/summer-survey-2024/. Accessed May 2024). It is perhaps unremarkable that in nature, this same phenomenon of survive through the capacity to adapt or die because of the inability to change occurs. In a century-long analysis of desert mammals and birds co-inhabiting the Mojave Desert, the effects of climate change were evident in the extreme temperatures (with increasing risk of heat radiation exposure and dehydration) evident. However, while the bird population dramatically declined, 'occupational levels' for the small mammal population remained "*unremarkably*" stable. As Riddell and colleagues explain, any modelling of climate change must consider the capacity of each species to "buffer" those changes. In this instance the mammals made changes to their microhabitat to buffer the additional heat they were exposed (Riddell et al. 2021). If anyone is fortunate to visit the famed lemon groves scattered throughout the Sorrento Coast in Italy and sample the locally produced *limoncello*, it is worth considering that the ubiquitous white nets surrounding the lemon trees are now more likely to be protecting them from the burning rays of the sun than damaging frosts; a problem that one of the locals (a florist whose family also have an inter-generational farm) explained is affecting all the plants they attempt to grow. Thus, we are replicating what is happening in nature. Although we may not be aware of doing so, but humans do the same thing for ourselves. The more adaptable we are, the more resilient we are likely to be—just like the small mammals living in the Mojave Desert.

## 2.5 The Bigger Picture—Who is Polluting the Planet and Choking in Return?

Thus far, this chapter has alternated between describing global events to how the human body responds to climatic extremes—thereby seeking to bypass any fatalistic thoughts around our future global climate. It does this by introducing and focusing on the concept that clinicians have the power to promote *climatic resilience* at the individual level. However, this doesn't abrogate our collective responsibility to highlight the drivers of anthropometric damage to our environment and advocate for policies that will (slowly) turn the tide—even if it is less strident as Greta Thunberg or climate activists on TikTok! At the very least, as health professionals who dedicate our lives to improving the lives of the people we care for, it is incumbent for us to at least understand the big picture. This includes who produces the most pollution (and therefore ultimately to blame for our planet's ill health) versus who is most affected by pollution? As will explored in later chapters, while health professionals are mainly ordinary citizens in the face of a global crisis, we do have the collective capacity to both influence and change our current trajectory towards higher rates of morbidity and premature mortality from a global to individual level. As often trusted health professionals and scientists (although this might be challenged given recent historical events in the US!) we can educate the public of the health implications of climate change.

In creating a world to accommodate us (humans) better, including the all-important *Industrial Revolution*, we've created a new problem—a hotter, more dynamic and unpredictable world with more climatic extremes that challenges thousands of years of bio-behavioural adaptation. The irony of the *Industrial Revolution* being almost exclusively about heat generated from carbon-based fossil fuels (from turning water into steam via steam engines and iron into steel from furnaces fed by copious amounts of black and brown coal) should not be lost on anyone. Fundamentally, those initial spot-fires of heat-generating machines spread rapidly around the world and ultimately led to us being able to both cool as well as heat our local environment and therefore thrive (in greater numbers) throughout the world.

As now widely acknowledged, human-derived carbon dioxide ($CO_2$) emissions (the most notorious pollutant collectively referred to as 'greenhouse gases') represents a key barometer of our contribution to a polluted atmosphere. Figure 2.2 (total emissions) and Fig. 2.3 (emissions per capita) provide a global picture of the differential impact of each country (>200 individual countries/principalities) in this regard (Worldometer 2024); highlighting the need to consider the importance of population density, life-styles and reliance/production of polluting technology, resources and industries in apportioning any blame in our achieving aspirational targets such as the Paris Accord and other climate goals (Beggs et al. 2024b). Overall, it is estimated that on average, each human produces 4.76 tonnes of $CO_2$ per annum. When considering the combination of average consumption with population dynamics, the top 10 polluting countries are as follows—(1) China, (2) United States, (3) India, (4) Russia, (5) Japan, (6) Germany, (7) Canada, (8) Iran, (9) South Korea and (10) Indonesia—producing a range of 530 million to 10.4 Billion tonnes of $CO_2$ tonnes per capita/annum. At the other end of the spectrum the least contributors are Greenland, Faeroe Islands, Saint Pierre and Miquelon, Saint Helena, Anguilla, Cook Islands, Palau, Falkland Islands, Kiribati and Sao Tome and Principe—producing between a total of 1530 to 56,195 tonnes of $CO_2$ per annum (Worldometer 2024). However, on a per capita basis, the top 10 polluters are as follows—(1) Qatar, (2) Montenegro, (3) Kuwait, (4) United Arab Emirates, (5) Trinidad and Tobago, (6) Oman, (7) Canada, (8) Brunei, (9) Gibraltar and (10) Luxemburg—individually producing

Tons

10432751400

1530

**Fig. 2.2** Global pattern of $CO_2$ emissions per country (2016 figures). This figure was generated from a website that tracks the levels of carbon production produced on a global scale with 209 individual countries/principalities represented. As per the scale (far bottom left corner), the darker the colour, the higher the total of $CO_2$ emissions (in 2016) produced by that country, As per the shading, a few very populous countries (notably China, India and the United States of America) are major $CO_2$ emitters (Worldometer 2024)

between 17.4 and 38.1 tonnes of $CO_2$ per capita/annum. At the other end of the spectrum the least contributors in this regard are Greenland, the Faeroe Islands, the Democratic Republic of Congo, Mali, Somalia, Ethiopia, Niger, Malawi, Burundi and Chad—producing 0.03–0.11 tonnes of $CO_2$ per capita/annum (Worldometer 2024).

On any given day, depending on the topography and climatic/environmental conditions (e.g. fires in the forests of Indonesia causing smoke and ash plumes to drift over the city nation Singapore located in the South China Sea), the population of a major urban area will be living in a cloud of pollution that will have both dramatic short and long-term effects on their health. Perhaps the most famous city to demonstrate a confluence of human excess, topography/geography and climate is the *City of Angels*—Los Angeles in the United States (Senn 1948). As indicated above, the United States is number two, for the most polluting country in the world (second only to China) and the near 4 million people living in a complex of valleys

and to a coastal plain (all bordered by a series of low, but steep mountains) of Los Angeles are famous for their cars and freeways. As explained in Chap. 6, these factors conspire to produce the perennial layer of yellow/orange smog that lays over Los Angeles unless favourable winds blow them away. So, for the local population, pollution (along with the threat posed by the San Andreas Faultline) poses a constant health hazard. Los Angeles, and more recently the citizens of Beijing, China (given the constant media stories now being reported) are not alone in this regard. As reported by the *World Air Quality Index Project*, the Air Quality Index (AQI) is derived from the regular (hourly) measurement of "*particulate matter ($PM_{2.5}$ and $PM_{10}$), Ozone ($O_3$), Nitrogen Dioxide ($NO_2$), Sulphur Dioxide ($SO_2$) and Carbon Monoxide (CO) emissions*" (World Air Quality Index 2024). Depending on the relative concentration of these particles, bands of AQI (from good to hazardous) are generated. There are various thresholds of poor air quality published across

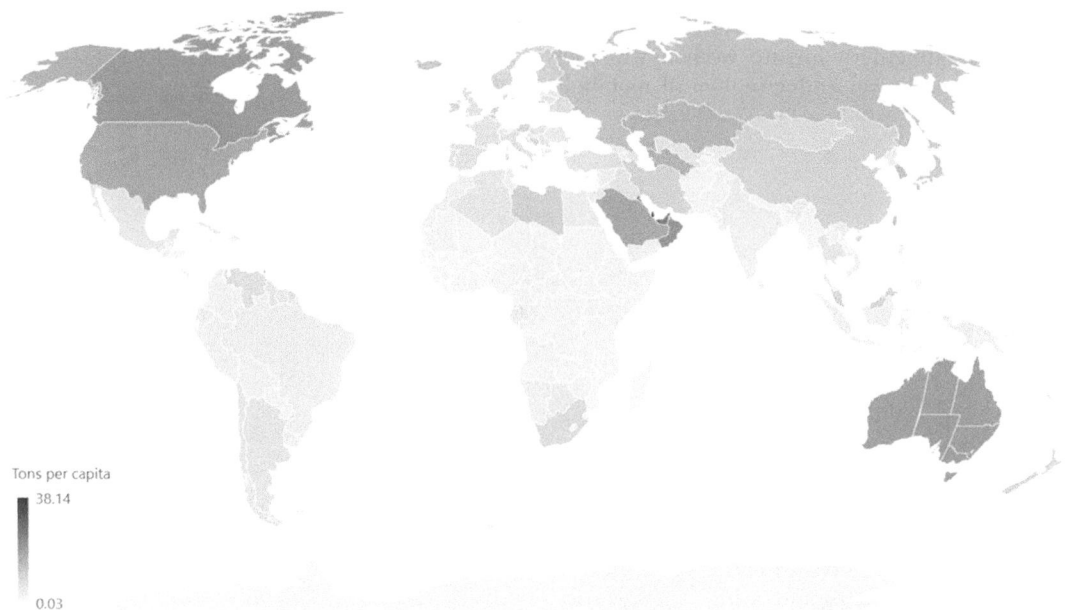

Tons per capita

38.14

0.03

**Fig. 2.3** Global pattern of $CO_2$ emissions per capita (2016 figures). This figure was generated from the same website used to generate Fig. 2.2 that tracks the levels of carbon production produced on a global scale with 209 individual countries/principalities represented. As per the scale (far bottom left corner), the darker the colour, the higher the total of $CO_2$ emissions per person (in 2016) as opposed to total volume, produced by the residents of that country, As per the more "mottled" shading, on a per capita basis, there are many more countries contributing to $CO_2$ emissions with the vast continents of Africa and South America noticeable lighter (Worldometer 2024)

different websites. However, an AQI above 200 is uniformly identified as 'unhealthy' to everyone. Below that, anything above 100 poses a threat to anyone with medical conditions (such as those with respiratory conditions) that can be provoked by airborne pollutants/particles. As shown in Fig. 2.4, in one randomly selected day in April 2024, China featured heavily in the most polluted cities with hazardous air conditions (indicated by the red dots), with numerous other Chinese cities affected by severe (blue dots) and unhealthy (green dots) air pollution. However, China was not alone in this regard. Other countries (who don't appear in the "highest" polluters list from a country to individual perspective) including Chad on the African continent and Kazakhstan in Central Asia appeared on the list (Real-time 100 Most Polluted Cities in the World. https://www.aqi.in/au/real-time-most-polluted-city-ranking). The distribution of cities with 'unhealthy' to 'severely unhealthy' air quality, on that one day, extended from Argentina to Mongolia and from the poorest (Bangladesh and Nepal) to wealthiest countries (United Kingdom and Sweden). Notably, as per increasing media reports, India was also a hotspot for air pollution with 13 cities (compared to a total of 12 Chinese cities) living in highly polluted atmospheres. These population giants (China/India) with 2.8 billion people combined and with aspirations to become world's largest economies to match, play and will continue to play a critical role in any response to climate change, given their relative contributions to pollution. Unlike many of the 'top' polluters named above, their air quality is providing a clear signal of the future. As will be highlighted throughout this book, those living in low-to-middle income countries who are yet to substantially contribute to climate change (see below) still face the consequences for economic development—noting that Nigeria and South Africa (historically, the economic powerhouses of sub-Saharan Africa) are represented in Fig. 2.4.

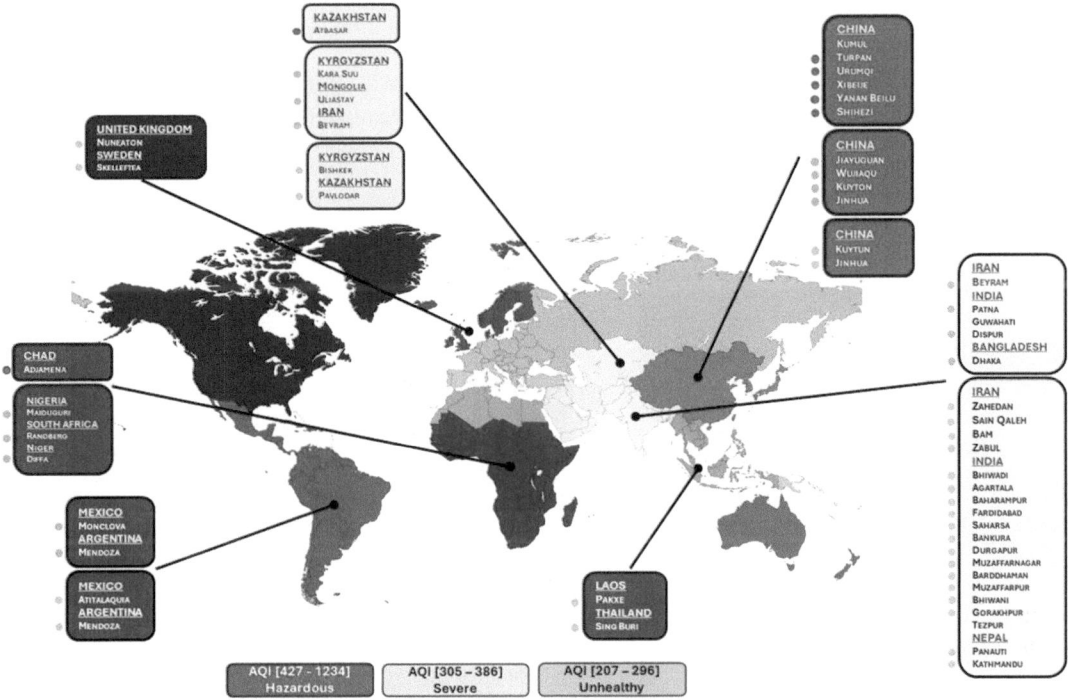

**Fig. 2.4** The most polluted cities in the world. This figure was generated from a website that tracks the air quality (as determined by measured by the Air Quality Index [AQI]) as published by monitoring agencies daily. It shows those cities with the poorest air quality (red dots being hazardous, blue dots being severely unhealthy and green dots unhealthy to everyone) on a global scale. While the biggest "polluters" are represented, there are may lower-to-middle income countries, with under-developed economies (historically at least) that are now living with the consequence of pollution at the local level (Real-time 100 Most Polluted Cities in the World. https://www.aqi.in/au/real-time-most-polluted-city-ranking)

## 2.6  An Evolutionary and Cultural Upheaval—Climate Change

As will be expanded upon in the next chapter, the price for our mastery of the elements and our environment, wherever we choose to apply our human footprint, is a global one—climate change. We started with more efficient ways to produce and then extinguish extreme heat in a factory and have ended up initiating a complex set of global warming phenomena/climate changes that will likely take centuries to reverse. Belying its specific title, it is beyond the scope of this book (and indeed author) to fully explain the mechanisms and nuances of the climate. However, as Mora and colleagues explain in a high-level review of the multiple threats posed by climate change (Mora et al. 2018), by us altering the delicate balance between incoming solar radiation and the subsequent loss of infra-red radiation, with a shift in the latter the most sensitive to human activity, the earth is now retaining more heat and $CO_2$ than it has in thousands/hundreds of thousands of years. The time-scale is unprecedented and, perhaps, running too fast for human evolution and inter-generational adaptations to keep-up. We are the most adaptable species on the planet, but even we are dwarfed by planet-wide changes to the ecology and environment.

At its fundamental level, the release of green-house gases creates an insulating blanket from which radiated heat cannot escape into space (solar energy). Multiplicative effects including the alteration of the convection- and salinity-driven mixing of oxygen-rich and oxygen-poor seawater only exacerbates the problem with acidification of oceans and reduced oxygen levels degrading the sea ecology and, ultimately, likely to impact on our ability to harvest seafood. Although heat increases the rate of water evaporation (predominantly from the planet's vast oceans) and the capacity of the atmosphere to hold moisture (thereby increasing humidity and cloud coverage), this doesn't mean that there is more rain and less drought. Simply put, as shown by the counterpoints of a La Niña (Australian Bureau of Meterology 2024) versus El Niño (Cordero et al. 2024) event—with more rain, colder temperatures and more dynamic weather conditions versus less rain, warmer temperatures/more extreme **bush/wildfire** conditions likely depending on what side of the Pacific Ocean you are located, climate change is inherently fickle and difficult to predict.

**Key Point:** As will be reinforced repeatedly in this book, simply focusing on the 'heating' element of climate change means we are in danger of missing the full spectrum of health provocations (particularly in regards to the cardiovascular system) and preventable events if we fail to understand the complex interaction between health and climatic/environmental conditions that have probably existed for thousands of years and how climate change will be more provocative across the full spectrum of temperature and weather extremes—not just provoke more heatwaves.

Critically, climate change will inevitably bring about more dynamic weather events because of the increased energy (due to more heat, moisture and static energy) and temperature differentials (with increasing wind speeds producing more rapid weather fronts). As dramatically suggested in Fig. 2.1, the oceans are the planet's heat conveyor belts—absorbing and retaining

heat more effectively than the atmosphere and therefore intensifying evaporation, precipitation and windspeeds (noting the critical importance of both ocean currents and the 'rivers of air', otherwise known as jet streams, that redistribute heat and atmospheric pressure on a global scale). Thus, beyond the prospect of more devastating weather events, including floods, hurricanes and even asthma storms (Thien 2018; Price et al. 2023; Beggs 2024), the frequency of more mundane weather changes will also increase. If these were benign from a cardiopulmonary perspective, they would be a welcome relief from hot conditions. Such dynamics are difficult to quantify. Alternatively, the loss of sea ice coverage and the rapid retreat of glaciers, along with degradation of the massive ice sheets covering Greenland and the Antarctic are far more obvious. This loss represents one of the major positive feedback loops underlying climate changes—the Ice-albedo feedback loop, whereby the high albedo of ice reflects incoming solar radiation back into the space. In reverse (i.e., when the planet cools for whatever reason, this mechanism is thought to trigger and maintain an ice-age) (https://www.unep.org/facts-about-climate-emergency?gad_source=1&gclid=Cj0KCQjwq86wBhDiARIsAJhuphmq6lFXvMLipqQIaPKDeBSHuiFG47pbdzgQNuOvItQxSHf29LoerhsaAsF9EALw_wcB). Other natural phenomena that reflect incoming solar radiation includes cloud cover (up to one third of solar energy). This represents a 'negative feedback loop', if, as expected, warmer temperatures increase evaporation and greater cloud cover. According to the *Stefan-Boltzmann Law*, the higher global temperatures reach, the higher the temperature differential with space and heat transfer—this represents yet another 'negative feedback' mechanism that will be provoked by climate change. Over time, both chemical weathering of rocks and increasing absorption of $CO_2$ into the oceans (to form carbonic acid) will likely sequester much of the excess $CO_2$ we've released as humans (https://www.unep.org/facts-about-climate-emergency?gad_source=1&gclid=Cj0KCQjwq86wBhDiARIsAJhuphmq6lFXvMLipqQIaPKDeBSHuiFG47pbdzgQNuOvI

tQxSHf29LoerhsaAsF9EALw_wcB). However, these negative feedback mechanisms work on geological timescales (millions of years) *not* human lifespans. More dramatic (positive) feedback mechanisms, including the melting of permafrost (thereby releasing the greenhouse gas Methane), loss of vegetation due to fire and drought and sea level rises, have already become evident in our lifetimes (https://www.unep.org/facts-about-climate-emergency?gad_source=↖1&gclid=Cj0KCQjwq86wBhDiARIsAJhuphmq6lFXvMLipqQIaPKDeBSHuiFG47pbdzgQNuOvItQxSHf29LoerhsaAsF9EALw_wcB). It is perhaps why the allegory of King Canute is so compelling (to me at least)!

As further identified by Mora and colleagues (Mora et al. 2018), the price to pay in terms of human suffering and loss of life, is almost too numerous to contemplate. However, beyond those mechanisms directly linking climate change leading to more/exacerbated heart disease (Khraishah et al. 2022; Kapp and McGuire 1960; Clearfield et al. 2014; Solomon and Landrigan 2024) (most of which is covered in this book) there are many identified health consequences, that will undoubtedly lead to more heart disease and, or exacerbate the impact of concurrent heart disease. This includes (with key examples) the following:

- Increasingly poor air quality [there are direct pathophysiological mechanisms leading from respiratory disease to pulmonary heart disease (Mocumbi et al. 2024)].
- More pathogenic diseases [rheumatic heart disease with its pathological process of cardiac inflammation and auto-immune induced scarring of the cardiac valves from group A streptococci infection (Karthikeyan et al. 2023; Marijon et al. 2021; Madeira et al. 2017; Zuhlke et al. 2015)].
- Movement of Tropical Zones/More vector-borne diseases [populations at risk from Chagas' disease, a parasitic infection that causes millions of cases of Chagas Cardiomyopathy in South America (Gomez-Ochoa et al. 2024; Silva et al. 2024), and human African trypanosomiasis (Barrett et al.

2024; Franco et al. 2024) in Sub-Saharan Africa are likely to increase].
- More Food Insecurity/Childhood Malnutrition [resulting in impaired renal function and endothelial early in life and then higher blood pressure levels and greater risk of hypertension later in life (Ogah et al. 2023)].
- More anxiety/depression [it is increasingly recognised, through the study of conditions such as Tako-Tsubo Cardiomyopathy, that major events, such as major storm, can trigger a stress response that is detrimental to the heart from both a structural (Smeijers et al. 2015) and autonomic perspective (Sado et al. 2018)].

In later chapters, these and the more obvious adverse effects of climate change on heart health/disease are discussed from an individual to population perspective, along with some preliminary ideas on how clinicians/health professionals can alter their perspective and practice to mitigate the impact of climate change among the people they care for. As will also be noted, however, we may be surprised by some unexpected benefits from climate change—something any scientist should be willing to contemplate given that the future remains challenging to predict with certainty.

## 2.7  Climate Change—Time for a 'Healthy' Response

This chapter has provided a 'opus' view of heart disease within an evolving picture/framework of climate change from a personal to global perspective. It's a truism that one cannot address a problem until that problem is recognised. For any health professional/scientist seeking to understand the depth of scientific evidence underpinning reports of human-induced climate change, one doesn't have to look much further than the United Nation's Environment Programme knowledge repository to recognise there is a problem (Fig. 2.5). Even if you remain a '*climate sceptic*' the issue of human illnesses being influenced by prevailing environmental

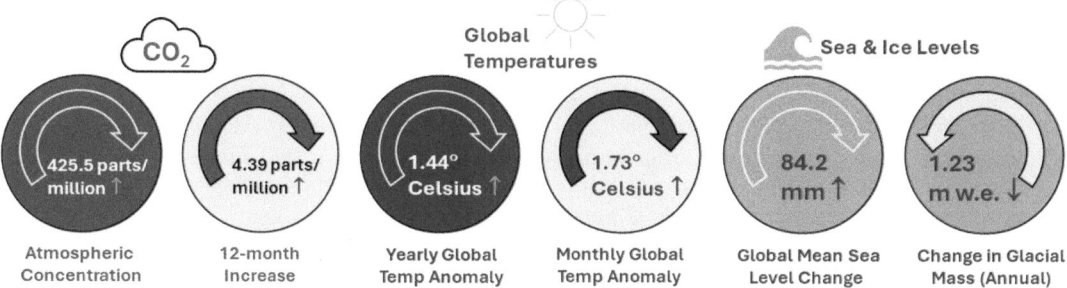

**Fig. 2.5** Climate health monitoring panel. The United Nation's Environment Programme's "*World Environment Situation Room*" provides, a running tally of the key metrics of human-induced climate change. Data from the website (for April 2024) are represented in this figure (https://wesr.unep.org/climate/. Accessed April 2024)

and climatic conditions (particularly in respect to ebbs and flows in hospital episodes and mortality) is one hard to ignore. This particularly applies to most forms of heart and cerebrovascular disease. The next chapter explores how this phenomenon occurs and how it relates to clinical management, rather than being a statistical curiosity.

# References

Abovich A, Matasic DS, Cardoso R, Ndumele CE, Blumenthal RS, Blankstein R, Gulati M. The AHA/ACC/HFSA 2022 heart failure guidelines: changing the focus to heart failure prevention. Am J Prev Cardiol. 2023;15:100527.

Ahlbom A. Seasonal variations in the incidence of acute myocardial infarction in Stockholm. Scand J Soc Med. 1979;7:127–30.

Ahn J, Uhm T, Han J, Won KM, Choe JC, Shin JY, Park JS, Lee HW, Oh JH, Choi JH, Lee HC, Cha KS, Hong TJ, Kim YH. Meteorological factors and air pollutants contributing to seasonal variation of acute exacerbation of atrial fibrillation: a population-based study. J Occup Environ Med. 2018;60:1082–6.

Akioka H, Yufu K, Teshima Y, Kawano K, Ishii Y, Abe I, Kondo H, Saito S, Fukui A, Okada N, Nagano Y, Shinohara T, Nakagawa M, Hara M, Takahashi N. Seasonal variations of weather conditions on acute myocardial infarction onset: Oita AMI registry. Heart Vessels. 2019;34:9–18.

Ansa VO, Ekott JU, Essien IO, Bassey EO. Seasonal variation in admission for heart failure, hypertension and stroke in Uyo, South-Eastern Nigeria. Ann Afr Med. 2008;7:62–6.

Arellano-Nava B, Halloran PR, Boulton CA, Scourse J, Butler PG, Reynolds DJ, Lenton TM. Destabilisation of the subpolar North Atlantic prior to the little ice age. Nat Commun. 2022;13:5008.

Asher E, Todt M, Rosenlof K, Thornberry T, Gao R, Taha G, Walter P, Alvarez S, Flynn J, Davis S, Evan S, Brioude J, Metzger J-M, Hurst DF, Hall E, Xiong E. Unexpected rapid aerosol formation in the Hunga Tonga plume. Proceed Natl Acad Sci. 2023;5:471.

Atherton JJ, Sindone A, De Pasquale CG, Driscoll A, MacDonald PS, Hopper I, Kistler P, Briffa TG, Wong J, Abhayaratna WP, Thomas L, Audehm R, Newton PJ, O'Loughlin J, Connell C, Branagan M. National heart foundation of Australia and cardiac society of Australia and New Zealand: Australian clinical guidelines for the management of heart failure 2018. Med J Aust. 2018;209:363–9.

Australian Bureau of Meterology. What is La Nina and does it impact Australia; 2024. http://www.bom.gov.au/climate/updates/articles/a020.shtml. Accessed May 2024.

Barrett MP, Priotto G, Franco JR, Lejon V, Lindner AK. Elimination of human African trypanosomiasis: the long last mile. PLoS Negl Trop Dis. 2024;18:e0012091.

Beggs PJ. Thunderstorm asthma and climate change. JAMA. 2024;331:878–9.

Beggs PJ, Trueck S, Linnenluecke MK, Bambrick H, Capon AG, Hanigan IC, Arriagada NB, Cross TJ, Friel S, Green D, Heenan M, Jay O, Kennard H, Malik A, McMichael C, Stevenson M, et al. The 2023 report of the MJA-lancet countdown on health and climate change: sustainability needed in Australia's health care sector. Med J Aust. 2024a;220:282–303.

Beggs PJ, Oliveira C, Giudice C. The United Nations framework convention on climate change (UNFCCC) 28th conference of the parties, Dubai (COP28): implications for lung disease. Respirology. 2024b;30:4871.

Boldor N, Bar-Dayan Y, Rosenbloom T, Shemer J, Bar-Dayan Y. Optimism of health care workers during a disaster: a review of the literature. Emerg Health Threats J. 2012;5:323.

Boulay F, Berthier F, Sisteron O, Gendreike Y, Gibelin P. Seasonal variation in chronic heart failure hospitalizations and mortality in France. Circulation. 1999;100:280–6.

Brode P, Fiala D, Blazejczyk K, Holmer I, Jendritzky G, Kampmann B, Tinz B, Havenith G. Deriving the operational procedure for the Universal Thermal Climate Index (UTCI). Int J Biometeorol. 2012;56:481–94.

Clearfield M, Pearce M, Nibbe Y, Crotty D, Wagner A. The, "New Deadly Quartet" for cardiovascular disease in the 21st century: obesity, metabolic syndrome, inflammation and climate change: how does statin therapy fit into this equation? Curr Atheroscler Rep. 2014;16:380.

Cordero RR, Feron S, Damiani A, Carrasco J, Karas C, Wang C, Kraamwinkel CT, Beaulieu A. Extreme fire weather in Chile driven by climate change and El Nino-Southern Oscillation (ENSO). Sci Rep. 2024;14:1974.

Crompton D, Kohleis P, Shakespeare-Finch J, FitzGerald G, Young R. Opportunistic mental health screening: is there a role following a disaster? Lessons from the 2010–2011 Queensland (Australia) floods and cyclones. Prehosp Disaster Med. 2023;38:223–31.

Deng J, Dai A. Arctic sea ice-air interactions weaken El Nino-Southern Oscillation. Sci Adv. 2024;10:eadk3990.

Deshmukh AJ, Pant S, Kumar G, Hayes K, Badheka AO, Dabhadkar KC, Paydak H. Seasonal variations in atrial fibrillation related hospitalizations. Int J Cardiol. 2013;168:1555–6.

Do V, Chen C, Benmarhnia T, Casey JA. Spatial heterogeneity of the respiratory health impacts of wildfire smoke PM(2.5) in California. Geohealth. 2024;8:e2023GH000997.

Filho WL, editor. Handbook of climate change adaptation. Heidelberg: Springer; 2011. https://doi.org/10.1007/978-3-642-38670-1

Ford B, Dore M, Bartlett B. Management of heart failure: updated guidelines from the AHA/ACC. Am Fam Physician. 2023;108:315–20.

Franco JR, Priotto G, Paone M, Cecchi G, Ebeja AK, Simarro PP, Sankara D, Metwally SBA, Argaw DD. The elimination of human African trypanosomiasis: monitoring progress towards the 2021–2030 WHO road map targets. PLoS Negl Trop Dis. 2024;18:e0012111.

Frost L, Johnsen SP, Pedersen L, Husted S, Engholm G, Sorensen HT, Rothman KJ. Seasonal variation in hospital discharge diagnosis of atrial fibrillation: a population-based study. Epidemiology. 2002;13:211–5.

Fujii T, Arima H, Takashima N, Kita Y, Miyamatsu N, Tanaka-Mizuno S, Shitara S, Urushitani M, Miura K, Nozaki K. Seasonal variation in incidence of stroke in a general population of 1.4 million Japanese: the Shiga stroke registry. Cerebrovasc Dis. 2022;51:75–81.

Furukawa Y. Meteorological factors and seasonal variations in the risk of acute myocardial infarction. Int J Cardiol. 2019;294:13–4.

Gomez-Ochoa SA, Rojas LZ, Hernandez-Vargas JA, Trujillo-Caceres SJ, Hurtado-Ortiz A, Licht-Ardila M, et al. Myocardial fibrosis by magnetic resonance and outcomes in chagas disease: a systematic review and meta-analysis. JACC Cardiovasc Imag. 2024;17:552–5.

Gragnano F, De Sio V, Calabro P. What is new in the 2023 AHA/ACC multisociety guideline on chronic coronary disease? Eur Heart J Cardiovasc Pharmacother. 2023;9:673–8.

Grigorieva EA, Alexeev VA, Walsh JE. Universal thermal climate index in the Arctic in an era of climate change: Alaska and Chukotka as a case study. Int J Biometeorol. 2023;67:1703–21.

Grossardt BR, Bower JH, Geda YE, Colligan RC, Rocca WA. Pessimistic, anxious, and depressive personality traits predict all-cause mortality: the Mayo Clinic cohort study of personality and aging. Psychosom Med. 2009;71:491–500.

Han MH, Yi HJ, Kim YS, Kim YS. Effect of seasonal and monthly variation in weather and air pollution factors on stroke incidence in Seoul Korea. Stroke. 2015;46:927–35.

Hasnain MG, Garcia-Esperon C, Tomari YK, Walker R, Saluja T, Rahman MM, Boyle A, Levi CR, Naidu R, Filippelli G, Spratt NJ. Bushfire-smoke trigger hospital admissions with cerebrovascular diseases: evidence from 2019–2020 bushfire in Australia. Eur Stroke J. 2024;1:23969873231223308.

Heaney A, Stowell JD, Liu JC, Basu R, Marlier M, Kinney P. Impacts of fine particulate matter from wildfire smoke on respiratory and cardiovascular health in California. Geohealth. 2022;6:e2021GH000578.

https://www.epa.gov/climate-indicators/climate-change-indicators-sea-surface-temperature. Accessed July 2024.

https://www.wired.com/story/climate-change-tiktok-science-communication/2021. Accessed May 2024

Hu HM, Shen CC, Chiang JCH, Trouet V, Michel V, Tsai HC, Valensi P, Spotl C, Starnini E, Zunino M, Chien WY, Sung WH, Chien YT, Chang P, Korty R. Split westerlies over Europe in the early little ice age. Nat Commun. 2022;13:4898.

Humbert M, Kovacs G, Hoeper MM, Badagliacca R, Berger RMF, Brida M, Carlsen J, Coats AJS, Escribano-Subias P, Ferrari P, Ferreira DS, Ghofrani HA, Giannakoulas G, Kiely DG, Mayer E, et al. 2022 ESC/ERS guidelines for the diagnosis and treatment of pulmonary hypertension. Eur Heart J. 2022;43:3618–731.

Humbert M, Kovacs G, Hoeper MM, Badagliacca R, Berger RMF, Brida M, Carlsen J, Coats AJS, Escribano-Subias P, et al. 2022 ESC/ERS guidelines for the diagnosis and treatment of pulmonary hypertension. Eur Respir J. 2023;61:491.

Inglis SC, Clark RA, Shakib S, Wong DT, Molaee P, Wilkinson D, Stewart S. Hot summers and heart failure: seasonal variations in morbidity and mortality in Australian heart failure patients (1994–2005). Eur J Heart Fail. 2008;10:540–9.

Jakovljevic D, Salomaa V, Sivenius J, Tamminen M, Sarti C, Salmi K, Kaarsalo E, Narva V, Immonen-Raiha P, Torppa J, Tuomilehto J. Seasonal variation in the occurrence of stroke in a Finnish adult population The FINMONICA stroke register Finnish monitoring trends and determinants in cardiovascular disease. Stroke. 1996;27:1774–17749.

Jiang N, Zhu C, Hu ZZ, McPhaden MJ, Chen D, Liu B, Ma S, Yan Y, Zhou T, Qian W, Luo J, Yang X, Liu F, Zhu Y. Enhanced risk of record-breaking regional temperatures during the 2023–2024 El Nino. Sci Rep. 2024;14:2521.

Kapp LA, McGuire JK. The influence of climate on patients with cardiovascular disease. GP. 1960;22:88–99.

Karagiannis A, Tziomalos K, Mikhailidis DP, Semertzidis P, Kountana E, Kakafika AI, Pagourelias ED, Athyros VG. Seasonal variation in the occurrence of stroke in Northern Greece: a 10 year study in 8204 patients. Neurol Res. 2010;32:326–31.

Karthikeyan G, Watkins D, Bukhman G, Cunningham MW, Haller J, Masterson M, Mensah GA, Mocumbi A, Muhamed B, Okello E, Sotoodehnia N, Machipisa T, Ralph A, Wyber R, Beaton A. Research priorities for the secondary prevention and management of acute rheumatic fever and rheumatic heart disease: a National Heart, Lung, and Blood Institute workshop report. BMJ Glob Health. 2023;8:7464.

Keller K, Hobohm L, Munzel T, Ostad MA. Sex-specific differences regarding seasonal variations of incidence and mortality in patients with myocardial infarction in Germany. Int J Cardiol. 2019;287:132–8.

Khraishah H, Alahmad B, Ostergard RL, AlAshqar A, Albaghdadi M, Vellanki N, Chowdhury MM, Al-Kindi SG, Zanobetti A, Gasparrini A, Rajagopalan S. Climate change and cardiovascular disease: implications for global health. Nat Rev Cardiol. 2022;19:798–812.

Kirchhof P, Benussi S, Kotecha D, Ahlsson A, Atar D, Casadei B, Castella M, Diener HC, Heidbuchel H, Hendriks J, Hindricks G, Manolis AS, Oldgren J, Popescu BA, Schotten U, et al. 2016 ESC guidelines for the management of atrial fibrillation developed in collaboration with EACTS. Europace. 2016;18:1609–78.

Kirchhof P, Benussi S, Kotecha D, Ahlsson A, Atar D, Casadei B, Castella M, Diener HC, Heidbuchel H, Hendriks J, Hindricks G, Manolis AS, Oldgren J, et al. 2016 ESC guidelines for the management of atrial fibrillation developed in collaboration with EACTS. Rev Esp Cardiol. 2017;70:50.

Konemann H, Ellermann C, Zeppenfeld K, Eckardt L. Management of ventricular arrhythmias worldwide: comparison of the latest ESC, AHA/ACC/HRS, and CCS/CHRS guidelines. JACC Clin Electrophysiol. 2023;9:715–28.

Kriszbacher I, Boncz I, Koppan M, Bodis J. Seasonal variations in the occurrence of acute myocardial infarction in Hungary between 2000 and 2004. Int J Cardiol. 2008;129:251–4.

Krum H, Jelinek MV, Stewart S, Sindone A, Atherton JJ, et al. 2011 update to national heart foundation of Australia and cardiac society of Australia and New Zealand guidelines for the prevention, detection and management of chronic heart failure in Australia, 2006. Med J Aust. 2011;194:405–9.

Kupari M, Koskinen P. Seasonal variation in occurrence of acute atrial fibrillation and relation to air temperature and sale of alcohol. Am J Cardiol. 1990;66:1519–20.

Li Y, Zhou Z, Chen N, He L, Zhou M. Seasonal variation in the occurrence of ischemic stroke: a meta-analysis. Environ Geochem Health. 2019;41:2113–30.

Loomba RS. Seasonal variation in paroxysmal atrial fibrillation: a systematic review. J Atr Fibrillation. 2015;7:1201.

MacIntyre K, Stewart S, Capewell S, Chalmers JW, Pell JP, Boyd J, Finlayson A, Redpath A, Gilmour H, McMurray JJ. Gender and survival: a population-based study of 201,114 men and women following a first acute myocardial infarction. J Am Coll Cardiol. 2001;38:729–35.

Madeira G, Chicavel D, Munguambe A, Langa J, Mocumbi A. Streptococcal pharyngitis in children with painful throat: missed opportunities for rheumatic heart disease prevention in endemic area of Africa. Cardiovasc Diagn Ther. 2017;7:421–3.

Mann ME, Zhang Z, Rutherford S, Bradley RS, Hughes MK, Shindell D, Ammann C, Faluvegi G, Ni F. Global signatures and dynamical origins of the little ice age and medieval climate anomaly. Science. 2009;326:1256–60.

Marijon E, Mocumbi A, Narayanan K, Jouven X, Celermajer DS. Persisting burden and challenges of rheumatic heart disease. Eur Heart J. 2021;42:3338–48.

Martin SS, Aday AW, Almarzooq ZI, Anderson CAM, Arora P, Avery CL, Baker-Smith CM, et al. 2024 Heart disease and stroke statistics: a report of US and global data from the American heart association. Circulation. 2024;149:e347–913.

Martinez-Selles M, Garcia Robles JA, Prieto L, Serrano JA, Munoz R, Frades E, Almendral J. Annual rates of admission and seasonal variations in hospitalizations for heart failure. Eur J Heart Fail. 2002;4:779–86.

Masri S, Jin Y, Wu J. Compound risk of air pollution and heat days and the influence of wildfire by SES across California, 2018–2020: implications for environmental justice in the context of climate change. Climate. 2022;10:1332.

Matsuda H, Kuragaichi T, Sato Y. Investigating the seasonal variation of heart failure hospitalizations and in-hospital mortality risks in Japan using a nationwide database. J Cardiol. 2024;83:236–42.

Maynard NG, Conway GA. A view from above: use of satellite imagery to enhance our understanding of potential impacts of climate change on human health in the Arctic. Alaska Med. 2007;49:38–43.

McAlister FA, Murphy NF, Simpson CR, Stewart S, MacIntyre K, Kirkpatrick M, Chalmers J, Redpath A, Capewell S, McMurray JJ. Influence of socioeconomic deprivation on the primary care burden and treatment of patients with a diagnosis of heart failure in general practice in Scotland: population based study. BMJ. 2004;328:1110.

McDonagh TA, Metra M, Adamo M, Gardner RS, Baumbach A, Bohm M, Burri H, Butler J, Celutkiene J, Chioncel O, Cleland JGF, Coats AJS, Crespo-Leiro MG, Farmakis D, Gilard M, et al. 2021 ESC guidelines for the diagnosis and treatment of acute and chronic heart failure. Eur Heart J. 2021;42:3599–726.

Mellish S, Ryan JC, Litchfield CA. Short-term psychological outcomes of Australia's 2019/20 bushfire season. Psychol Trauma. 2024;16:292–302.

Mocumbi AO, Cebola B, Muloliwa A, Sebastiao F, Sitefane SJ, Manafe N, Dobe I, Lumbandali N, Keates A, Stickland N, Chan YK, Stewart S. Differential patterns of disease and injury in Mozambique: new perspectives from a pragmatic, multicenter, surveillance study of 7809 emergency presentations. PLoS ONE. 2019;14:e0219273.

Mocumbi A, Humbert M, Saxena A, Jing ZC, Sliwa K, Thienemann F, Archer SL, Stewart S. Pulmonary hypertension. Nat Rev Dis Primers. 2024;10:1.

Mokhtar K, Chuah LF, Abdullah MA, Oloruntobi O, Ruslan SMM, Albasher G, Ali A, Akhtar MS. Assessing coastal bathymetry and climate change impacts on coastal ecosystems using Landsat 8 and Sentinel-2 satellite imagery. Environ Res. 2023;239:117314.

Mora C, Spirandelli D, Franklin EC, et al. Broad threat to humanity from cumulative climate hazards intensified by greenhouse gas emissions. Nat Clim Change. 2018;8:1062–71.

Murphy NF, Stewart S, MacIntyre K, Capewell S, McMurray JJ. Seasonal variation in morbidity and mortality related to atrial fibrillation. Int J Cardiol. 2004;97:283–8.

Naqvi HR, Mutreja G, Shakeel A, Singh K, Abbas K, Naqvi DF, Chaudhary AA, Siddiqui MA, Gautam AS, Gautam S, Naqvi AR. Wildfire-induced pollution and its short-term impact on COVID-19 cases and mortality in California. Gondwana Res. 2023;114:30–9.

National Centers for Environmental Information (National Oceanic and Atmospheric Administration)—Southern Oscillation Index; 2024. https://www.ncei.noaa.gov/access/monitoring/enso/soi. Accessed June 2024.

National Hurricane Center; 2024. https://www.nhc.noaa.gov/. Accessed June 2024

National Oceanic and Atmospheric Administration. Volcano News; 2023. https://csl.noaa.gov/news/2023/393_1220.html. Accessed May 2024.

Nganou-Gnindjio CN, Awah-Epoupa RA, Wafeu-Sadeu G, Tchapmi-Njeunje DP, Endomba-Angong FT, Menanga AP. Seasonal variation of decompensated heart failure admissions and mortality rates in sub-Saharan Africa, Cameroon. Ann Cardiol Angeiol. 2021;70:148–52.

Norris JR, Allen RJ, Evan AT, Zelinka MD, O'Dell CW, Klein SA. Evidence for climate change in the satellite cloud record. Nature. 2016;536:72–5.

OECD. Doctors by age, sex and category; 2024. https://www.oecd-ilibrary.org/sites/aa9168f1-en/index.html?itemId=/content/component/aa9168f1-en. Accessed May 2024.

Ogah OS, Oguntade AS, Chukwuonye II, Onyeonoro UU, Madukwe OO, Asinobi A, Ogah F, Orimolade OA, Babatunde AO, Okeke MF, Attah OP, Ebengho IG, Sliwa K, Stewart S. Childhood and Infant exposure to famine in the Biafran war is associated with hypertension in later life: the Abia NCDS study. J Hum Hypertens. 2023;37:936–43.

Oida M, Suzuki S, Arita T, Yagi N, Otsuka T, Kishi M, Semba H, Kano H, Matsuno S, Kato Y, Uejima T, Oikawa Y, Hoshino S, Matsuhama M, Inoue T, Yajima J, Yamashita T. Seasonal variations in the incidence of ischemic stroke, extracranial and intracranial hemorrhage in atrial fibrillation patients. Circ J. 2020;84:1701–8.

Ommen SR, Ho CY, Asif IM, Balaji S, Burke MA, Day SM, Dearani JA, Epps KC, Evanovich L, Ferrari VA, Joglar JA, Khan SS, Kim JJ, Kittleson MM, Krittanawong C, Martinez MW, Mital S, Naidu SS, Saberi S, Semsarian C, Times S, Waldman CB. 2024 AHA/ACC/AMSSM/HRS/PACES/SCMR guideline for the management of hypertrophic cardiomyopathy: a report of the American heart association/American college of cardiology joint committee on clinical practice guidelines. Circulation. 2024;51:6541.

Ong GJ, Sellers A, Mahadavan G, Nguyen TH, Worthley MI, Chew DP, Horowitz JD. Bushfire Season' in Australia: determinants of increases in risk of acute coronary syndromes and takotsubo syndrome. Am J Med. 2023;136:88–95.

Palinkas LA. Global climate change, population displacement and public health. Cham: Springer; 2020. https://doi.org/10.1007/978-3-642-40455-9.

Peters A, Schneider A. Cardiovascular risks of climate change. Nat Rev Cardiol. 2021;18:1–2.

Price D, Hughes KM, Dona DW, Taylor PE, Morton DAV, Stevanovic S, Thien F, Choi J, Torre P, Suphioglu C. The perfect storm: temporal analysis of air during the world's most deadly epidemic thunderstorm asthma (ETSA) event in Melbourne. Ther Adv Respir Dis. 2023;17:17534666231186726.

Primeau C, Homoe P, Lynnerup N. Temporal changes in childhood health during the medieval little ice age in Denmark. Int J Paleopathol. 2019;27:80–7.

Real-time 100 Most Polluted Cities in the World; 2024. https://www.aqi.in/au/real-time-most-polluted-city-ranking. Accessed 12th April 2024.

Renner R. NASA's flagship satellite will revolutionize study of climate change. Environ Sci Technol. 1999;33:271A-A272.

Riddell EA, Iknayan KJ, Hargrove L, Tremor S, Patton JL, Ramirez R, Wolf BO, Beissinger SR. Exposure to climate change drives stability or collapse of desert mammal and bird communities. Science. 2021;371:633–6.

Roth GA, Mensah GA, Johnson CO, Addolorato G, Ammirati E, Baddour LM, Barengo NC, Beaton AZ, Benjamin EJ, Benziger CP, Bonny A, Brauer M, Brodmann M, Cahill TJ, Carapetis J, et al. Global burden of cardiovascular diseases and risk factors, 1990–2019: update from the GBD 2019 study. J Am Coll Cardiol. 2020;76:2982–3021.

Sado J, Kiyohara K, Iwami T, Kitamura Y, Ando E, Ohira T, Sobue T, Kitamura T. Three-year follow-up after the great east Japan earthquake in the incidence of out-of-hospital cardiac arrest with cardiac origin. Circ J. 2018;82:919–22.

Senn CL. General atmospheric pollution; Los Angeles smog. Am J Public Health Nations Health. 1948;38:962–5.

Shah M, Patnaik S, Patel B, Arora S, Patel N, Garg L, Agrawal S, Martinez MW, Figueredo VM. Regional and seasonal variations in heart failure admissions and mortality in the USA. Arch Cardiovasc Dis. 2018;111:297–301.

Sheehy S, Fonarow GC, Holmes DN, Lewis WR, Matsouaka RA, Piccini JP, Zhi L, Bhatt DL. Seasonal variation of atrial fibrillation admission and quality of care in the United States. J Am Heart Assoc. 2022;11:e023110.

Silva GGD, Lopez VM, Vilarinho AC, Datto-Liberato FH, Oliveira CJF, Poulin R, Guillermo-Ferreira R. Vector species richness predicts local mortality rates from Chagas disease. Int J Parasitol. 2024;54:139–45.

Sliwa K, Viljoen CA, Stewart S, Miller MR, Prabhakaran D, Kumar RK, Thienemann F, Piniero D, Prabhakaran P, Narula J, Pinto F. Cardiovascular disease in low- and middle-income countries associated with environmental factors. Eur J Prev Cardiol. 2024;31:688–97.

Smeijers L, Szabo BM, van Dammen L, Wonnink W, Jakobs BS, Bosch JA, Kop WJ. Emotional, neurohormonal, and hemodynamic responses to mental stress in Tako-Tsubo cardiomyopathy. Am J Cardiol. 2015;115:1580–6.

Smith WL, Weisz E, Knuteson R, Revercomb H, Feldman D. Retrieving decadal climate change from satellite radiance observations: a 100-year CO(2) doubling OSSE demonstration. Sensors. 2020;20:13352.

Smithsonian Institute NMoNH. Global vulcanism program. https://volcano.si.edu/volcano.cfm?vn=262000. Accessed May 2024.

Solomon CG, Landrigan PJ. Fossil fuels, climate change, and cardiovascular disease: a call to action. Circulation. 2024;149:1400–1.

Spengos K, Vemmos K, Tsivgoulis G, Manios E, Zakopoulos N, Mavrikakis M, Vassilopoulos D. Diurnal and seasonal variation of stroke incidence in patients with cardioembolic stroke due to atrial fibrillation. Neuroepidemiology. 2003a;22:204–10.

Spengos K, Vemmos KN, Tsivgoulis G, Synetos A, Zakopoulos N, Zis VP, Vassilopoulos D. Seasonal variation of hospital admissions caused by acute stroke in Athens, Greece. J Stroke Cerebrovasc Dis. 2003b;12:93–6.

Srinivas G, Vialard J, Liu F, Voldoire A, Izumo T, Guilyardi E, Lengaigne M. Dominant contribution of atmospheric nonlinearities to ENSO asymmetry and extreme El Nino events. Sci Rep. 2024;14:8122.

Stewart S, Hart CL, Hole DJ, McMurray JJ. Population prevalence, incidence, and predictors of atrial fibrillation in the Renfrew/Paisley study. Heart. 2001a;86:516–21.

Stewart S, MacIntyre K, Hole DJ, Capewell S, McMurray JJ. More 'malignant' than cancer? Five-year survival following a first admission for heart failure. Eur J Heart Fail. 2001b;3:315–22.

Stewart S, MacIntyre K, MacLeod MM, Bailey AE, Capewell S, McMurray JJ. Trends in hospital activity, morbidity and case fatality related to atrial fibrillation in Scotland, 1986–1996. Eur Heart J. 2001c;22:693–701.

Stewart S, MacIntyre K, MacLeod MM, Bailey AE, Capewell S, McMurray JJ. Trends in hospitalization for heart failure in Scotland, 1990–1996: an epidemic that has reached its peak? Eur Heart J. 2001d;22:209–17.

Stewart S, McIntyre K, Capewell S, McMurray JJ. Heart failure in a cold climate. Seasonal variation in heart failure-related morbidity and mortality. J Am Coll Cardiol. 2002;39:760–6.

Stewart S, Hart CL, Hole DJ, McMurray JJ. A population-based study of the long-term risks associated with atrial fibrillation: 20-year follow-up of the Renfrew/Paisley study. Am J Med. 2002a;113:359–64.

Stewart S, Demers C, Murdoch DR, McIntyre K, MacLeod ME, Kendrick S, Capewell S, McMurray JJ. Substantial between-hospital variation in outcome following first emergency admission for heart failure. Eur Heart J. 2002b;23:650–7.

Stewart S, Keates AK, Redfern A, McMurray JJV. Seasonal variations in cardiovascular disease. Nat Rev Cardiol. 2017;14:654–64.

Swampillai J, Wijesinghe N, Sebastian C, Devlin GP. Seasonal variations in hospital admissions for ST-elevation myocardial infarction in New Zealand. Cardiol Res. 2012;3:205–8.

Sweltering Cities Report. https://swelteringcities.org/2024/04/04/summer-survey-2024/. Accessed May 2024.

The ATOC Consortium. Ocean climate change: comparison of acoustic tomography, satellite altimetry, and modeling. Science. 1998;281:1327–32.

The Climate Clock. https://climateclock.world/. Accessed May 2024

The Doomsday Clock. https://time.com/6249856/dooms-day-clock-catastrophe-ukraine/. Accessed May 2024

Thien F. Melbourne epidemic thunderstorm asthma event 2016: lessons learnt from the perfect storm. Respirology. 2018;23:976–7.

Tian Y, Xiang M, Peng J, Duan Y, Wen Y, Huang S, Li L, Yu S, Cheng J, Zhang X, Wang P. Modification effects of seasonal and temperature variation on the association between exposure to nitrogen dioxide and ischemic stroke onset in Shenzhen, China. Int J Biometeorol. 2022;66:1747–58.

Tollefson J. Climate scientists push for access to world's biggest supercomputers to build better Earth models. Nature. 2023;23:2249. https://doi.org/10.1038/d41586-023-02249-6.

Tsuji T, Nagata M, Ogawa M. Epidemiologic studies on seasonal variation in mortality and morbidity of stroke. Nihon Eiseigaku Zasshi. 1975;30:185.

UK Climate Classification; 2024. https://en.climate-data.org/europe/united-kingdom/scotland-257/. Accessed May 2024.

United Nations. Synergy solutions for a world in crisis: tackling climate and SDG action together report on strengthening the evidence base. 2023. https://sdgs.un.org/sites/default/files/2023-09/UN%20Climate%20SDG%20Synergies%20Report-091223B_1.pdf. Accessed May 2024.

United Nations. United Nations framework convention on climate change. 2024. https://unfccc.int/resource/ccsites/zimbab/conven/text/art01.htm#:~:text=2.,3. Accessed May 2024.

United Nations Education Programme. Facts about climate change. https://www.unep.org/facts-about-climate-emergency?gad_source=1&gclid=Cj0-KCQjwq86wBhDiARIsAJhuphmq6lFXvMLipqQIaPKDeBSHuiFG47pbdzgQNuOvItQxSHf29LoerhsaAsF9EALw_wcB.

United Nations Environment Programme. World environment situation room; 2024. https://wesr.unep.org/climate/. Accessed June 2024.

Vaduganathan M, Mensah GA, Turco JV, Fuster V, Roth GA. The global burden of cardiovascular diseases and risk: a compass for future health. J Am Coll Cardiol. 2022;80:2361–71.

Virani SS, Newby LK, Arnold SV, Bittner V, Brewer LC, Demeter SH, Dixon DL, Fearon WF, Hess B, Johnson HM, Kazi DS, Kolte D, Kumbhani DJ, LoFaso J, et al. 2023 AHA/ACC/ACCP/ASPC/NLA/PCNA guideline for the management of patients with chronic coronary disease: a report of the American heart association/American college of cardiology joint committee on clinical practice guidelines. J Am Coll Cardiol. 2023;82:833–955.

Whitfield JB, Zhu G, Landers JG, Martin NG. Pessimism is associated with greater all-cause and cardiovascular mortality, but optimism is not protective. Sci Rep. 2020;10:12609.

World Air Quality Index. https://waqi.info/. Accessed May 2024.

World Heart Federation. World heart report 2023: full report; 2023. https://world-heart-federation.org/resource/world-heart-report-2023/. Accessed May 2024

World Meteorological Organization's. World weather and climate extremes archive; 2021. https://wmo.asu.edu/content/arctic-circle-highest-temperature. Accessed June 2024.

Worldometer. Carbon dioxide emissions, 2024. https://www.worldometers.info/co2-emissions/co2-emissions-per-capita/. Accessed May 2024.

Xie E, Howard C, Buchman S, Miller FA. Acting on climate change for a healthier future: critical role for primary care in Canada. Can Fam Phys. 2021;67:725–30.

Xue J, Liu P, Xia X, Qi X, Han S, Wang L, Li X. Seasonal variation in neurological severity and clinical outcomes in ischemic stroke patients: a 9-year study of 5238 patients. Circ J. 2023;87:1187–95.

Yang J, Zhou M, Ou CQ, Yin P, Li M, Tong S, Gasparrini A, Liu X, Li J, Cao L, Wu H, Liu Q. Seasonal variations of temperature-related mortality burden from cardiovascular disease and myocardial infarction in China. Environ Pollut. 2017;224:400–6.

Yang Y, Piper DJW, Xu M, Gao J, Jia J, Normandeau A, Chu D, Zhou L, Wang YP, Gao S. Northwestern Pacific tropical cyclone activity enhanced by increased Asian dust emissions during the little ice age. Nat Commun. 2022;13:1712.

Zhao S, Liu M, Tao M, Zhou W, Lu X, Xiong Y, Li F, Wang Q. The role of satellite remote sensing in mitigating and adapting to global climate change. Sci Total Environ. 2023;904:166820.

Zou CZ, Goldberg MD, Hao X. New generation of U.S. satellite microwave sounder achieves high radiometric stability performance for reliable climate change detection. Sci Adv. 2018;4:eaau0049.

Zuhlke L, Engel ME, Karthikeyan G, Rangarajan S, Mackie P, Cupido B, Mauff K, Islam S, Joachim A, Daniels R, Francis V, Ogendo S, Gitura B, Mondo C, Okello E, Lwabi P, et al. Characteristics, complications, and gaps in evidence-based interventions in rheumatic heart disease: the Global Rheumatic Heart Disease Registry (the REMEDY study). Eur Heart J. 2015;36:1115–1122a.

**Abstract**

Having described and explored the global to individual context of climate change and health, this chapter now explores the biological imperative of any organism to maintain homeostasis. As a key function of this imperative, in humans, the cardiovascular system plays a key role in maintaining homeostasis. Over time we (humans) have overcome many of the biological limitations/constraints of these protective mechanisms through a combination of behavioural, cultural and technological adaptations—thereby allowing us to migrate and thrive in nearly every corner of the world. Within this biological to historical context, rapid climate change (through exposure to more weather extremes) is now challenging the limits of our ability to maintain homeostasis. How the opposing forces of human adaptability, maintaining thermoregulation and increasing climatic provocations to health plays out in vulnerable individuals and communities (from those living in poverty to older individuals living with chronic heart disease) will provide important context to later chapters.

**Keywords**

Human adaptation · Human migration · Thermoregulation · Homeostasis · Heart response · Cold response · Human comfort

## 3.1 Understanding Biological Imperatives

The dominant species on the planet, humans are remarkably adaptable and resilient. It is this reason that some of us can walk across desolated and wind-swept ice-packs with the polar bears in the Arctic Circle. It also explains how others can transverse the stiflingly hot and humid rainforests alongside giant anacondas in the Amazon without rapidly expiring! However, for most of us, our environments and behavioural patterns are carefully controlled to avoid extreme challenges to our preferred (in terms of comfort) and optimal (from a biological perspective) homeostasis (Billman 2020). The cardiovascular system including the micro- to macro-vasculature and the major organ at the centre of it (the all-important heart) takes the brunt of any decision we make to test our limits in this regard.

As first described in previous Chapters, increasingly, we (humanity) rely on advanced technology to master any environment and maintain our individual equilibrium. From lightweight 'breathing' clothes to portable air conditioning units, this technology continues to evolve. We've also pursued better ways to travel, communicate, socialise and entertain ourselves, commercialising many aspects of our lives in the process. But have we upset the equilibrium of a complex network of global ecosystems in the process? From an economy of effort to achieve

© The Author(s) 2024
S. Stewart, *Heart Disease and Climate Change*, Sustainable Development Goals Series,
https://doi.org/10.1007/978-3-031-73106-8_3

a specific goal [an inherent aspect of biological efficiency within every organism (Brett and Groves 1979)], we've transitioned and embraced an era of societal excess to enhance our lives.

It is this excess that may well, if we are to believe the doomsayers, bring about our downfall—at least in terms of an advanced civilisation striving for peace and seeking to ensure that every individual is well-fed and achieves their full potential (United Nations Department of Economic and Social Affairs 2024). This downfall comes in the form of human-provoked *climate change*—noting once again, how the terminology has shifted from a highly simplistic description of *global warming* and a bleak future of unending heatwaves to a more nuanced and sophisticated description of radically altered weather patterns (United Nations Education Programme 2024). Thus, if this book were written 20–30 years ago, it might well focus on heat stress alone as the predominant provocateur of increasing cardiac events worldwide. Instead, despite the persistent focus of media outlets and major research groups alike (who undoubtedly profit from the attention they gain) on the threat posed extreme heat, this book considers the full spectrum of current to future climatic conditions that play a fundamental role in determining our 'heart health' and when we might experience a potentially fatal cardiovascular event.

To fully understand what's ahead of us (humanity), it is critical to reflect on how we have populated nearly every corner of the world through a combination of environmental control, adaptive behaviours/cultural practices and physiological adaptations. It is also critical to accept that we have never truly conquered our environment. If we had, then the chances of us experiencing an acute myocardial infarction (AMI) or cerebrovascular event would be truly random and anchored to 'when' we develop clinically significant atherosclerosis and/or reach the point of no return where an inflamed atherosclerotic lesion provokes a deadly cascade of thrombosis formation and vascular occlusion (Gherasie et al. 2023). Instead, as will be subsequently described, this has likely never been the case. As mentioned previously while major

reports such as those presented by the American Heart Association (Virani et al. 2023; Ford et al. 2023; Abovich et al. 2023), European Society of Cardiology (Konemann et al. 2023; Gersh and Packer 2007; Kirchhof et al. 2016; McDonagh et al. 2021, 2023) and Global Burden of Disease consortium (and other groups) (Vaduganathan et al. 2022; Roth et al. 2020) routinely provide updated statistics on cardiovascular morbidity and mortality as if events occurred on a uniform basis throughout the calendar year and regardless of climatic conditions, this is simply not the case. As an extension of this oversight (noting the increasing recognition of the role of environmental pollution in provoking cardiovascular events—see Chap. 6), expert clinical guidelines for the primary prevention to secondary management of heart disease remain a desert (pun intended) in respect to assessing an individual's susceptibility to climatic provocations to their health and the need to modulate their management accordingly.

Within the context of seeking to understand the fundamental limits of human health, the following sections provide an overview (noting this is neither a book on the complex mechanics of climate change or complex human physiology) of the key considerations needed, by any health professional, to understand how our environmental conditions (including climatic conditions) shape our behaviours and health status.

## 3.2  The Limits of Human Adaptation

As already described, humans are remarkably resilient and adaptable. But what if we were forced to live without modern clothing and the support of advanced technology to maintain homeostasis? How would we fare if we had to live in the most extreme environments, from unrelenting hot and humid conditions to below freezing temperatures without these essential items? The simple answer is that beyond those still living and practicing generations of culturally adaptive traditions and behaviours, most of us wouldn't. Indeed, even the fittest of persons, would struggle to cope physically (and perhaps mentally) with any prolonged discomfort imposed by temperatures at either spectrum of human tolerance. It is worth repeating (see Chap. 2) here that in places like Scandinavia there is a tradition of doing exactly that—combining a red-hot sauna with temperature rising to 100° Fahrenheit followed by an ice-cold plunge (preferably in a beautiful fjord!) (Heinonen and Laukkanen 2018). Clearly, there is a level of tolerance that will develop over time with this practice (especially if growing up with that tradition), but does this mean that everyone in Norway, Finland, Sweden, or Finland walks around in winter without warm and protective clothing? The obvious answer is a resounding no! In terms of global longevity Sweden (rank 12, with average life expectancy for males and females of 85.2 and 82.1 years in men and women, respectively, in 2023), Norway (rank 13, 85.2 and 81.9 years) and Finland (rank 85, 85.0 and 80.0 years) are impressive on a global scale without being dramatically better than those without such traditions (World Bank 2024). It has been suggested that instead, this practice results in less depression and greater happiness. As such, according to the 'World Happiness Report', Finland is the "*happiest country in the world*" according to a combination of six parameters—"*gross domestic product per capita, social support, healthy life expectancy, freedom to make your own life choices, generosity of the general population, and perceptions of internal and external corruption levels*" (World Population Review 2024)—most of which, one could argue, has nothing to do with exposure to extreme heat and cold!

For the first timer at least (noting again my ability to speak from personal experience), the transition from a scalding-hot sauna into a freezing-cold fjord in sub-zero temperatures is both exhilarating and painful in equal measures. At both ends of the temperature extremes, the body is forced into dynamic, autonomic changes to preserve vital bodily functions that defy any attempts to apply mind-over matter—a highly over-rated concept when your body is applying years of evolutionary processes and adaptations!

From sweating to extreme vasoconstriction, there are fundamental reasons why the body reacts and responds the way it does in such a dramatic scenario. This includes an overwhelming urge to reduce that extreme exposure as soon as humanly possible, with physical pain and mental shock a key component to survival. No matter how much we might think that the pain and discomfort can be tolerated, the inevitable conclusion from prolonged exposure to extreme heat or extreme cold is death. So, what is the ideal, 'goldilocks' temperature for a human and at what point does the body begin to expire? These are some of the questions that will be explored in this Chapter.

## 3.3 Home is Where the Heart Is

As succinctly described by the *Smithsonian* (Smithsonian Institute 2008), while there are many gaps in our knowledge around human migration, it is believed that modern humans first migrated from sub-Saharan Africa between 60,000 and 80,000 years ago—their first migratory path being Asia. To understand how and why we know so much (or little) about early human development and our African origins, one only has to visit the wonderful, World Heritage listed *Cradle of Humankind* site just north of Johannesburg in South Africa—where the story of Hominids who first emerged around 7 million years before their ancestors [us!] is told so eloquently (https://www.maropeng.co.za/content/page/introduction-to-your-visit-to-the-cradle-of-humankind-world-heritage-site). From there, humans followed natural land bridges and used boats to migrate to Indonesia, Papua New Guinea and Australia. At a later stage (estimated to be no earlier than 40,000 years ago) humans then migrated from Africa into Europe via two main routes (along the Danube River via Turkey and the Mediterranean coast). In the midst of this migration, the Neanderthals migrated to the less hospitable mountainous regions of modern Croatia, the Iberian Peninsula and the Crimea to, most probably, avoid conflict/competition with our direct ancestors. Despite this, they became extinct around 25,000 years ago (Smithsonian Institute 2008). Around 10,000 years later, humans finally bridged the last frontier (at least in terms of habitability) and migrated from Asia into Northern America and then into Southern America (Smithsonian Institute 2008).

In considering these historical patterns of migration, it is critical to consider the driving forces that ensured humans [noting the extinction of the Neanderthals forced into higher climes and therefore more climatic extremes (Smithsonian Institute 2008)] followed a predictable pathway, whereby they could survive and thrive. Remarkably, despite all our technological advances, this pattern still exists today, when one considers the distribution and density of the global population according to latitude (https://www.weforum.org/agenda/2022/05/mapped-the-worlds-population-density-by-latitude/#:~:text=In%20 particular%2C%20the%2025th%20and,States% 2C%20Mexico%2C%20and%20others). Much as the Earth sits in a 'Goldilocks Zone' (just right in terms of the warming derived from the solar radiation generated by the Sun without the atmosphere being 'stripped' by solar winds in the process), humans have largely settled in the those climates that feel the most comfortable and the least challenging to homeostasis. Specifically, as described by a visually compelling report (hosted by an online publisher dedicated to generating 'data-driven' visuals), the 25th and 26th parallels adjacent to the equator (the warmest, most climatically stable, water-rich and bio-diverse band on the planet) are the most densely populated latitudes. Along these latitude bands, around 280 million people reside in a broad range of countries such as India, Bangladesh, China, the United States and Mexico. The specific pattern of human habitation within these bands is/was driven by additional factors such as the topography of the land (from mountainous barriers to large river systems), soil fertility and micro-climatic conditions (https://www.weforum.org/agenda/2022/05/mapped-the-world-s-population-density-by-latitude/#:~: text=In%20particular%2C%20the%2025th%20 and,States%2C%20Mexico%2C%20and%20others). If one were to focus on the single continent of Australia (Fig. 3.1), one can see these factors

in play. As such, much of the continent is 'uninhabitable' from a human perspective—a major exception to this statement, being the Indigenous peoples who originally migrated from South-Eastern Asia and then established a sustainable population founded on a complex hunter and gatherer/nomadic lifestyle that is/was ecologically sensitive, thousands of years before European wave of migration/occupation occurred (Aboriginal Heritage Office 2020). Nevertheless, if one were to consider the location of three of the five continental cities that make-up a good proportion of Australia's population from east-to-west (Sydney, Adelaide and Perth), it is immediately clear that not only are they adjacent to the sea (at major junctions of major rivers/harbours) but are located within a geographically narrow (in planetary terms) band across the continent. Fundamentally, these broad patterns of human inhabitation reflect our physiological limitations, beyond a 'Goldilocks' equivalent of a 'thermoneutral' air temperature range of 25–31 °C, whereby, our bodies do not spend excessive amounts of energy to either limit or over-express temperature loss to the external environment; noting that our psychological comfort zone operates in a different temperature range (Kingma et al. 2014).

## 3.4 Thermoregulation—The Key to Our Survival!

As described in many reports (Osilla et al. 2024; Kingma et al. 2022; Hardy 1961; Burton 1956), thermoregulation is the processes by which we maintain a stable core body temperature that is optimal for all physiological and metabolic activity. It is achieved through the balance of heat generation (via internal metabolic activity) with heat loss (to the external environment). Healthy individuals typically require an internal temperature between 36.5 and 37.0 °C in this regard. The body's 'thermostat' resides in the preoptic region of the hypothalamus, with central thermoreceptors located in the viscera, spinal cord and the hypothalamus itself and monitor core temperatures. Peripheral thermoreceptors are also located in the skin and these

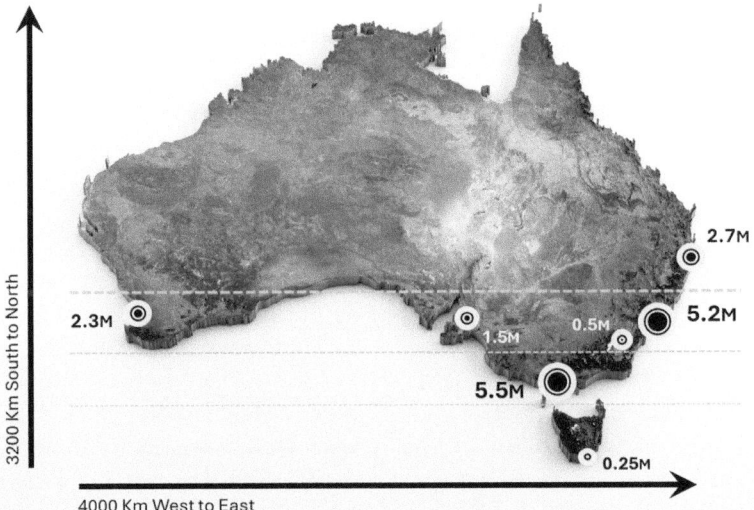

**Fig. 3.1** Where the Europeans decided to settle in a vast continent. This figure shows the location of the major population centres in the southern regions of Australia (all but one [Canberra] represent the historical locations chosen by European settlers as the best place to live in the eighteenth and nineteenth centuries). These cities and surrounding areas continue to grow, although Australia remains a very small country by population size—total of 26.8 million people. (https://www.abs.gov. au/statistics/people/population). This contrasts with the much more dispersed/continent-wide, Aboriginal peoples (estimated to number ~750,000 in 1788) (Aboriginal Heritage Office 2020). The arrows indicate the approximate depth and width of the main continent (comprising 7.7 million Km²)—noting the substantive size and biodiversity of the southern "island" of Tasmania with large swathes of uninhabited wilderness and a much cooler/more dynamic climate than the mainland

sense peripheral temperatures—hence the disconnect between voluntary control mechanisms/behaviours based on comfort versus involuntary control mechanisms essentially guarding our biological need for homeostasis (Osilla et al. 2024; Kingma et al. 2022; Hardy 1961; Burton 1956).

## 3.5  The Limits of Human Adaptation—Maintaining Homeostasis

As elegantly described by Billman (2020), homeostasis can be defined as a *"a self-regulating process by which an organism can maintain internal stability while adjusting to changing external conditions"*. As further explained, this is not a static process. As reflected in human behaviour and adaptation, it is *"a dynamic process that can change internal conditions as required to survive external challenges"*. Of particular relevance to heart and broader cardiopulmonary health, the relative success or failure to maintain homeostasis is reflected in the health and vitality of the organism (Billman 2020). Critically, there are multiple and complex feedback mechanisms (some of which will be discussed below) in humans that automatically adjust our internal environment to maintain homeostasis—with a high degree of fine-control and adaptability that enables us to adapt (probably better than any other species on the planet) to variable environmental conditions. However, these mechanisms and the role they play, cannot be disconnected from our conscious behaviours. They also cannot be divorced from our increasing reliance on technology to push the limits where we can live comfortably in a health state. In his excellent treatise on the *"Central Organizing Principle of Physiology"*, Billman, reflects on our historical understanding of homeostasis through the words of the Greek physician/philosopher Alcmaeon of Croton. In circa 500 BC, he wrote about a balance of opposites—"*Health is the equality of rights of the functions, wet-dry, cold-hot, bitter-sweet and the rest; but single rule of either pair is deleterious*" (Freman 1948). Hippocrates expanded upon this concept when he wrote "*Health is primarily that state in which these constituent substances are in correct proportion to each other, both in strength and quantity and are well mixed*" (Chadwick 2007).

Thus, although collectively we may have long forgotten its importance (with the growing threat of climate change, likely to reverse such collective amnesia), our ability to control our environment to maintain physiological homeostasis, should be a major consideration when assessing anyone presenting with a major illness/chronic disease. Before providing a holistic framework of how that can be considered through the lens of climatic vulnerability, it's worth considering the physiological challenges posed by the extremes of cold versus hot conditions.

## 3.6  The Challenge of Maintaining Homeostasis in the Cold

As outlined by Yurkevicius and colleagues in their excellent review of "*human cold adaptation*" (Yurkevicius et al. 2022), in response to cold, humans exhibit/apply three main adaptations—(1) A hypermetabolic state, (2) Insulative physiology/behaviours and (3) Habituated behavioural responses (Aj 1996; Castellani and Young 2016). At the fundamental level, cold exposure triggers acute and chronic physiological responses of varying degrees depending on our age and sex, as well as cardio-respiratory fitness and the metabolic reserves/capacity to overcome thermoregulatory fatigue (Rammsayer et al. 1993; Therminarias et al. 1988; Thompson 1977). As will be explained, these responses can have detrimental consequences to the cardiovascular system—thereby triggering a cardiac event/crisis in those with minimal tolerance to any additional physiological demands to

maintain homeostasis. The broad risk of compromising core temperature via exposing skin to cold temperatures is modulated by the ambient air temperature, the windchill factor and humidity levels (Xu et al. 2023). As most people would know, exposure to cold, wet, wintry conditions can be as challenging as being exposed to freezing dry temperatures on a still day given the enhanced capacity of water, whether it be on the skin or in the air, to act as a conduit to heat exchange. Although the relative humidity of cold air is much lower than warm air, water (whether it be in the form of rain, snow and mist/condensation) along with factors like windchill will always compromise our comfort levels/thermoregulation—hence the imperative to stay dry and insulated! (Montreal Science Centre 2023).

The primary thermoregulatory responses to lower peripheral skin temperatures is peripheral vasoconstriction and shivering. The former reduces heat loss/thermal conduction to the external environment by removing blood vessels from the skin surface, while the latter increases (metabolic) heat production via muscular activity—see Fig. 3.2. Largely mediated via sympathetic activation, both result in an elevated heart rate and blood pressure (Castellani and Young 2016). More prolonged exposure to cold has been linked to hyper-platelet activation (Zhang et al. 2004) and elevated inflammatory markers (Halonen et al. 2010). Ironically, the degree of heat loss and the potential adverse effects/response to cold exposure may be exacerbated following exercise (Castellani et al. 1999).

In considering these physiological responses to hot and cold climatic conditions, beyond the raw figures indicating that due to the phenomenon known as epidemiological transition (Blacher et al. 2016; Omran 1998), the global burden of disease associated with hypertension (Kario et al. 2024; Lu et al. 2024), type II diabetes (Xie et al. 2024; Luo et al. 2024; Yang et al. 2024; Forray et al. 2023; Jiang et al. 2024) and established forms of cardiovascular disease [including ischaemic heart disease, heart failure (Salerno et al. 2024; Shahim et al. 2023; Kularatna et al. 2023) and atrial fibrillation

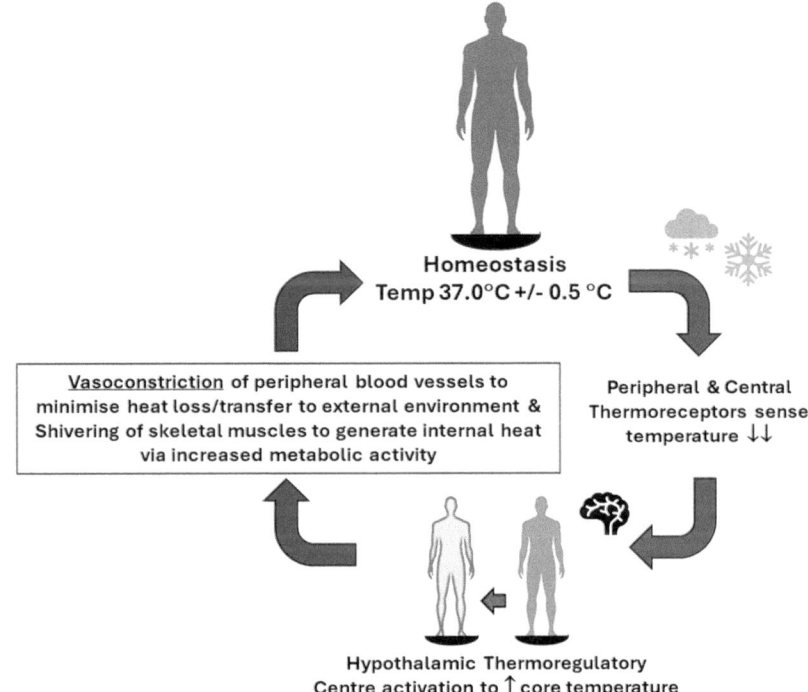

**Fig. 3.2**  Cold responses to maintain thermoregulatory control. This figure summarises the main physiological responses (in men and women) to cold

(Kularatna et al. 2023; Ball et al. 2013, 2015; Ma et al. 2024; Ohlrogge et al. 2023)] continues to grow, critically, global obesity rates also continue to rise (Gona et al. 2021; Belancic et al. 2020; Seidell and Halberstadt 2015; Kelly et al. 2008). This is important, because, despite postulations that 'cold-exposure therapy' might be beneficial to them (Ivanova and Blondin 2021; Falk et al. 1994), obese people have a higher skin surface area. Much like a bigger sail catching the wind, their potential thermal loss through exposed skin is much higher and their physiological responses are likely to be less robust than if they maintained a much healthier weight (Falk et al. 1994). Unfortunately, obesity is not confined to adults. Childhood obesity rates are also rising, along with a seemingly paradoxical rise in malnutrition (both of which have a serious impact on a child's ability to maintain homeostasis) (Chong et al. 2023; Malta et al. 2024).

Of course, the issue of cold exposure in the short-term, needs to be distinguished in those who do and don't live in cold conditions. Such discrimination becomes relevant in both the context of seasonal changes and the likely impact of climate change on the frequency of unexpected exposure to cold spells/events. Previous studies suggest that those living in colder climes exhibit the following (main) adaptive responses—insulative (in the form of more cutaneous fat and less peripheral circulation) and metabolic (shivering or non-shivering thermogenesis) (Makinen 2010). As described by Makinen, such adaptation is dependent on the type and intensity of cold exposure (with various levels/thresholds of provocation) and, as noted above, the age and sex, anthropometric profile and cardio-respiratory fitness of the individual (with varying levels of physiological responses) (Bittel 1992). Studies suggest that '*cold adaptation*' (the increased threshold at which a lower temperature triggers a physiological response to maintain homeostasis) is mediated by norepinephrine, the levels of which has been shown to decrease with cold exposure in a controlled environment (Leppaluoto et al. 2001).

Concurrently, it has been suggested brown adipose tissue is an important mediator in thermoregulatory control and its deposition is even triggered by cold exposures, thereby representing an important component of cold adaptation. However, more research is needed in this area (Tabei et al. 2024).

As Makinen, further explains, it is difficult to determine if observed thermoregulatory adaptations (to both cold and hot climates) are genotypic or phenotypic, given the inability to control for environmental/climatic conditions provoking such changes (Lambert and Dugas 2008). Nevertheless, there is likely to be intergenerational adaptations to cold (and heat) that will be challenged by (more) rapid climate change.

## 3.7 The Challenge of Maintaining Homeostasis in the Heat

At the other end of the spectrum (exposure to hot conditions), there is corollary activation of thermoregulatory mechanisms designed to prevent overheating and metabolic stress. As can be imagined, given the focus on *"global warming"* the physiological effects and perception of rising temperatures due to climate change is an area of active research (Khudaiberdiev et al. 1992; Lim 2020; Kenney et al. 2014). Compared to the 'cold response' outlined above (Castellani and Young 2016), these are generally more systemic than peripheral in nature and often result in more immediate, adverse responses at an individual to population level. One of the immediate responses to heat (see Fig. 3.3), is peripheral vasodilatation and sweating (Taylor 2014). Once again, this automated response to the environment is modulated by the air temperature and humidity in terms of the efficiency in radiating heat to lower the core body temperature (Green et al. 2006). In response to vasodilation (resulting in a consequential reduction in vascular resistance), sympathetic activation prevents a critical loss of blood pressure. This is achieved through compensatory increases in

heart rate and stroke volume—both of which induce a state of cardiac stress (Greaney et al. 2015, 2016). As described in a previous review, if required, *"blood flow to the skin can be dramatically increased from a baseline of 0.3 to 7.5 l/min"* (Holowatz et al. 2010). With just 1% loss of body weight [noting there are 2–4 million sweat glands in the body and up to 3 l/h of bodily fluid is readily mobilised as sweat to provide highly efficient evaporative cooling (Greaney et al. 2015, 2016)], the risk of hyperthermia rises markedly in the setting of dehydration as both fluid and electrolyte depletion occurs (Kenney et al. 2014). With more advanced hyperthermia, compromised preload due that combined loss, can trigger hypotension and tachycardia. If not reversed, this provokes a cascade of low/highly compromised cerebral and cardiac with consequential syncope and myocardial ischaemia. Heat stress is also associated with a prothrombotic state. This is due to haemoconcentration/high blood viscosity, not only due to a decrease in plasma volume but also an increasing number of circulating blood cells—including all-important platelets (Keatinge and Donaldson 2004; Keatinge 1991). When the bodies' thermoregulatory control is completely overwhelmed, rhabdomyolysis (Kruijt et al. 2023) and the associated release of cellular contents (including potassium, phosphate, myoglobin, creatinine kinase and lactate dehydrogenase) are likely to induce a disseminated intravascular coagulation. The consequence of this pathological process is increasing peripheral and end-organ ischaemia (Burgess 2022). In the most extreme cases, therefore, along with the risk of a fatal cardiac arrhythmia (either extreme bradycardia or ventricular tachycardia due to hyperkalaemia), prolonged and severe heat stress will trigger end-organ failure, with the production of heat-shock proteins [as a protective mechanism (Mellati 2006; Sun and MacRae 2005)] as a precursors to cardiogenic shock and death (Hanna and Tait 2015; Kaltsatou et al. 2020; Fujii et al. 2019).

Once again, the issue of a rising number of obese people worldwide needs to be highlighted given that physiological studies demonstrate that people with higher proportions

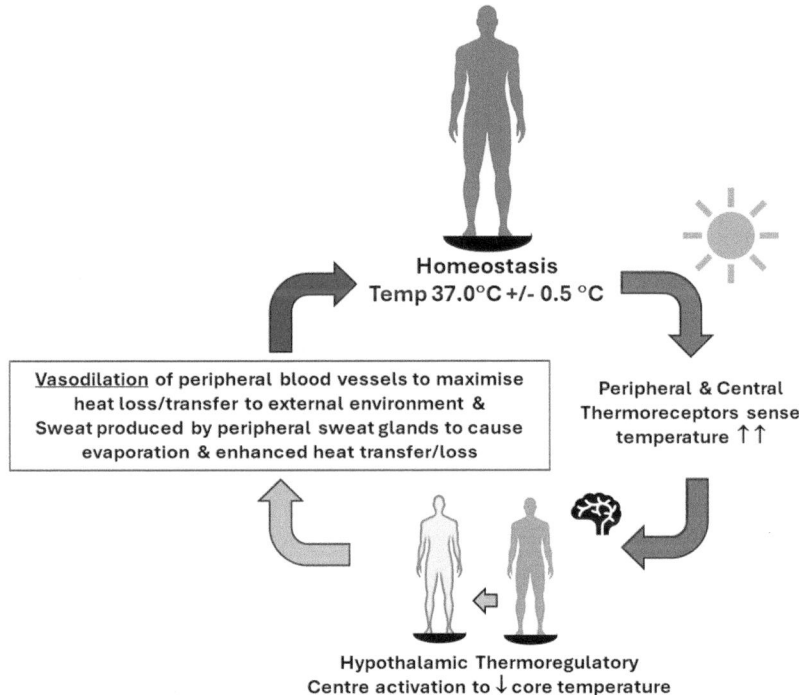

**Fig. 3.3** Heat response to maintain thermoregulatory control. This figure summarises the main physiological responses (in men and women) to cold

of body fat (i.e. women and obese individuals) have a lower heat tolerance because of a reduced capacity to store heat (Cheung et al. 2000). At the other end of the spectrum, despite typically having the advantage of a greater body surface area-to-mass ratio than adults (one study showing this advantage was 64% when comparing young children aged 9 months to 4.5 years to their mothers) for thermoregulation, children are disadvantaged in terms of their sweating and skin temperature response—their body temperature rising higher under the same conditions (Tsuzuki-Hayakawa et al. 1995). This has prompted public warnings in North America, that children are much more likely to be adversely affected by heat stress as the number and severity of heat waves increase (Scientific America 2022). When one considers the age profile of highly populous and poverty-stricken countries such as Mozambique, with a predominance of children aged under 10 years living in a country with hot, tropical

conditions (Mocumbi et al. 2019) these findings and warnings reinforce the discussion around the progress of UNSDGs and what will happen if climate change is not addressed from multiple perspectives (United Nations Department of Economic and Social Affairs 2024; United Nations 2023). The scope of the problem (i.e. number of young children to adults at risk) is somewhat tempered by evidence of, as with the potential for intergenerational cold acclimatisation/adaptation, physiological acclimatisation to heat beyond an ability to tolerate warmer temperatures (Cheung and McLellan 1998). Such individuals demonstrate more rapid and effective sweating and differential vasodilatation, particularly if provoked by exercising in warmer temperatures (Murray et al. 2021; Horowitz and Robinson 2007). Highlighting the physiological differences between cold versus hot provocations to maintaining haemostasis, it's important to note the less profound 'protection' from heat stress among young, aerobically fit people when compared to that conferred against cold temperatures (Selkirk and McLellan 2001; McLellan 2001).

## 3.8   Non-modifiable Factors to Consider—Age and Sex

Concurrent to historically high rates of obesity in many parts of the world along with an ever-expanding human population/foot-print, it's important to note the rapidly ageing populations of many higher-income countries. This is important from a climate and heart health perspective, because older individuals exhibit a blunted (physiological) response to cold in order to maintain thermoregulation (Smolander 2002; Inoue et al. 1992). Moreover, although there is an inherent bias towards studying responses of older people to heat events (Ratwatte et al. 2022), advancing age blunts perceptions of cold requiring a behavioural/adaptive response (Frank et al. 2000; Fujii et al. 2017; Taylor et al. 1995)—leaving them at increased risk when, inevitable wintry conditions and/or an unseasonal cold-spell arrives, for those living in most countries. From a heat perspective, endocrine-activated sweat glands are activated less and produce less progressively in older people (Keatinge and Donaldson 2004; Keatinge 1991). The circulatory response to elevated temperatures is also blunted, while cardiac reserve is more compromised when trying to compensate for a combination of vasodilatation and lower plasma volume; particularly among typically older people with common chronic diseases such as diabetes and heart disease (Carrillo et al. 2016; Kenny et al. 2010). A vital therapeutic target in the management of people with diabetes, hypertension and left ventricular systolic dysfunction/heart failure, the renin–angiotensin system is also increasingly compromised in older age (Wang et al. 2021; Werner et al. 2014; Pedone et al. 2006; Bevan 2001). Significantly, many agents deliberately blunt the response of renin-angiotensin system as a therapeutic strategy (Castiglione et al. 2023; Silva et al. 2022; Silva-Cardoso et al. 2019; Teerlink 2002).

Nevertheless, the renin-angiotensin system plays an integral role in any heat response. Thus, any compromise, will likely blunt its response to both cold and heat stressors (Kosunen et al. 1976; Hiramatsu et al. 1984). From a biological perspective, there are compelling reasons why women and men respond differently in the presence of temperature extremes and these are explained in relation the risk posed to the cardiovascular system from climate change in Chap. 4. For example, as touched upon earlier, while women tend to have a greater peripheral fat content (greater heat insulation negated by less capacity to generate heat) than men, they also tend to have larger surface areas, potentially resulting in a net increase in convective heat loss under the same climatic conditions (Makinen 2010; Lim 2020). Any consideration of the differential impact of climatic challenges to the health of men and women, therefore, has to consider the biological to behavioural (many based on gender/identify) factors that influence the historical pattern of heart disease on a sex-specific basis (see Fig. 3.4) (Khraishah et al. 2022). The importance of these factors will be more fully explored in the next Chapter.

## 3.9   Beyond the Physiological —Perceptions of Human Comfort

As already noted above, age blunts our perceptions. Moreover, anyone who has landed in a different climate and season and wondered why their clothing (either deciding to wear light clothing or wearing a heavy insulated coat) is discordant with that of the local population will realise that acclimatisation is not just about physiological adaptation, it is a combination of physiology and perception. Trying to quantify the concept of thermal comfort is difficult when one considers how many times two people in the same household will argue over the heating or cooling settings! Nevertheless, the *Universal Thermal Climate Index* was developed to quantify thermal comfort in relation to the ability of the body to maintain homeostasis through

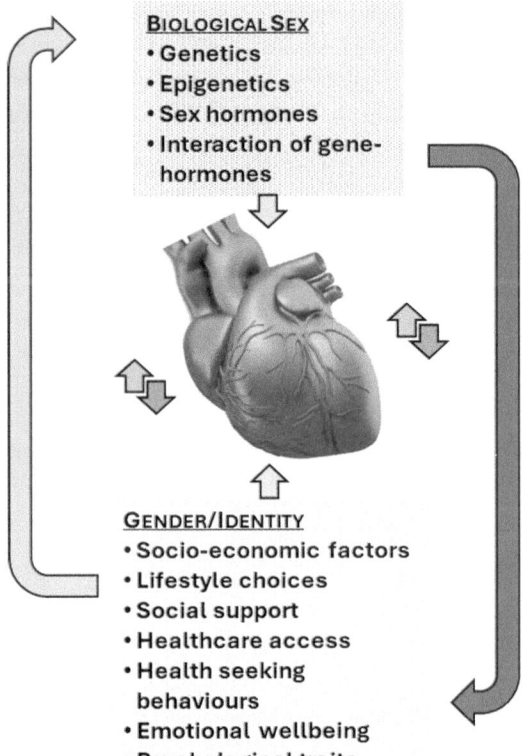

**BIOLOGICAL SEX**
- **Genetics**
- **Epigenetics**
- **Sex hormones**
- **Interaction of gene-hormones**

**GENDER/IDENTITY**
- **Socio-economic factors**
- **Lifestyle choices**
- **Social support**
- **Healthcare access**
- **Health seeking behaviours**
- **Emotional wellbeing**
- **Psychological traits**

**Fig. 3.4** Cardiac health—biological and gender-related factors. This figure summarises the main (inter-related) biological and gender-identity influenced factors that might positively or negatively influence cardiac health—resulting in the divergent pattern of heart disease seen in men and women that has the potential to widen under the influence of climate change (see Chap. 4)

self-thermoregulation. Expressed in degrees °C, it considers air temperature, solar and thermal radiation, humidity levels and wind speed, while considering *"human heat balance"* models in respect to thermoregulation (Brode et al. 2012). As recently concluded by Grigoriev and colleagues, in considering comfort levels, not all climate change is 'bad' when one considers the propensity for otherwise inhospitable regions of the world to become more comfortable (at least according to this index)—*"Overall, a decrease in the UTCI categories of extremely cold stress is coupled with an increase in the comfortable range in both Alaska and Chukotka. The salient conclusion is that, from the point of view of comfort and safety, global warming has a positive impact on the climatology of thermal stress in the Arctic, providing advantages for the development of tourism and recreation"* (Grigorieva et al. 2023). As will be explored further, whether these observed shifts result in improved heart health and protection from cardiac events provoked by inevitable corrections/shifts towards colder temperatures, is very much open to debate. The European COPERNICUS initiative (https://climate.copernicus.eu) is mapping global bands based on the UTCI, noting that thermal comfort can rapidly change to thermal discomfort (COPERNICUS Group 2024). To interpret this statement further, one could argue that while such change can occur on a day-to-day basis it can be argued that rapid climate change can rapidly alter/reset the threshold at which comfort transitions to discomfort. As commented on the COPERNICUS website— *"Understanding how UTCI may evolve in the future is important as global temperature rise is expected to exacerbate thermal stress across the globe"* (COPERNICUS Group 2024).

## 3.10   The Importance of Behaviours and Cultural Adaptations

Much of the factors explored above (from climatic conditions to key functions of the autonomic system) are beyond the control of the individual. One could argue that, despite standard indices of comfort/tolerability such as the UTCI (Grigorieva et al. 2023), that even comfort levels have a component of inevitability when one considers cultural norms and expectations around expressing discomfort. More, but not completely, controllable, is the voluntary choices humans make in terms of placing themselves (conscious or otherwise) at risk or in danger. Fundamentally, in the context of our interactions with the environment, a person's cognitively informed decisions and actions when confronted by low or high temperatures that threaten their comfort/homeostasis, form an integral part of a composite autonomic and behavioural thermo-effector response (Stewart et al. 2019, 2017). Reflecting the importance of how communities living in distinctive geographic and climatic regions specifically adapt to their environment over the longer-term, a range of studies have described sociocultural variances in preserving heat (indoor heating, wearing more thermo-protective clothing and modulating exercise levels), that explain why individuals living in one region of the world will be more protected/resilient when exposed to the effects of both predictable and unpredictable cold extremes (Donaldson et al. 1998). Alternatively, there are unique risks attached to 'normal' behaviours such as exercising or shovelling snow in cold extremes for those living in colder climates, that may not be immediately obvious to the affected individual (Glass and Zack 1979; Baranchuk 2017; Sauter et al. 2015).

While behaviour can range from the impulsive to ingrained/habitual, studies consistently reveal that many important behaviours that increase the risk of a cardiovascular event [e.g. consuming a diet richer in saturated fats during winter (Ockene et al. 2004)] vary according to predictable (at least prior to climate change), seasonal transitions in the weather. For example, the SEASONS Study, a large prospective cohort study conducted in Massachusetts, USA, revealed potentially important differences in dietary patterns (such as an autumn peak for both saturated and unsaturated fat consumption) and physical activity (lowest levels in winter) on a seasonal basis—noting a higher degree of variation was found in non-white, middle-aged men with low education levels (Ma et al. 2006). One of the most compelling studies describing this phenomenon (behavioural change linked to seasonal climatic transitions) involved an analysis of cross-sectional data from 24 population-based studies from 15 countries from both hemispheres (Marti-Soler et al. 2014). Overall, pooled data on common cardiovascular risk-factors in 237,979 people according to the season in which they were profiled were analysed. Regardless of the hemispheres, risk factor levels were higher in winter and lower in summer (these data broadly correlating with 'when' cardiovascular events occur—see Chap. 5). For those living in the

Northern Hemisphere, estimated seasonal variations were $0.26 \, \mathrm{kg/m^2}$ for BMI, $0.6 \, \mathrm{cm}$ for waist circumference, $2.9/1.4 \, \mathrm{mmHg}$ for systolic/diastolic BP, $0.02 \, \mathrm{mmol/L}$ for triglycerides, $0.10 \, \mathrm{mmol/L}$ for total cholesterol, $0.01 \, \mathrm{mmol/L}$ for HDL cholesterol, $0.11 \, \mathrm{mmol/L}$ for LDL cholesterol and $0.07 \, \mathrm{mmol/L}$ for glycaemia. Broadly similar results (with some variations) were observed in those living in the southern hemisphere (Marti-Soler et al. 2014). In a study derived from the FINRISK in Finland (comprising 4690 participants) conducted in 2012, 70% displayed seasonal variations in sleep duration, social activity, mood and energy levels and 40% variations in their weight and appetite (Basnet et al. 2016). As measured by the *Global Seasonality Score* [a tool used to assess seasonal affective disorder (Murray 2004; Magnusson 1996)—see below], the occurrence of angina and depression along with hypertension, dyslipidaemia, diabetes and other non-communicable diseases correlated with 'higher' seasonality (Basnet et al. 2016). As with most historical studies, it's unclear how climate change will alter seasonal/climatic-driven behaviours. However, it is important to note the climate change won't eradicate winters/colder temperatures, nor will there be dramatic variations in the duration of sunlight.

## 3.11  Other Important Considerations

As will be described in more detail, the increasing role of communicable, parasitic and vector-borne infectious diseases leading to heart disease in the context of seasonal variations in risk exposures and climate change cannot be ignored (see Chap. 7). Nor can the role of indoor to outdoor pollution (in all its forms—see Chap. 6). Beyond these factors, it also important to consider, more closely two factors related to mental health and sunlight exposure –

- *Seasonal affective disorder* is characterised by major depression (with associated insomnia and nightmares) and, logically, is reported to occur more (~10% of the population)

according to strict criteria in higher latitude countries/regions (Rosen et al. 1990) such as Greenland (Kegel et al. 2009) and Alaska (Booker and Hellekson 1992) and less in countries like Australia (0.3%) where relatively long days are preserved even in winter (Murray 2004). Many causes of seasonal affective disorder have been postulated. This includes sunlight-sensitive variations in serotonin regulation (via serotonin transporter binding) postulated in some individuals (Tyrer et al. 2016a, b; Gupta et al. 2013). Associated changes in behaviour, including reduced exercise/activity levels and increased alcohol intake, have the potentially to markedly elevate an individual's cardiovascular risk (Neshumova et al. 1994).

• *Deficiency of Vitamin D* (fat-soluble secosteroids responsible for increasing intestinal absorption of calcium, magnesium and phosphate) is highly prevalent among those living in high-latitude countries where sunlight is markedly reduced during winter months. As reported by Cui et al. (2000), in a systematic review and meta-analysis of 308 studies with nearly 8 million participants from 81 countries, globally, 15.7% (95%CI 13.7–17.8%), 47.9% (95%CI 44.9–50.9%) and 76.6% (95% CI74.0–79.1%) of people had serum 25-hydroxyvitamin D levels less than 30, 50 and 75 nmol/l, respectively (Cui et al. 2000). Overall, those living in higher latitudes had a higher prevalence of Vitamin D deficiency, while this figure was 1.7-fold (95%CI 1.4–2.0) higher in winter-spring compared to summer-autumn (Cui et al. 2000). Of note, women and those living in the Eastern Mediterranean region and/or a low-to-middle-income country were more likely to display Vitamin D deficiency. A consistent inverse relationship between low Vitamin D levels and increasing levels of cardiovascular risk and all CVD subtypes has been demonstrated in a range of countries (Frentusca et al. 2023; Verdoia and De Luca 2023; Zhou et al. 2022). Overall, despite a large body research, the acute versus chronic (as well as the indirect versus direct cardiovascular) effects of Vitamin D deficiency and its therapeutic treatment are not yet fully elucidated (Frentusca et al. 2023; Tousoulis 2018).

As will be further explained in the next Chapter, all the above factors come into play when considering the potential consequences of cold and heat extremes in provoking cardiovascular events, but for different reasons—see Fig. 3.5.

## 3.12   Thinking Outside the Clinic or Hospital

Although clinicians are used to monitoring a person's temperature when they are acutely ill and clinics/hospitals (at least in middle-to-high income countries) in which they are managed, are environmentally controlled and comfortable, they rarely consider the circumstances in which a person lives and works. This truism first struck me on my first visit to the township of Soweto (the acronym of south-west township under the South African apartheid regime) on the outskirts of Johannesburg. I had just spent time in the busy Cardiology Unit of the Baragwanath Hospital which was (at the time in the early twenty-first century), reportedly the largest hospital in the world. I was there to design and implement the *Heart of Soweto Study* [a study that revealed much of the transition from communicable to non-communicable forms of heart disease that will be influenced/modulated by climate change in that part of the world (Sliwa et al. 2008; Stewart 2009; Stewart et al. 2006, 2008, 2011; Tibazarwa et al. 2009)] with my host (and Co-Principal Investigator) Professor Karen Sliwa. With many residents unable to easily return to the hospital clinicians were forced to go to them for important follow-up assessments. As I attended one of these mobile follows-up, we were forced several times (due to the lack of formal streets and numbers) to ask for directions. It was there I took a picture of three generations of Sowetans who kindly showed me why they were sitting outside their small concrete home—although the picture can't be presented in this book, it became an iconic image for the

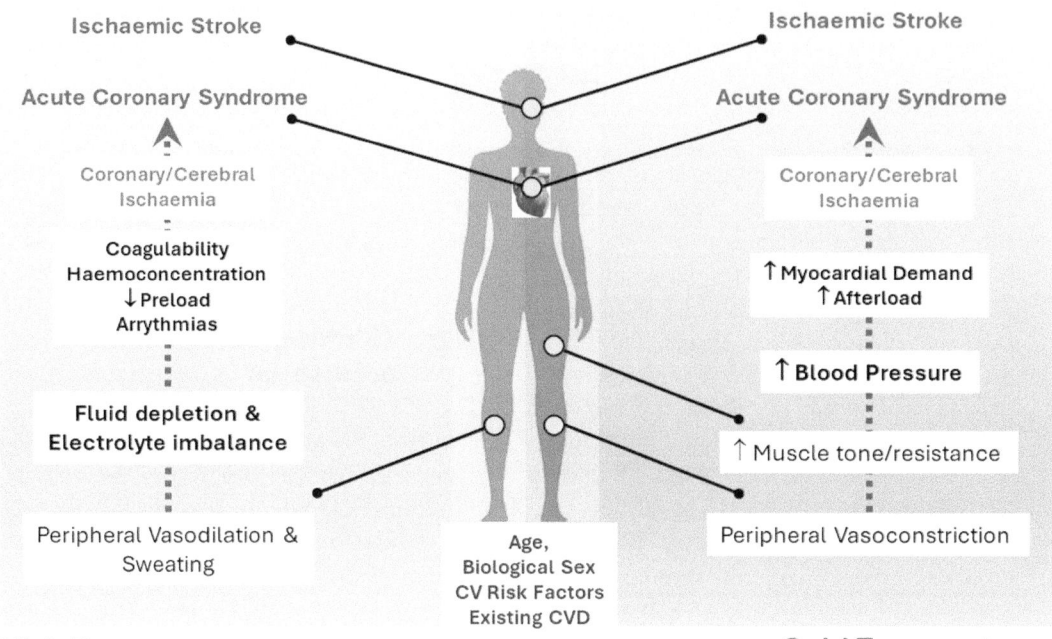

**High Temperatures**                                                **Cold Temperatures**

**Fig. 3.5** Differential pathways to cardiovascular events from hot and cold extremes. This figure summarises the differential pathways that extreme/challenging heat (left) and cold (right) exposures can provoke myocardial ischaemia/acute coronary syndrome and, consequently, potential death (Khraishah et al. 2022)

study because it encapsulated much of challenges of living in Soweto. In very simple terms, at the time of the picture (nearing the midday sun), their modest home was too unbearably hot and still heavy with the smoke from the charcoal they had used to cook their lunch. Without an air conditioner or fan extractor they had no other choice but sit outside. It was then I was struck [once again given my commitment to home-based heart failure management (Stewart et al. 1999)] that we can only truly understand how best to support someone's health, if we understand '*where*' they spend most of their time. This will become more imperative with climatic change and more provocative climatic conditions. However, we can't visit everyone. Instead, we need to prioritise who needs a more in-depth understanding of their circumstances and how that arises. The next chapter provides a holistic framework for deciding who is most climatically vulnerable and how me might move them towards a more '*resilient*' state and therefore protect their cardiovascular health.

## References

Aboriginal Heritage Office. A brief aboriginal history; 2020. https://www.aboriginalheritage.org/history/history/. Accessed May 2024.

Abovich A, Matasic DS, Cardoso R, Ndumele CE, Blumenthal RS, Blankstein R, Gulati M. The AHA/ACC/HFSA 2022 heart failure guidelines: changing the focus to heart failure prevention. Am J Prev Cardiol. 2023;15:100527.

Aj Y. In: Fregly MJ, Cm B, editors. Handbook of physiology: environmental physiology. Bethesda: American Physiological Society; 1996. pp. 419–438.

Ball J, Carrington MJ, McMurray JJ, Stewart S. Atrial fibrillation: profile and burden of an evolving epidemic in the 21st century. Int J Cardiol. 2013;167:1807–24.

Ball J, Thompson DR, Ski CF, Carrington MJ, Gerber T, Stewart S. Estimating the current and future prevalence of atrial fibrillation in the Australian adult population. Med J Aust. 2015;202:32–5.

Baranchuk A. The cardiovascular risk of snowfall and snow shovelling in Canada. CMAJ. 2017;189:E545.

Basnet S, Merikanto I, Lahti T, Mannisto S, Laatikainen T, Vartiainen E, Partonen T. Seasonal variations in mood and behavior associate with common chronic diseases and symptoms in a population-based study. Psychiatry Res. 2016;238:181–8.

Belancic A, Klobucar Majanovic S, Stimac D. The escalating global burden of obesity following the COVID-19 times: are we ready? Clin Obes. 2020;10:e12410.

Bevan M. Assessing renal function in older people. Nurs Older People. 2001;13:27–8.

Billman GE. Homeostasis: the underappreciated and far too often ignored central organizing principle of physiology. Front Physiol. 2020;11:200.

Bittel J. The different types of general cold adaptation in man. Int J Sports Med. 1992;13(Suppl 1):S172–6.

Blacher J, Levy BI, Mourad JJ, Safar ME, Bakris G. From epidemiological transition to modern cardiovascular epidemiology: hypertension in the 21st century. Lancet. 2016;388:530–2.

Booker JM, Hellekson CJ. Prevalence of seasonal affective disorder in Alaska. Am J Psychiatry. 1992;149:1176–82.

Brett JR, Groves TDD. Physiological energetics. London: Academic Press Inc.; 1979.

Brode P, Fiala D, Blazejczyk K, Holmer I, Jendritzky G, Kampmann B, Tinz B, Havenith G. Deriving the operational procedure for the universal thermal climate index (UTCI). Int J Biometeorol. 2012;56:481–94.

Burgess S. Rhabdomyolysis: an evidence-based approach. J Intensive Care Soc. 2022;23:513–7.

Burton AC. The clinical importance of the physiology of temperature regulation. Can Med Assoc J. 1956;75:715–20.

Carrillo AE, Flouris AD, Herry CL, Poirier MP, Boulay P, Dervis S, Friesen BJ, Malcolm J, Sigal RJ, Seely AJE, Kenny GP. Heart rate variability during high heat stress: a comparison between young and older adults with and without Type 2 diabetes. Am J Physiol Regul Integr Comp Physiol. 2016;311:R669–75.

Castellani JW, Young AJ. Human physiological responses to cold exposure: acute responses and acclimatization to prolonged exposure. Auton Neurosci. 2016;196:63–74.

Castellani JW, Young AJ, Kain JE, Rouse A, Sawka MN. Thermoregulation during cold exposure: effects of prior exercise. J Appl Physiol. 1999;87:247–52.

Castiglione V, Gentile F, Ghionzoli N, Chiriaco M, Panichella G, Aimo A, Vergaro G, Giannoni A, Passino C, Emdin M. Pathophysiological rationale and clinical evidence for neurohormonal modulation in heart failure with preserved ejection fraction. Card Fail Rev. 2023;9:e09.

Chadwick JMWN. The medical works of hippocrates: a new translation from the original greek made especially for english readers. Oxford: Blackwell Publishers; 2007.

Cheung SS, McLellan TM. Heat acclimation, aerobic fitness, and hydration effects on tolerance during uncompensable heat stress. J Appl Physiol. 1998;84:1731–9.

Cheung SS, McLellan TM, Tenaglia S. The thermophysiology of uncompensable heat stress. Physiological manipulations and individual characteristics. Sports Med. 2000;29:329–59.

Chong B, Jayabaskaran J, Kong G, Chan YH, Chin YH, Goh R, Kannan S, Ng CH, Loong S, Kueh MTW, Lin C, Anand VV, Lee ECZ, Chew HSJ, Tan DJH, Chan KE, Wang JW, Muthiah M, Dimitriadis GK, Hausenloy DJ, Mehta AJ, Foo R, Lip G, Chan MY, Mamas MA, et al. Trends and predictions of malnutrition and obesity in 204 countries and territories: an analysis of the global burden of disease study 2019. EClinicalMedicine. 2023;57:101850.

COPERNICUS Group. How changes in thermal stress will impact lives in the future; 2024. https://climate.copernicus.eu/how-changes-thermal-stress-will-impact-lives-future. Accessed April 2024.

Cradle of Humankind. https://www.maropeng.co.za/content/page/introduction-to-your-visit-to-the-cradle-of-humankind-world-heritage-site. Accessed May 2024.

Cui A, Zhang T, Xiao P, Fan Z, Wang H, Zhuang Y. Global and regional prevalence of vitamin D deficiency in population-based studies from 2000 to 2022: a pooled analysis of 7.9 million participants. Front Nutr. 2023;10:1070808.

Donaldson GC, Ermakov SP, Komarov YM, McDonald CP, Keatinge WR. Cold related mortalities and protection against cold in Yakutsk, eastern Siberia: observation and interview study. BMJ. 1998;317:978–82.

Falk B, Bar-Or O, Smolander J, Frost G. Response to rest and exercise in the cold: effects of age and aerobic fitness. J Appl Physiol. 1994;76:72–8.

Ford B, Dore M, Bartlett B. Management of heart failure: updated guidelines from the AHA/ACC. Am Fam Phys. 2023;108:315–20.

Forray AI, Coman MA, Simonescu-Colan R, Mazga AI, Chereches RM, Borzan CM. The global burden of type 2 diabetes attributable to dietary risks: insights from the global burden of disease study 2019. Nutrients. 2023;15:5292.

Frank SM, Raja SN, Bulcao C, Goldstein DS. Age-related thermoregulatory differences during core cooling in humans. Am J Physiol Regul Integr Comp Physiol. 2000;279:R349–54.

Freman K. Ancilla to the pre-socratic philosophers: a complete translation of the fragments diels fragmente der Vorsokratiker. Cambridge: Harvard University Press; 1948.

Frentusca CF, Babes K, Galusca DI. Vitamin D deficiency as an independent predictor of cardiovascular disease. Acta Endocrinol. 2023;19:319–25.

Fujii N, Aoki-Murakami E, Tsuji B, Kenny GP, Nagashima K, Kondo N, Nishiyasu T. Body temperature and cold sensation during and following exercise under temperate room conditions in cold-sensitive young trained females. Physiol Rep. 2017;5:133.

Fujii N, McGarr GW, Hatam K, Chandran N, Muia CM, Nishiyasu T, Boulay P, Ghassa R, Kenny GP. Heat shock protein 90 does not contribute to cutaneous vasodilatation in older adults during heat stress. Microcirculation. 2019;26:e12541.

Gersh BJ, Packer D. AHA/ACC/ESC 2006 atrial fibrillation guidelines: looking towards the future. Nat Clin Pract Cardiovasc Med. 2007;4:59.

Gherasie FA, Popescu MR, Bartos D. Acute coronary syndrome: disparities of pathophysiology and mortality with and without peripheral artery disease. J Pers Med. 2023;13:944.

Glass RI, Zack MM. Increase in deaths from ischaemic heart-disease after blizzards. Lancet. 1979;1:485–7.

Gona PN, Kimokoti RW, Gona CM, Ballout S, Rao SR, Mapoma CC, Lo J, Mokdad AH. Changes in body mass index, obesity, and overweight in Southern Africa development countries, 1990 to 2019: findings from the global burden of disease, injuries, and risk factors study. Obes Sci Pract. 2021;7:509–24.

Greaney JL, Alexander LM, Kenney WL. Sympathetic control of reflex cutaneous vasoconstriction in human aging. J Appl Physiol. 2015;119:771–82.

Greaney JL, Kenney WL, Alexander LM. Sympathetic regulation during thermal stress in human aging and disease. Auton Neurosci. 2016;196:81–90.

Green DJ, Maiorana AJ, Siong JH, Burke V, Erickson M, Minson CT, Bilsborough W, O'Driscoll G. Impaired skin blood flow response to environmental heating in chronic heart failure. Eur Heart J. 2006;27:338–43.

Grigorieva EA, Alexeev VA, Walsh JE. Universal thermal climate index in the arctic in an era of climate change: Alaska and Chukotka as a case study. Int J Biometeorol. 2023;67:1703–21.

Gupta A, Sharma PK, Garg VK, Singh AK, Mondal SC. Role of serotonin in seasonal affective disorder. Eur Rev Med Pharmacol Sci. 2013;17:49–55.

Halonen JI, Zanobetti A, Sparrow D, Vokonas PS, Schwartz J. Associations between outdoor temperature and markers of inflammation: a cohort study. Environ Health. 2010;9:42.

Hanna EG, Tait PW. Limitations to thermoregulation and acclimatization challenge human adaptation to global warming. Int J Environ Res Public Health. 2015;12:8034–74.

Hardy JD. Physiology of temperature regulation. Physiol Rev. 1961;41:521–606.

Heinonen I, Laukkanen JA. Effects of heat and cold on health, with special reference to Finnish sauna bathing. Am J Physiol Regul Integr Comp Physiol. 2018;314:R629–38.

Hiramatsu K, Yamada T, Katakura M. Acute effects of cold on blood pressure, renin-angiotensin-aldosterone system, catecholamines and adrenal steroids in man. Clin Exp Pharmacol Physiol. 1984;11:171–9.

Holowatz LA, Thompson-Torgerson C, Kenney WL. Aging and the control of human skin blood flow. Front Biosci. 2010;15:718–39.

Horowitz M, Robinson SD. Heat shock proteins and the heat shock response during hyperthermia and its modulation by altered physiological conditions. Prog Brain Res. 2007;162:433–46.

Inoue Y, Nakao M, Araki T, Ueda H. Thermoregulatory responses of young and older men to cold exposure. Eur J Appl Physiol Occup Physiol. 1992;65:492–8.

Ivanova YM, Blondin DP. Examining the benefits of cold exposure as a therapeutic strategy for obesity and type 2 diabetes. J Appl Physiol. 2021;130:1448–59.

Jiang S, Yu T, Di D, Wang Y, Li W. Worldwide burden and trends of diabetes among people aged 70 years and older, 1990–2019: a systematic analysis for the global burden of disease study 2019. Diabetes Metab Res Rev. 2024;40:e3745.

Kaltsatou A, Notley SR, Kenny GP. Effects of exercise-heat stress on circulating stress hormones and interleukin-6 in young and older men. Temperature. 2020;7:389–93.

Kario K, Okura A, Hoshide S, Mogi M. The WHO global report 2023 on hypertension warning the emerging hypertension burden in globe and its treatment strategy. Hypertens Res. 2024;47:1099–102.

Keatinge WR. Global warming and health. BMJ. 1991;302:965–6.

Keatinge WR, Donaldson GC. The impact of global warming on health and mortality. South Med J. 2004;97:1093–9.

Kegel M, Dam H, Ali F, Bjerregaard P. The prevalence of seasonal affective disorder (SAD) in Greenland is related to latitude. Nord J Psychiatry. 2009;63:331–5.

Kelly T, Yang W, Chen CS, Reynolds K, He J. Global burden of obesity in 2005 and projections to 2030. Int J Obes. 2008;32:1431–7.

Kenney WL, Craighead DH, Alexander LM. Heat waves, aging, and human cardiovascular health. Med Sci Sports Exerc. 2014;46:1891–9.

Kenny GP, Yardley J, Brown C, Sigal RJ, Jay O. Heat stress in older individuals and patients with common chronic diseases. CMAJ. 2010;182:1053–60.

Khraishah H, Alahmad B, Ostergard RL, AlAshqar A, Albaghdadi M, Vellanki N, Chowdhury MM, Al-Kindi SG, Zanobetti A, Gasparrini A, Rajagopalan S. Climate change and cardiovascular disease: implications for global health. Nat Rev Cardiol. 2022;19:798–812.

Khudaiberdiev MD, Sultanov FF, Pokormyakha LM. Perception of temperature elevation in human seasonal heat adaptation. Neurosci Behav Physiol. 1992;22:236–40.

Kingma BR, Frijns AJ, Schellen L, van Marken Lichtenbelt WD. Beyond the classic thermoneutral zone: Including thermal comfort. Temperature. 2014;1:142–9.

Kingma BRM, Charkoudian N, White MD. Physiology and pharmacology of temperature regulation: from basic to applied and across environments. Temperature. 2022;9:115–8.

Kirchhof P, Benussi S, Kotecha D, Ahlsson A, Atar D, Casadei B, Castella M, Diener HC, Heidbuchel H, Hendriks J, Hindricks G, Manolis AS, Oldgren J, Popescu BA, Schotten U, Van Putte B, Vardas P. 2016 ESC guidelines for the management of atrial fibrillation developed in collaboration with EACTS. Kardiol Pol. 2016;74:1359–469.

Konemann H, Ellermann C, Zeppenfeld K, Eckardt L. Management of ventricular arrhythmias worldwide: comparison of the latest ESC, AHA/ACC/HRS, and CCS/CHRS guidelines. JACC Clin Electrophysiol. 2023;9:715–28.

Kosunen KJ, Pakarinen AJ, Kuoppasalmi K, Adlercreutz H. Plasma renin activity, angiotensin II, and aldosterone during intense heat stress. J Appl Physiol. 1976;41:323–7.

Kruijt N, van den Bersselaar LR, Hopman MTE, Snoeck MMJ, van Rijswick M, Wiggers TGH, Jungbluth H, Bongers C, Voermans NC. Exertional heat stroke and rhabdomyolysis: a medical record review and patient perspective on management and long-term symptoms. Sports Med Open. 2023;9:33.

Kularatna S, Jadambaa A, Hewage S, Brain D, McPhail S, Parsonage W. Global, regional, and national burden of heart failure associated with atrial fibrillation. BMC Cardiovasc Disord. 2023;23:345.

Lambert MTM, Dugas J. Ethnicity and thermoregulation. In: Fe M, editor. Thermoregulation and human performance physiological and biological aspects basel. Switzerland: Med Sport Sci; 2008.

Leppaluoto J, Korhonen I, Hassi J. Habituation of thermal sensations, skin temperatures, and norepinephrine in men exposed to cold air. J Appl Physiol. 2001;90:1211–8.

Lim CL. Fundamental concepts of human thermoregulation and adaptation to heat: a review in the context of global warming. Int J Environ Res Public Health. 2020;17:5468.

Lu M, Li D, Hu Y, Zhang L, Li Y, Zhang Z, Li C. Persistence of severe global inequalities in the burden of hypertension heart disease from 1990 to 2019: findings from the global burden of disease study 2019. BMC Public Health. 2024;24:110.

Luo J, Zhao X, Li Q, Zou B, Xie W, Lei Y, Yi J, Zhang C. Evaluating the global impact of low physical activity on type 2 diabetes: insights from the global burden of disease 2019 study. Diabetes Obes Metab. 2024;26:2456–65.

Ma Y, Olendzki BC, Li W, Hafner AR, Chiriboga D, Hebert JR, Campbell M, Sarnie M, Ockene IS. Seasonal variation in food intake, physical activity, and body weight in a predominantly overweight population. Eur J Clin Nutr. 2006;60:519–28.

Ma Q, Zhu J, Zheng P, Zhang J, Xia X, Zhao Y, Cheng Q, Zhang N. Global burden of atrial fibrillation/flutter: trends from 1990 to 2019 and projections until 2044. Heliyon. 2024;10:e24052.

Magnusson A. Validation of the seasonal pattern assessment questionnaire (SPAQ). J Affect Disord. 1996;40:121–9.

Makinen TM. Different types of cold adaptation in humans. Front Biosci. 2010;2:1047–67.

Malta DC, Gomes CS, Felisbino-Mendes MS, Veloso GA, Machado IE, Cardoso LO, Azeredo RT, Jaime PC, Vasconcelos LLC, Naghavi M, Ribeiro ALP. Undernutrition, and overweight and obesity: the two faces of malnutrition in Brazil, analysis of the Global Burden of Disease, 1990 to 2019. Public Health. 2024;229:176–84.

Marti-Soler H, Gubelmann C, Aeschbacher S, Alves L, Bobak M, Bongard V, Clays E, de Gaetano G, Di Castelnuovo A, Elosua R, Ferrieres J, Guessous I, Igland J, et al. Seasonality of cardiovascular risk factors: an analysis including over 230 000 participants in 15 countries. Heart. 2014;100:1517–23.

McDonagh TA, Metra M, Adamo M, Gardner RS, Baumbach A, Bohm M, Burri H, Butler J, Celutkiene J, Chioncel O, Cleland JGF, Coats AJS, Crespo-Leiro MG, Farmakis D, Gilard M, Heymans S, Hoes AW, Jaarsma T, Jankowska EA, Lainscak M, Lam CSP, et al. 2021 ESC guidelines for the diagnosis and treatment of acute and chronic heart failure. Eur Heart J. 2021;42:3599–726.

McDonagh TA, Metra M, Adamo M, Gardner RS, Baumbach A, Bohm M, Burri H, Butler J, Celutkiene J, Chioncel O, Cleland JGF, Crespo-Leiro MG, Farmakis D, Gilard M, Heymans S, Hoes AW, Jaarsma T, Jankowska EA, Lainscak M, et al. 2023 Focused update of the 2021 ESC guidelines for the diagnosis and treatment of acute and chronic heart failure. Eur Heart J. 2023;44:3627–39.

McLellan TM. The importance of aerobic fitness in determining tolerance to uncompensable heat stress. Comp Biochem Physiol A Mol Integr Physiol. 2001;128:691–700.

Mellati AA. The role of heat shock proteins as chaperones on several human diseases. Saudi Med J. 2006;27:1302–5.

Mocumbi AO, Cebola B, Muloliwa A, Sebastiao F, Sitefane SJ, Manafe N, Dobe I, Lumbandali N, Keates A, Stickland N, Chan YK, Stewart S. Differential patterns of disease and injury in Mozambique: New perspectives from a pragmatic, multicenter, surveillance study of 7809 emergency presentations. PLoS ONE. 2019;14:e0219273.

Montreal Science Centre. Dry cold, damp cold… is winter weather colder when humidity is higher? 2023. https://www.montrealsciencecentre.com/blog/dry-cold-damp-cold-winter-weather-colder-when-humidity-higher. Accessed May 2024.

Murray G. How common is seasonal affective disorder in temperate Australia? A comparison of BDI and SPAQ estimates. J Affect Disord. 2004;81:23–8.

Murray KO, Brant JO, Iwaniec JD, Sheikh LH, de Carvalho L, Garcia CK, Robinson GP, Alzahrani JM, Riva A, Laitano O, Kladde MP, Clanton TL. Exertional heat stroke leads to concurrent long-term epigenetic memory, immunosuppression and altered heat shock response in female mice. J Physiol. 2021;599:119–41.

Neshumova TV, Danilenko KV, Putilov AA. Response of the cardiovascular system during seasonal affective disorder and phototherapy. Fiziol Cheloveka. 1994;20:83–8.

Ockene IS, Chiriboga DE, Stanek EJ, Harmatz MG, Nicolosi R, Saperia G, Well AD, Freedson P, Merriam PA, Reed G, Ma Y, Matthews CE, Hebert JR. Seasonal variation in serum cholesterol levels: treatment implications and possible mechanisms. Arch Intern Med. 2004;164:863–70.

Ohlrogge AH, Brederecke J, Schnabel RB. Global burden of atrial fibrillation and flutter by national income: results from the global burden of disease 2019 database. J Am Heart Assoc. 2023;12:e030438.

Omran A. The epidemiological transition theory revisited thirty years later. 1998. https://iris.who.int/bitstream/handle/10665/330604/WHSQ-1998-51-n2-3-4-eng.pdf. Accessed May 2024.

Osilla EV, Marsidi JL, Shumway KR, Sharma S. Physiology, temperature regulation StatPearls treasure island (FL) ineligible companies. Disclosure: Jennifer Marsidi declares no relevant financial relationships with ineligible companies. Disclosure: Karlie Shumway declares no relevant financial relationships with ineligible companies. Disclosure: Sandeep Sharma declares no relevant financial relationships with ineligible companies; 2024.

Pedone C, Corsonello A, Incalzi RA, Investigators G. Estimating renal function in older people: a comparison of three formulas. Age Ageing. 2006;35:121–6.

Rammsayer T, Hennig J, Bahner E, von Georgi R, Opper C, Fett C, Wesemann W, Netter P. Lowering of body core temperature by exposure to a cold environment and by a 5-HT1A agonist: effects on physiological and psychological variables and blood serotonin levels. Neuropsychobiology. 1993;28:37–42.

Ratwatte P, Wehling H, Kovats S, Landeg O, Weston D. Factors associated with older adults' perception of health risks of hot and cold weather event exposure: a scoping review. Front Public Health. 2022;10:939859.

Rosen LN, Targum SD, Terman M, Bryant MJ, Hoffman H, Kasper SF, Hamovit JR, Docherty JP, Welch B, Rosenthal NE. Prevalence of seasonal affective disorder at four latitudes. Psychiatry Res. 1990;31:131–44.

Roth GA, Mensah GA, Johnson CO, Addolorato G, Ammirati E, Baddour LM, Barengo NC, Beaton AZ, Benjamin EJ, Benziger CP, Bonny A, Brauer M, Brodmann M, Cahill TJ, Carapetis J, Catapano AL, Chugh SS, Cooper LT, Coresh J, Criqui M, DeCleene N, Eagle KA, et al. Global burden of cardiovascular diseases and risk factors, 1990–2019: update from the GBD 2019 study. J Am Coll Cardiol. 2020;76:2982–3021.

Salerno PR, Chen Z, Wass S, Motairek I, Elamm C, Salerno LM, Hassani NS, Deo SV, Al-Kindi SG. Sex-specific heart failure burden across the United States: global burden of disease 1990–2019. Am Heart J. 2024;269:35–44.

Sauter T, Haider DG, Ricklin ME, Exadaktylos AK. The snow, the men, the shovel, the risk? ER admissions after snow shovelling: 13 winters in Bern. Swiss Med Wkly. 2015;145:w14104.

Scientific America. Heat waves affect children more severely; 2022. https://www.scientificamerican.com/article/heat-waves-affect-children-more-severely/. Accessed May 2024.

Seidell JC, Halberstadt J. The global burden of obesity and the challenges of prevention. Ann Nutr Metab. 2015;66(Suppl 2):7–12.

Selkirk GA, McLellan TM. Influence of aerobic fitness and body fatness on tolerance to uncompensable heat stress. J Appl Physiol. 2001;91:2055–63.

Shahim B, Kapelios CJ, Savarese G, Lund LH. Global public health burden of heart failure: an updated review. Card Fail Rev. 2023;9:e11.

Silva JE, Melo N, Ferreira AI, Silva C, Oliveira D, Lume MJ, Pereira J, Araujo JP, Lourenco P. Prognostic impact of neurohormonal modulation in very old patients with chronic heart failure. Age Ageing. 2022;51:598.

Silva-Cardoso J, Bras D, Canario-Almeida F, Andrade A, Oliveira L, Padua F, Fonseca C, Braganca N, Carvalho S, Soares R, Santos JF. Neurohormonal modulation: the new paradigm of pharmacological treatment of heart failure. Rev Port Cardiol. 2019;38:175–85.

Sliwa K, Wilkinson D, Hansen C, Ntyintyane L, Tibazarwa K, Becker A, Stewart S. Spectrum of heart disease and risk factors in a black urban population in South Africa (the Heart of Soweto Study): a cohort study. Lancet. 2008;371:915–22.

Smithsonian Institute. The great human migration; 2008. https://www.smithsonianmag.com/history/the-great-human-migration-13561/. Accessed June 2024.

Smolander J. Effect of cold exposure on older humans. Int J Sports Med. 2002;23:86–92.

Stewart S. Population screening for heart disease in vulnerable populations: lessons from the Heart of Soweto study. Heart Lung Circ. 2009;18:104–6.

Stewart S, Marley JE, Horowitz JD. Effects of a multidisciplinary, home-based intervention on unplanned readmissions and survival among patients with chronic congestive heart failure: a randomised controlled study. Lancet. 1999;354:1077–83.

Stewart S, Wilkinson D, Becker A, Askew D, Ntyintyane L, McMurray JJ, Sliwa K. Mapping the emergence of heart disease in a black, urban population in Africa: the Heart of Soweto Study. Int J Cardiol. 2006;108:101–8.

Stewart S, Wilkinson D, Hansen C, Vaghela V, Mvungi R, McMurray J, Sliwa K. Predominance of heart failure in the heart of Soweto Study cohort: emerging challenges for urban African communities. Circulation. 2008;118:2360–7.

Stewart S, Carrington M, Pretorius S, Methusi P, Sliwa K. Standing at the crossroads between new and historically prevalent heart disease: effects of migration and socio-economic factors in the Heart of Soweto cohort study. Eur Heart J. 2011;32:492–9.

Stewart S, Keates AK, Redfern A, McMurray JJV. Seasonal variations in cardiovascular disease. Nat Rev Cardiol. 2017;14:654–64.

Stewart S, Moholdt TT, Burrell LM, Sliwa K, Mocumbi AO, McMurray JJ, Keates AK, Hawley JA. Winter peaks in heart failure: an inevitable or preventable consequence of seasonal vulnerability? Card Fail Rev. 2019;5:83–5.

Sun Y, MacRae TH. The small heat shock proteins and their role in human disease. FEBS J. 2005;272:2613–27.

Tabei S, Chamorro R, Meyhofer SM, Wilms B. Metabolic effects of brown adipose tissue activity due to cold exposure in humans: a systematic review and meta-analysis of RCTs and non-RCTs. Biomedicines. 2024;12:1133.

Taylor NA. Human heat adaptation. Compr Physiol. 2014;4:325–65.

Taylor NA, Allsopp NK, Parkes DG. Preferred room temperature of young versus aged males: the influence of thermal sensation, thermal comfort, and affect. J Gerontol A Biol Sci Med Sci. 1995;50:M216–21.

Teerlink JR. Recent heart failure trials of neurohormonal modulation (OVERTURE and ENABLE): approaching the asymptote of efficacy? J Card Fail. 2002;8:124–7.

Therminarias A, Quirion A, Pellerei E, Laurencelle L. Effects of acute cold exposure on physiological responses obtained during a short exhaustive exercise. J Sports Med Phys Fitness. 1988;28:45–50.

Thompson GE. Physiological effects of cold exposure. Int Rev Physiol. 1977;15:29–69.

Tibazarwa K, Ntyintyane L, Sliwa K, Gerntholtz T, Carrington M, Wilkinson D, Stewart S. A time bomb of cardiovascular risk factors in South Africa: results from the heart of Soweto Study "heart awareness days." Int J Cardiol. 2009;132:233–9.

Tousoulis D. Vitamin D deficiency and cardiovascular disease: fact or fiction? Hellenic J Cardiol. 2018;59:69–71.

Tsuzuki-Hayakawa K, Tochihara Y, Ohnaka T. Thermoregulation during heat exposure of young children compared to their mothers. Eur J Appl Physiol Occup Physiol. 1995;72:12–7.

Tyrer AE, Levitan RD, Houle S, Wilson AA, Nobrega JN, Rusjan PM, Meyer JH. Serotonin transporter binding is reduced in seasonal affective disorder following light therapy. Acta Psychiatr Scand. 2016a;134:410–9.

Tyrer AE, Levitan RD, Houle S, Wilson AA, Nobrega JN, Meyer JH. Increased seasonal variation in serotonin transporter binding in seasonal affective disorder. Neuropsychopharmacology. 2016b;41:2447–54.

United Nations. Synergy solutions for a world in crisis: tackling climate and SDG action together Report On Strengthening The Evidence Base; 2023. https://sdgs.un.org/sites/default/files/2023-09/UN%20Climate%20SDG%20Synergies%20Report-091223B_1.pdf. Accessed May 2024.

United Nations Department of Economic and Social Affairs. Sustainable development (the 17 goals); 2024. https://sdgs.un.org/goals. Accessed June 2024

United Nations Education Programme. Facts about climate change; 2024. https://www.unep.org/facts-about-climate-emergency?gad_source=1&gclid=Cj0KCQjwq86wBhDiARIsAJhuphmq6lFXvMLipqQIaPKDeBSHuiFG47pbdzgQNuOvItQxSHf29LoerhsAsF9EALw_wcB. Accessed May 2024

Vaduganathan M, Mensah GA, Turco JV, Fuster V, Roth GA. The global burden of cardiovascular diseases and risk: a compass for future health. J Am Coll Cardiol. 2022;80:2361–71.

Verdoia M, De Luca G. Is there an actual link between vitamin D deficiency, cardiovascular disease, and glycemic control in patients with type 2 diabetes mellitus? Pol Arch Intern Med. 2023;133:57452.

Virani SS, Newby LK, Arnold SV, Bittner V, Brewer LC, Demeter SH, Dixon DL, Fearon WF, Hess B, Johnson HM, Kazi DS, Kolte D, Kumbhani DJ, LoFaso J, Mahtta D, Mark DB, Minissian M, Navar AM, Patel AR, et al. 2023 AHA/ACC/ACCP/ASPC/NLA/PCNA guideline for the management of patients with chronic coronary disease: a report of the American heart association/American college of cardiology joint committee on clinical practice guidelines. J Am Coll Cardiol. 2023;82:833–955.

Wang Y, Zhang W, Qian T, Sun H, Xu Q, Hou X, Hu W, Zhang G, Drummond GR, Sobey CG, Charchar FJ, Golledge J, Yang G. Reduced renal function may explain the higher prevalence of hyperuricemia in older people. Sci Rep. 2021;11:1302.

Werner KB, Elmstahl S, Christensson A, Pihlsgard M. Male sex and vascular risk factors affect cystatin C-derived renal function in older people without diabetes or overt vascular disease. Age Ageing. 2014;43:411–7.

World Bank. Population indicators; 2024. https://databank.worldbank.org/indicator/SP.DYN.LE00.IN/1ff4a498/Popular-Indicators. Accessed May 2024.

World Economic Forum. https://www.weforum.org/agenda/2022/05/mapped-the-world-s-population-density-by-latitude/#:~:text=In%20particular%2C%20the%2025th%20and,States%2C%20Mexico%2C%20and%20others. Accessed June 2024.

World Population Review. https://worldpopulationreview.com/country-rankings/happiest-countries-in-the-world. Accessed May 2024.

Xie J, Lin X, Fan X, Wang X, Pan D, Li J, Hao Y, Jie Y, Zhang L, Gu J. Global burden and trends of primary liver cancer attributable to comorbid type 2 diabetes mellitus among people living with hepatitis b: an observational trend study from 1990 to 2019. J Epidemiol Glob Health. 2024;32:5642.

Xu X, Rioux T, Friedl K, Gonzalez J, Castellani J. Development of interactive guidance for cold exposure using a thermoregulatory model. Int J Circumpolar Health. 2023;82:2190485.

Yang X, Sun J, Zhang W. Global trends in burden of type 2 diabetes attributable to physical inactivity across 204 countries and territories, 1990–2019. Front Endocrinol. 2024;15:1343002.

Yurkevicius BR, Alba BK, Seeley AD, Castellani JW. Human cold habituation: physiology, timeline, and modifiers. Temperature. 2022;9:122–57.

Zhang JN, Wood J, Bergeron AL, McBride L, Ball C, Yu Q, Pusiteri AE, Holcomb JB, Dong JF. Effects of low temperature on shear-induced platelet aggregation and activation. J Trauma. 2004;57:216–23.

Zhou A, Selvanayagam JB, Hypponen E. Non-linear Mendelian randomization analyses support a role for vitamin D deficiency in cardiovascular disease risk. Eur Heart J. 2022;43:1731–9.

# From Climatic Resilience to Vulnerability

## Abstract

In the last chapter, the biological limits of human adaptation in the face of climatic conditions outside of our mandated physiological and personal comfort zones were explored—with a major focus on what happens to the heart and cardiovascular system when confronted with cold to hot extremes. It also introduced the concept of non-modifiable versus modifiable factors that modulate the biological impact of climatic provocations to an individual. As an extension of this concept, this chapter now presents a holistic, interdisciplinary framework/model that helps to explain why certain people (with consideration of sex-based differences) are '*climatically vulnerable*', while others are more '*climatically resilient*'—identifying the key characteristics and attributes that might be altered to prevent climatically provoked cardiac events and premature mortality at the individual level.

## Keywords

Bio-behavioural model · Sex-specific differences · Physiological traits · Modulating factors · Vulnerable phenotype · Resilient phenotype · Winter · Summer

## 4.1 A Spectrum of Resilience to Vulnerability

If one looks up the meaning of 'vulnerable' (e.g. the online Cambridge dictionary), one of the key examples of it being used in a sentence immediately stands out—"*Older people are especially vulnerable to cold temperatures even inside their homes*" (https://dictionary.cambridge.org/dictionary/english/vulnerable). The opposite to vulnerability, a word that is increasingly used in an increasingly stressful world [but not often in terms of the climate and our response to climatic provocations) is 'resilient'. From a climate change perspective, the example provided by the Cambridge dictionary is very apt ("*This rubber ball is very resilient and immediately springs back into shape*" (https://dictionary.cambridge.org/dictionary/english/resilient)]. This is probably how we would want to describe the global climate once our increasingly harmful human footprint (described in Chap. 1) is modulated for the better.

As remarked by the Massachusetts Institute of Technology's Adam Schlosser (their Deputy Director of the MIT Joint Program on the Science and Policy of Global Change) in relation to reports from the prestigious Intergovernmental Panel on Climate Change (Intergovernmental Panel on Climate Change 2023a; b)—"*If I had to rate odds, I would say*

© The Author(s) 2024
S. Stewart, *Heart Disease and Climate Change*, Sustainable Development Goals Series,
https://doi.org/10.1007/978-3-031-73106-8_4

*the chances of climate change driving us to the point of human extinction are very low, if not zero*" (MIT Climate Portal 2024). If this perspective (from a highly informed climate scientist) is correct, the relevant question is not "*will humans become extinct?*" but "*will humans become sicker, have poorer quality of life and live shorter lives in the face to climate changes?*". While previous Chapters have already laid the foundations for answering this question, this Chapter will describe the characteristics of those who are likely to still thrive (i.e. be '*climatically resilient*') versus those who are most likely (without therapeutic intervention) to be '*climatically vulnerable*' and suffer accordingly. For the specific purposes of this book, these two opposing states/phenotypes are best defined as follows:

- *Climatically Resilient*—an inherent ability (typically strengthened during childhood exposures and cultural reinforcement/practices) to buffer and rapidly recover from exposures to challenging and extreme climatic (and other environmental) conditions from a physiological (physical and psychological) perspective—thereby decreasing the probability of cardiovascular instability and/or a cardiac-specific or broader cardiovascular-related event occurring at certain times of the day/year.
- *Climatically Vulnerability*—an impaired ability to 'buffer' and rapidly recover from exposures to challenging and extreme climatic (and other environmental) conditions from a physiological (physical and psychological) perspective—thereby increasing the risk of cardiovascular-related clinical instability, with a higher probability of a cardiac-specific event or cardiovascular-related crisis occurring at certain times of the day/year.

As will be explained in more detail in the next chapter, while we cannot reliably predict when marked variations in climatic conditions (e.g. heatwaves and cold spells) will occur—noting the major influence of hemisphere-wide phenomena such as La Niña (Australian Bureau of Meterology 2024) and El Niño (Srinivas et al.

2024; Deng and Dai 2024; Jiang et al. 2024; Cordero et al. 2024), there is a reliable indicator of climatic provocations to the health of nearly every person on the planet (noting that even the tropics have subtle transitions from relatively 'dry' to 'wet' conditions)—'seasonal transitions' in weather conditions. It is this phenomenon that acts as a natural and pervasive 'control' phenomenon that influences many aspects of our health and behaviours without us consciously being aware of it. It is also the basis for local to global models seeking to understand the impact of the weather and longer-term climatic conditions on human health. As succinctly put—"*Climate is what you expect and weather is what you get!*" (National Centers for Environmental Information 2024). This is an important distinction, given that while acute weather changes are dynamic from every perspective, they can still have an accumulative effect over a period of time—the seasons being characterised by specific weather patterns (typically mean temperature, rainfall, wind, cloud cover and, of course, the number of daylight hours). It is these predictable and more prolonged weather conditions that are pivotal to our understanding/modelling what will happen with climate change at the whole population level. Before considering this, it is important to explore what 'climatic resilience' versus 'climatic vulnerability' looks like at the individual level.

## 4.2   Sex-Specific Differences in Cardiovascular Vulnerability

As originally reported from the Framingham Heart Study (Lloyd-Jones et al. 1999), the lifetime risk of developing coronary heart disease is approximately one in two men and one in three men from the age of 40 years onwards. Specifically, the Framingham Investigators reported that at age 40 and 70 years, respectively, the lifetime risk was 48.6% (95% CI 45.8–51.3) for men and 31.7% (29.2–34.2) for women and then 34.9% (31.2–38.7) for men and 24.2% (21.4–27.0) for women (Lloyd-Jones et al. 1999). However, when considering the specific location (a relatively wealthy and homogenous community in North America), it is important to consider that when one considers other communities and other cardiovascular conditions, let alone changing risk factors based on sex/gender, these differences may not be as stark as first seems. For example, as reported by Wang and Hoy (2013), in an Australian study of a remote aboriginal community followed up for 20-years, the average age at which men and women developed coronary heart disease was very similar (48 and 49 years, respectively). Overall, the competing risk was 70.7% for men and 63.8% for women. When adjusting for competing risk of death from non-cardiac causes, the lifetime risk remained similar at 52.6% for men and 49.2% for women (Wang and Hoy 2013). When one considers the pattern of hypertension and heart failure in men and women in regions such as sub-Saharan Africa, there is a complete reversal, with more women than men affected (Keates et al. 2017). Such differences appear at an early age [most likely around the age of puberty and the transition from adolescence to adulthood (Chen et al. 2023; Melaku et al. 2019)]. In high-income countries, there is a predominance of heart failure/left ventricular dysfunction characterised by more systolic impairment (reduced ejection fraction) in men and more preserved systolic/impaired diastolic function (preserved ejection fraction) in women (McDonagh et al. 2021). Moreover, when the pattern of ejection fractions reported in the *National Echo Database of Australia* [a large national, clinical cohort with individually linked mortality outcomes (Strange et al. 2021, 2019a, b; Playford et al. 2023, 2021)] was examined it confirmed two important things. Firstly, as with previous studies, overall, there are roughly equal numbers of men and women affected by heart failure with women more likely to develop the syndrome at an older age. Secondly, when adjusting for their left ventricular ejection fraction (LVEF), there was a different threshold of mortality. Specifically, the adjusted risk for cardiovascular-related mortality for a LVEF of 55.0–59.9% was 1.36 (95% CI 1.16–1.59) in women compared to 1.21 (95% CI 1.05–1.39). In women

only, a LVEF of 60.0–64.9% was still associated with a significant risk of cardiovascular mortality (1.33-fold higher, 95% CI 1.16–1.52). These findings strengthened among those aged <65 years (Stewart et al. 2021). From the perspective of acute coronary syndromes and events leading to wall motion abnormalities of the left ventricular wall (detected on echocardiography) that severely compromise systolic function and therefore a person's LVEF, the *National Echo Database of Australia*, confirmed a male bias is such cardiac pathology (Playford et al. 2023). Specifically, in a study of 492,338 men and women aged $61.9 \pm 17.9$ years undergoing routine investigation with echocardiography, a total of 39, 346 of 255,697 men (15.4%) versus 17,834 of 236,641 women (7.5%) had evidence of a *wall motion abnormality of the left ventricle* (often without clear evidence that they had been treated for the cardiac event that caused it). Thus, the ratio of men and women detected with this abnormality was approximately 2:1. Moreover, where the abnormality was found (indicating a differential pattern of coronary artery disease and subsequent survival was different on a sex-specific basis, with the a defect found in the left ventricular inferior wall versus anterior wall being the most and least common wall motion abnormality in men (8.0 and 2.5%) and women (3.3 and 1.1%), respectively. Any wall motion abnormality was associated with increased 5-year mortality. In men, it rose from 17.5% (no defect) to 29.7% (any defect), while in women it rose from 14.9% (no defect) to 30.8% (any defect) in women. Known myocardial infarction (0.86-fold [95% CI, 0.80–0.93] less likely) or coronary revascularisation procedure including bypass or angioplasty (0.87-fold [95% CI, 0.82–0.92] less likely) was independently associated with a better prognosis. Alternatively, men (1.22-fold increase versus women) and those with greater systolic/diastolic dysfunction had a worse prognosis. Indicative of the importance of sex-specific pattern of wall abnormalities detected, among those with any wall motion abnormality found on echo, apical (1.08-fold [95% CI, 1.02–1.13] more likely) or inferior (1.09-fold [95% CI, 1.04–1.15] more likely) akinesis, dyskinesis or aneurysm, or a wall motion score index > 3.0 conveyed the worst prognosis (Playford et al. 2023).

As will be discussed in Chap. 5, the timing of the cardiac 'insult' (almost certainly a non-fatal acute myocardial infarction) that resulted in the defects detected in this study is undoubtedly influenced by climatic conditions. Moreover, people who develop heart failure are also at the mercy of the weather. What we don't know, is whether certain cardiac abnormalities are more susceptible to climatic provocations. However, differential responses to the weather and extreme climatic conditions on a sex-specific basis, suggest this might be the case, beyond specific physiological differences between biological males and females. As reported in a previous Scottish report, even if a man and woman suffer the same event (e.g. an acute coronary event) the consequences are often different in respect to—an out-of-hospital cardiac arrest, survival to acute hospital care, initial survival/mortality when being actively managed (within 7-days) (Ruane et al. 2017) and then within 28-days and 1-year of the event

(MacIntyre et al. 2001). Overall, the rate of men and women developing the potentially fatal syndrome of heart failure following an acute myocardial infarction appears to be declining (Docherty et al. 2023). However, regardless of the likely disproportionate impact of climate change, improving the care of women affected by all forms of heart disease, represents a particularly important clinical and public health priority (Lam and Myhre 2023; Ravera et al. 2022; DeFilippis et al. 2022; Zaman et al. 2021; McMurray et al. 2020; Beale et al. 2018).

Without delving into the exact reasons why men and women differ in their risk of developing heart disease (and indeed many other forms of cardiovascular disease), with markedly different trajectories and natural histories depending on their life-stage, as initially suggested in Chap. 3, the section above highlights an important point when assessing the threat of climate change on the heart health based on biological sex and even gender identify. As recently commented by Suman et al. (2023)—"*Cardiovascular death in women is major concern which is still under-recognition*

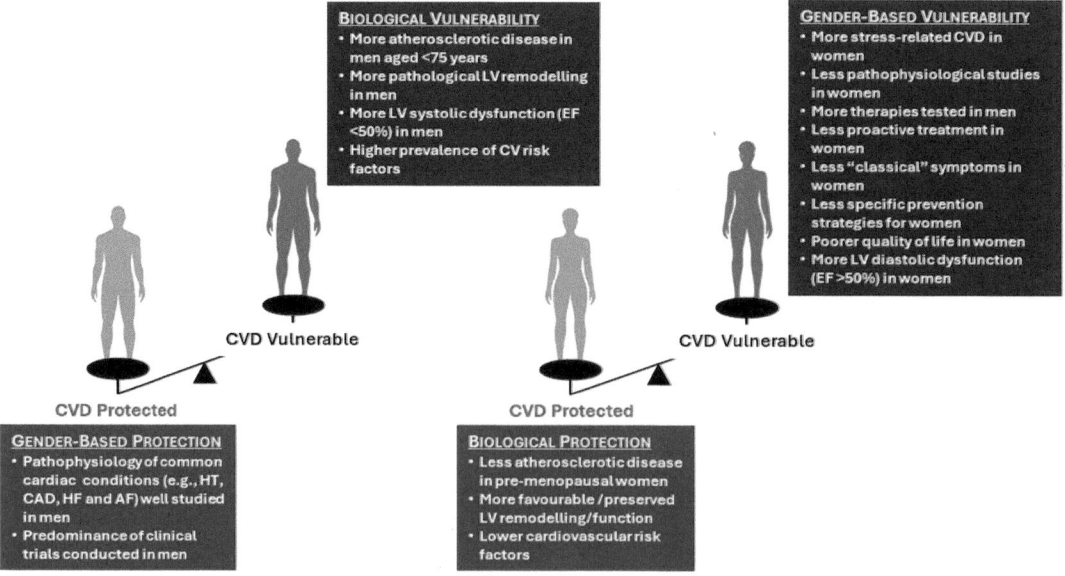

**Fig. 4.1** Sex-specific differences in the development and outcomes from cardiovascular disease. Building on the sex- and gender-based influences on cardiac health, this figure summaries the key factors and characteristics of men and women from a cardio-protective to vulnerability to cardiovascular disease perspective (irrespective of climatic provocations) that will drive a 'basal' level of cardiovascular events with any population

*and untreated. Consideration of gender differences is important for prevention, diagnosis, treatment, and management of CVD".*

In this context, Fig. 4.1 summarises some of the key characteristics that are of major importance when assessing climatic vulnerability in men and women—noting that the age profile and indeed, cardiovascular pathology according to biological sex is likely to vary, as will a person's lifestyle/decision-making be influenced by their gender/identity. As will be explained below, this becomes more important when considering the characteristics of who is more likely to experience clinical deterioration/instability and/or major cardiac event in response to external climatic/weather conditions, as opposed to the "internal" mechanisms of cardiovascular pathology. For example, as reported by Gifford et al., who undertook a systematic review (36 studies) and meta-analysis (22 studies) of the relevant literature, the reported rate of heat-related illnesses (note this was not a cardiac-specific report) among women was consistently lower than men across the lifespan. Specifically, they found that males were 2.28-fold (95% CI 1.66–3.16) more likely to suffer an event and observed differences persisted when correcting for the severity the event and their occupation (noting the expectation that manual occupations have more heat exposure) (Gifford et al. 2019).

When writing this section of the book, I attended the funeral of a person who was the inspiration for a Parliamentary Event in Canberra, Australia in 2017—*Women and Heart Disease*. She bravely talked in front of many important people from the Federal Parliament/Senate and Department of Health, uttering the same, powerful words we used to run a media campaign to highlight the forgotten component (something very close to my own family's heart) of heart disease in many countries—"*my husband doesn't have heart disease I do!*" (Sydney Morning Herald 2017). **Vale—Ruth Marie Budge (April 2024).**

Unfortunately, the ratio of women being included in major therapeutic trials over the past three decades has been poor (Vogel et al. 2021; McMurray and Berry 2000; McMurray et al. 2012; Packer et al. 2015; Jin et al. 2020). This phenomenon (with women being typically older and excluded from trials because of multimorbidity) is something not replicated to the same degree in multidisciplinary management trials (McAlister et al. 2004). Encouragingly, with increased recognition of the role of diastolic dysfunction/preserved ejection fraction in driving poor health outcomes, and the application of new therapeutics such as sacubitril-valsartan and the much-vaunted SGLT-2 inhibitors, female participation in clinical trials has recently increased (Mehran et al. 2022; Pfeffer et al. 2022; Vaduganathan et al. 2022; Yeoh et al. 2022; Dewan et al. 2023; Ern Yeoh et al. 2023; Malik et al. 2023; McDonagh et al. 2023). Moreover, there are efforts to better understand [informed by big-data surveillance studies (Stewart et al. 2021)], to identify sex-specific thresholds for therapeutic intervention, rather than rely on clinical data mainly derived from younger men (McMurray et al. 2020; Gohar et al. 2017). It is certainly postulated within this book, that we need to understand climate change and its impact on the health of men and women through the lens of likely differential 'triggers and responses', while appreciating the fundamentals of how the body reacts to environmental stimuli and extremes and the most common forms of heart disease affecting both men and women alike across different regions of the world.

## 4.3   A Holistic Framework to Understanding Climatic Vulnerability

Created by a panel of interdisciplinary experts to explain why hospitalisations and deaths in Australia and globally peak at certain times of the year (Stewart et al. 2017) (see Chap. 5) and then refined and validated in a series of observational and a prospective trial (Stewart et al.

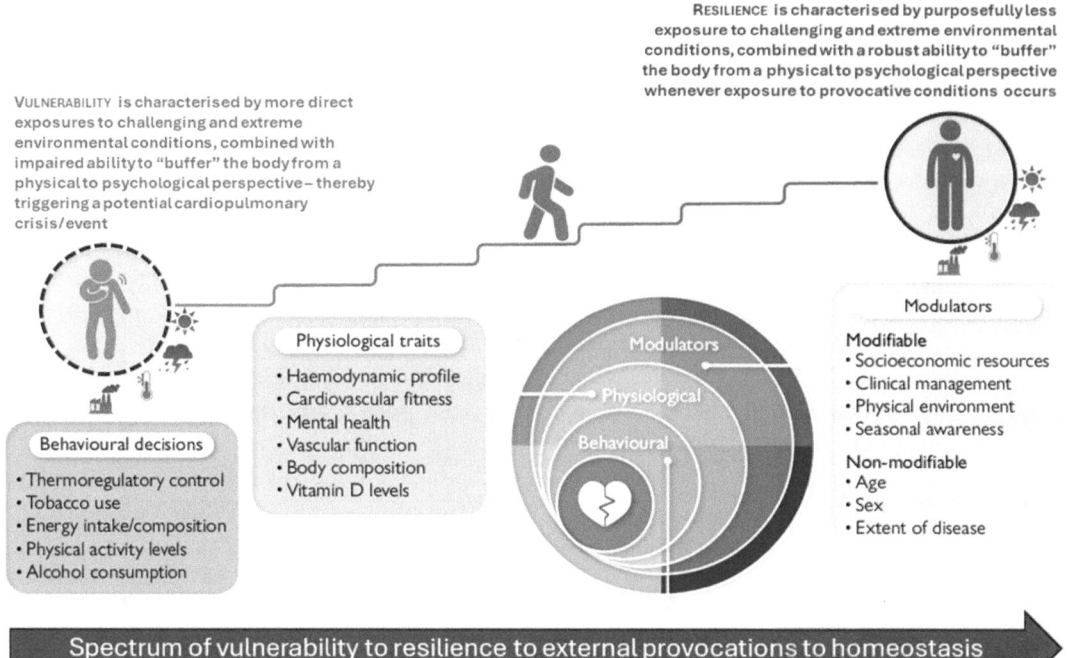

**Fig. 4.2** A holistic model of climatic vulnerability. This (hopefully) self-explanatory framework/model introduces the concept of a spectrum of climatic vulnerability to resilience with a range of modifiable to non-modifiable factors that determine where someone lies on that spectrum—adapted from original published figure (Stewart et al. 2017)

2024, 2019; Loader et al. 2019a), Fig. 4.2 is a framework/model to explain why some people are more likely to have a cardiac event at certain point in time. Specifically, it shows a spectrum of vulnerability-to-resilience to climatic provocations to the heart and cardiovascular system. In doing so, it describes a number of critical components/factors that help to explain why any intervention to address vulnerability/promote resilience requires a multifaceted approach (Stewart et al. 2024). Moreover, it explains why so much more research is needed to fully understand/explain the non-randomness of cardiovascular-related outcomes outlined in the next chapter. Nevertheless, there are key elements that can explain much of what we observe when tracking hospitalisations and deaths according to the prevailing weather/climatic conditions—from a population health to clinical cohort perspective. These are explained in more detail below.

### 4.3.1  Behavioural Decisions

When thinking about conscious decisions to behave in a certain way and suffer the consequences of being exposed to potentially deadly climatic conditions/events, it's hard to go past deciding to run a competitive race, or even marathon, in the heat (Bouscaren et al. 2019; Gill et al. 2015; Trubee et al. 2014). As highlighted by Bouscaren et al. (2019)—"*high running intensity (especially for the fastest runners), the urban context with high albedo effect materials, and the hot self-generated microclimate in mass-participation events (especially for the average to slow runners) are specific risk factors associated with marathon running in hot environments*". If race organisers are particularly wise, they might organise the race earlier in the day to avoid extreme temperatures—something being increasingly seen in major events (Cheuvront et al. 2021). Unsurprisingly, this seems to work.

For example, Cheuvront et al. (2021) reported that—"*The 2007 decision to make the Boston Marathon start time earlier by 2 h has reduced by ~1.4 times the odds that runners will be exposed to environmental conditions associated with exertional heat illness*". As reported by Thorsson et al. (2021), applying indices such as the *Wet-Bulb Globe Temperature, Physiological Equivalent Temperature* or *Universal Thermal Climate Index* help decide when a runner is likely to suffer from heat stress during sporting events. However, these are hardly indices/measures that an ordinary individual considers when, for example, they decide to do the gardening or walk to the shops on a hot day! At the other end of spectrum, consider the phenomenon of "*snow-shovelling mortality*" (Baranchuk 2017; Sauter et al. 2015; Whittington 1977) that encompasses the long-known (reported) risk of coronary plaque rupture (Hammoudeh and Haft 1996). At face-value the coronary risks of this activity is not immediately obvious. However, in a study of 500 patients (aged $65.7 \pm 13.4$ years and 67% male) presenting with an acute coronary syndrome in Canada, 7% were related to snow-shovelling. Adjusted analyses suggested that the strongest predictor of such an event (versus non-snow-shovelling presentations) was a familial history of premature cardiovascular disease (3.6-fold increase) and being male (4.8-fold more likely) (Nichols et al. 2012).

These two phenomena, while extreme, highlight the fact that an individual's cognitive decisions and actions (especially in the context or setting of temperatures extremes) form an integral part of the composite autonomic and behavioural thermo-effector response reflected in Fig. 4.2 and described previously in Chap. 3. Such behaviour and consequent actions crossover to the more mundane—particularly in respect to thermal protection across different climates. Consequently, a range of studies have highlighted regional and sociocultural differences in preserving heat (indoor heating, wearing more thermoprotective clothing, and modulating exercise levels), that help to explain why individuals living in/being exposed to typically warm-hot to relatively milder climates are more exposed to colder weather/unexpected cold spells and vice versa in respect to those being exposed to unexpected hot condtions (Donaldson et al. 1998; Terblanche-Greeff et al. 2018; Banwell et al. 2012; Blinman 2008). As described in Chap. 2, this is likely in an era of unpredictable climate change as opposed to global warming.

It is important to note that responsive behaviour (to the weather and climate) can be both immediate and habitual. A range of studies have identified seasonal variations in behaviour that might contribute to increased risk of cardiovascular events during certain parts of the year and, by extension in response to climate change. As described in a previous chapter, the SEASONS Study (Ma et al. 2021) revealed significant differences in dietary patterns (e.g. autumn peaks for both saturated and unsaturated fat consumption) and physical activity (with lowest levels of exercise in the colder winter months) among a large prospective cohort of patients in the USA according to the seasonal transitions they were exposed to. Consistent with Fig. 4.2 and the importance of 'empowerment', the greatest variations occurred in those were non-white (i.e. racial minorities), middle-aged men and those with the lowest education levels (Ma et al. 2021). As also noted, this phenomenon is not confined to the USA. A composite study of cardiovascular risk in >230,000 participants from 15 countries demonstrated a very similar pattern of risk levels tracking higher in colder winter months that declinined in warmer summer months (Marti-Soler et al. 2014). It is tempting of course, to wonder if these negative trends in relation to body mass index, waist circumference, blood pressure, and lipid and glucose levels (i.e. all the major cardiovascular risk factors) will be attenuated if climate change moderates winter temperatures. However, this is unlikely to be the case if there is more weather extremes and climate volatility. Nevertheless, in countries like Finland in Northern Europe (parts of which are in the Arctic Circle), even if climate change modulates the weather, the transition to short days/eternal night in many parts of the country are unlikely to modulate strong variations in social activity, energy level, mood, sleep duration and the risk of seasonal affective disorder evident on a population basis (Ma et al. 2006).

Clearly, without knowing their specific 'risk', an individual or society as a whole will not modulate their behaviour(s). One of the problems, as will be discussed in more detail in Chap. 9, is the lack of specific information (based on a lack of research), beyond general weather alerts around obvious events such as heatwaves and cold spells/cold-snaps (Nicholls et al. 2008; Masato et al. 2015), around what to tell people to inform them better about climatic provocations to their health. This represents an ongoing challenge when considering the uncertainties around what climate change will deliver year-by-year to different communities. Nevertheless, the power of social media to inform the populace about climate change (with potential to inform the messaging) has at least been demonstrated in recent study that analysed the effect of social network site usage on climate change awareness in 18 Latin American countries (Gomez-Casillas and Gomez 2023). As will be discussed below, even if people are made aware of the dangers posed to their health (by the weather/climatic conditions) it's another thing to enact them. However, it's appropriate to consider who is most in need of education and behavioural support to reduce their exposure to climate provocations to their health.

## 4.3.2   Physiological Traits

It is perhaps unsurprising that anyone with the following conditions should be considered at 'high-risk' of *climatic vulnerability* given that the autonomic mechanisms used to achieve thermoregulation and homeostasis, while centrally mediated, relies on the cardiovascular system and often subtle changes in the distribution of blood supply via the peripheral circulation with arterial compliance, vasoconstriction/dilatation and arterial pressure being important components in this regard:

- Uncorrected congenital heart disease
- Hypertension
- Obesity/metabolic disease

- Type 2 diabetes
- Left ventricular hypertrophy.
- Coronary artery disease
- Any form left ventricular dysfunction/heart failure
- Any form of right ventricular dysfunction/cor-pulmonale
- Pulmonary hypertension
- Cardiac arrhythmias (including atrial fibrillation and conditions such as heart block requiring a Cardiac Pacemaker and/or Defibrillator)
- Valvular Heart Disease (including aortic stenosis, mitral valve disease, tricuspid disease and the mixture of valvular dysfunction caused by conditions such as Rheumatic Heart Disease).
- Cardiomyopathies (including hypertrophic cardiomyopathy, idiopathic cardiomyopathy and cardiac amyloidosis)

Not unsurprisingly, given their physiological connections (and indeed) to, and control of the circulatory system, this list extends to:

- Any form of cerebrovascular disease
- Renal dysfunction/disease

For the reasons outlined in Chap. 3, any person with a combination of these conditions is vulnerable by any 'stimulus' that provokes potentially large fluctuations in blood pressure, afterload, preload and cardiac output (which is the product of heart rate and stroke volume) (King and Lowery 2024). To place cardiac output into context, a normal cardiac output is 5–6 L/min, but elite athletes can push their output to 35 L/min (Bhatt et al. 2021). Why is this important? Athletes have the capacity to maximise their oxygen intake (via maximal lung function and haemoglobin mediated oxygen transport), cardiac function and utilisation of oxygen/removal of lactic acid through optimal metabolic processes, because they are fit and healthy. They are unlikely to have the profile of impaired cardiac function atherosclerosis producing ischaemia and impaired endothelial function that is characteristic of the conditions listed above

(Gonzalez and Selwyn 2003; Hinderliter and Caughey 2003; Vita and Keaney 2002; Paterick and Fletcher 2001; Linke et al. 2008; Butt et al. 2009; Okada et al. 2010; Reriani et al. 2010; Tanaka et al. 2020; Ding et al. 2021; Bianconi et al. 2023; Van De Maele and Bruyndonckx 2024). Adding to this list (for indirect reasons) is cognitive dysfunction. Cognitive dysfunction is closely linked to cardiovascular disease (Ball and Carrington 2013; Cameron et al. 2013; McLennan et al. 2011; Pennington et al. 2018; Tadic et al. 2016; Weintraub et al. 2017) and is a major impediment to conveying information and therefore modulating positive behaviours in any health context (Ball et al. 2018; McLennan et al. 2006; Stewart et al. 2000)—let alone in respect to explaining the potential danger from climatic conditions and future change. Unfortunately, within many ageing populations living in high-income countries multimorbidity (comprising many malignant combinations of the conditions listed above) is becoming increasingly common (Dewan et al. 2023; Chen et al. 2019; Forman et al. 2018; Stewart et al. 2016).

### 4.3.3 Modulating Factors

Unsurprisingly, even if we are 'healthy', as we age endothelial function and cardiorespiratory fitness declines (Loader et al. 2019b) with a combination of factors likely to accelerate this process (Konigstein et al. 2022; Jujic et al. 2023; Babcock et al. 2022; Martinez-Majander et al. 2021; Froldi and Dorigo 2020). The inherent benefits of maintaining cardiorespiratory fitness (Alvares et al. 2024; Fosstveit et al. 2024; Cadenas-Sanchez et al. 2024; Men et al. 2023; McAlister et al. 2023; Mateo-Gallego et al. 2022) to prevent cardiovascular-related events and premature mortality (Bolam et al. 2024; Duggan et al. 2024; Kodama et al. 2009; Lang et al. 2024; Sparks et al. 2024; Ung et al. 2024) will be a major point of discussion later on in the book. However, for many people (including those who have already become frail or suffer from a musculoskeletal injury/condition) attaining the physical training required to improve cardiorespiratory fitness is impossible. A study reported by Madrigano et al. conducted in New York City in the USA, is instrumental in highlighting the fact that even with the best knowledge and intentions, there are often barriers that are difficult to overcome. They conducted a random digit dial telephone survey of 801 adults aged 18 and older during the "Fall" of 2015. Overall, 13% did not possess an air conditioner and another 15% had one used it infrequently or not at all. On adjusted basis, those least likely to possess an air conditioner were non-Hispanic blacks (twofold less likely) and those with a low annual household income (3.1-fold less likely). Only 12% of these individuals actively sought air conditioning in a public place. Overall, while low-income individuals were less likely to be aware of heat warnings, they were 1.6-fold more likely to be concerned that heat could make them ill (this included concern around the impact of climate change in this regard) (Madrigano et al. 2018). Unpacking the lessons from these instructive data is difficult, given the potentially mixed messages, but they do emphasise that having the resources to make change/positive health decisions along with having an education to improve health literacy (the discussion of which informs "how" we can promote climatic resilience later in this book) is a key reason why social inequality results in poor health outcomes (including communicable and non-communicable forms of heart disease) (Capewell et al. 2001; Kure et al. 2016; Deo et al. 2023; Gwatkin et al. 1999; Klassen et al. 2024; Bukhman et al. 2023, 2020, 2015; Coates et al. 2021; Jarvis and Townsend 2021; Kwan et al. 2016). As highlighted in Fig. 4.2, in framing the potential to address vulnerability in people with existing heart disease and multimorbidity (Stewart et al. 2024) (see Chap. 3 and the section below), the key modifiable factors that can "restore" the inherent resilience most of us are born with are as follows:

- Socio-economic resources to enact positive changes.
- Clinical management to restore clinical stability/optimal cardiovascular function.

- The physical environment in which we reside (and take shelter in if needed)
- Seasonal/climatic awareness to make decisions that protect us from weather extremes and the need to rely on autonomic thermoregulatory responses to restore homeostasis.

These factors will be explored more fully, when considering how we can future-proof our individual to collective heart health in the face of climate change in Chaps. 8 and 9.

## 4.4    The Vulnerable Versus Resilient Phenotype

As indicated by Fig. 4.2, if one accepts that there is a spectrum of vulnerability to resilience in respect to climatic/weather-induced provocations to the cardiovascular system (as it constantly tries to maintain homeostasis), it should be possible to identify individuals at each end of the spectrum—and even within the diagnostic groups listed above. As such, on the left-hand side of Fig. 4.2, one can see a *'porous'* interaction between the figure and their external environment with challenges from pollution (see Chap. 6), temperature variations, storms, and heatwaves most likely to provoke an adverse health response (most particularly from a cardiovascular perspective, but also more generally from a cardiopulmonary perspective). Based on the explanations provided in the above sections, beyond the inability to control for advancing age, extent of (cardiac/cardiovascular and common multimorbid) disease and biological sex, there are both behavioural decisions and physiological traits that enable us to identify the 'climatically vulnerable' phenotype. This includes behaviours such as smoking, sedentary behaviours, dietary choices, and excessive alcohol intake. From a physiologic perspective, this includes indicators such as hypertension or hypotension, tachycardia/arrhythmias, anxiety/depression, obesity, and low Vitamin D levels—the specific profiling of which will be discussed later in the book—once again noting that the clinical research around how we profile

individuals at risk of current and future climatic conditions remains in its infancy (Stewart et al. 2024). As has, and will be reinforced repeatedly throughout this book, this requires all clinicians, health professionals and policy makers to add a new dimension to how they view the 'how' and 'who' Is affected by different forms of cardiovascular disease and related disorders and consider 'when' events occur. In essence, those advocating for better clinical management (i.e. Expert Guideline Panels) and those reporting of event rates should embrace this paradigm change (even in the face of climate change scepticism) to improve health resource and health service delivery planning. Ironically, the COVID-19 pandemic, while not acting anything like a seasonally infectious illness yet (due to varying level of immune protection, climatic conditions, and human behaviours), has at least provoked more careful thinking around 'waves' of infection and associated deaths in every continent (Yang et al. 2024; Nogareda et al. 2023; Vikstrom et al. 2023; Abe et al. 2022; Carbonell et al. 2021; Hoogenboom et al. 2021; Jassat et al. 2021; Contreras and Priesemann 2021; Prabhakaran et al. 2022; Ogah et al. 2021).

This careful interpretation of when someone with a cardiovascular condition is inherently more vulnerabile to a cardiac event is important when further delineating between vulnerable phenotypes—noting that the 'climatically resilient' phenotype represented on the right-hand side of Fig. 4.2, is premised on a universal protection to all climatic/weather provocations. As will be explored in Chap. 6, it seems clear that certain individuals are more likely to be readmitted or die at either ends of climatic conditions/extremes (Janos et al. 2024). Indeed a recent European report by the Lancet Countdown on Health and Climate Change states—*"Heat-related deaths are estimated to have risen across most of Europe, with an average increase of 17.2 deaths per 100,000 inhabitants between the periods of 2003–2012 and 2013–2022"*—with concerning trends around when people are exercising. It is on this basis that Fig. 4.3 (*Winter Vulnerable Phenotype*) and Fig. 4.4 (*Summer Vulnerable Phenotype*) demonstrates that, while

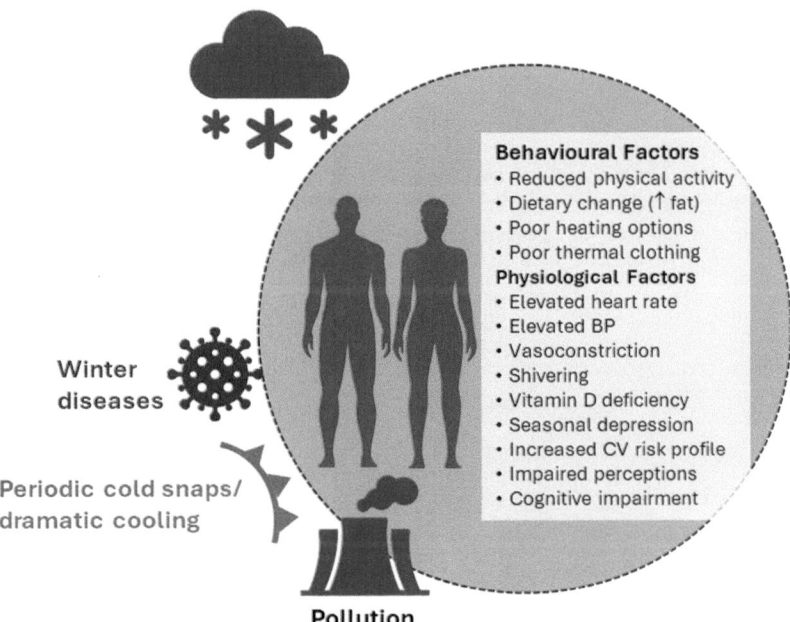

**Fig. 4.3** The vulnerable phenotype in winter. Building on Fig. 4.2, this figure shows the predominant characteristics of men and women who demonstrate vulnerability to climatic conditions (and pollution) typically occurring in the winter/cold weather—adapted from the original published figure (Stewart et al. 2017)

there are likely to be similar factors involved (e.g. compromised ability to afford either heating or cooling in winter or summer) there is a distinction in such vulnerability that has relevance at the clinical level. Once again, this will be explored more in later chapters—particularly around profiling high-risk people, mitigating vulnerability to promote resilience and considering whether it is climatic volatility (rather than heat or cold per se) that is most provocative to the heart and broader cardiovascular health of vulnerable people worldwide.

## 4.5   A More 'Resilient' Future

This Chapter has explored the concepts of climatic vulnerability and resilience, presenting vulnerability to resilience not as a dichotomy, but more as a spectrum that infers (if we are to intervene in positive ways) it is possible to move towards a more resilient state and collective future. This predicates that we steadily lose the natural resilience that we are born with and strengthen in our childhood via learned behaviours and even cultural practices in modifiable ways. However, this degradation is neither linear nor inevitable. Thus, beyond non-modifiable factors such as our biological sex and the way that shapes the natural history of cardiovascular disease and other related disorders, advancing age and the underlying extent of pre-existing heart disease, this chapter identifies what can be potentially modified to maintain/regain climatic resilience. As further explored, this doesn't necessarily mean there is an easy fix. As will be discussed in later chapters there is much more research to be performed in this space. Nor do we have a lexicon of validated tools to identify who is vulnerable and who is resilient—although their typical characteristics arising from the research my group has undertaken, is presented as evidence that this is (and should be) routinely possible in clinical practice. Much of the book thus far has been presented on the idea that the weather conditions and climate we experience are indeed important considerations when providing clinical care to people with heart disease

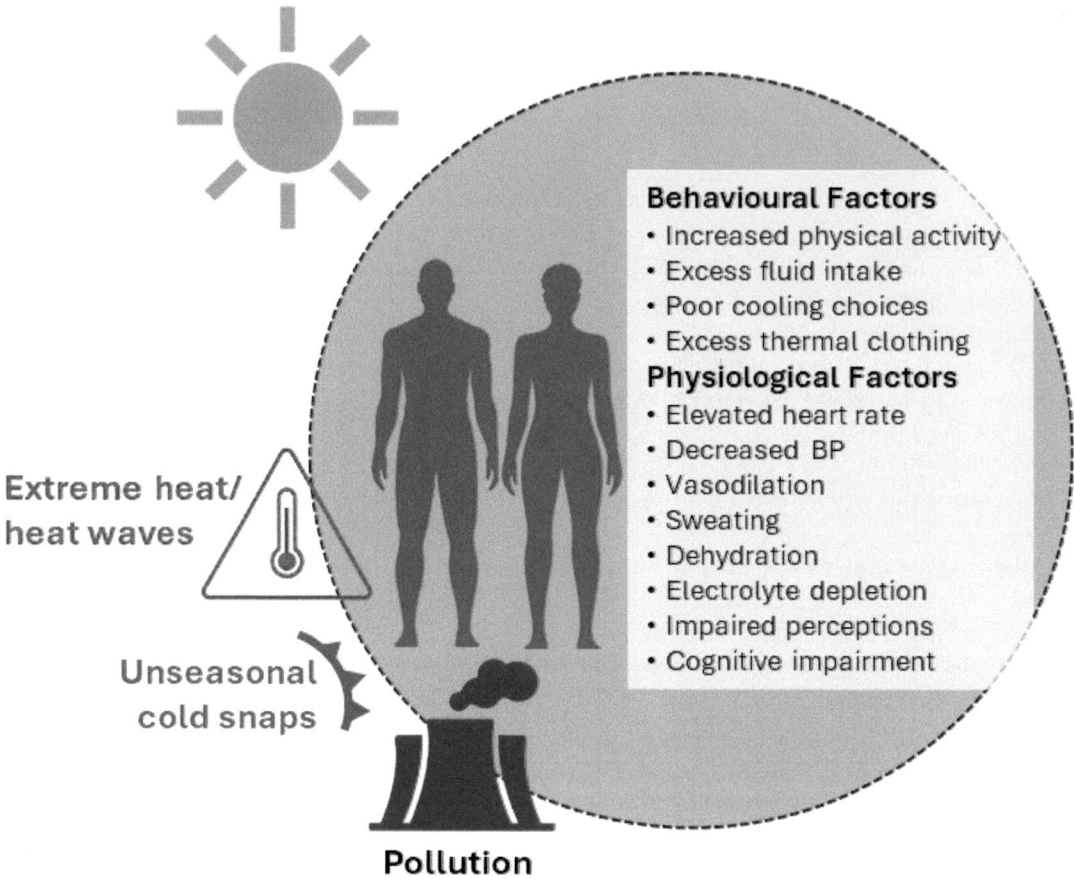

**Fig. 4.4** The vulnerable phenotype in summer. Building on Fig. 4.2 (and in contrast to Fig. 4.3), this figure shows the predominant characteristics of men and women who demonstrate vulnerability to climatic conditions (and pollution) typically occurring in the summer/warmer weather—adapted from the original published figure (Stewart et al. 2017)

and other common forms of cardiovascular disease (most notably cerebrovascular disease); thereby refuting the passive view that most cardiac events are random in the sense of pathological triggers. The next chapter proffers the hard evidence that this is not the case from a population to clinical cohort perspective—demonstrating that we (as health professionals/clinicians) should be concerned about 'when' and 'how' the people we care for and treat are at risk during certain times of the day and year. It also presents a case for why we should be concerned that climate change will likely (some might say inevitably) change these dynamics for the worse; especially among disadvantaged communities.

## References

Abe H, Ushijima Y, Bikangui R, Ondo GN, Lell B, Adegnika AA, Yasuda J. Delays in the arrival of the waves of COVID-19: a comparison between Gabon and the African continent. Lancet Microbe. 2022;3:e476.

Alvares TS, de Souza LVM, Soares RN, Lessard SJ. Cardiorespiratory fitness is impaired in type 1 and type 2 diabetes: a systematic review, meta-analysis, and meta-regression. Med Sci Sports Exerc. 2024;82:6458.

Australian Bureau of Meterology. What is La Nina and does it impact Australia; 2024. http://www.bom.gov.au/climate/updates/articles/a020.shtml. Accessed May 2024.

Babcock MC, DuBose LE, Witten TL, Stauffer BL, Hildreth KL, Schwartz RS, Kohrt WM, Moreau

KL. Oxidative stress and inflammation are associated with age-related endothelial dysfunction in men with low testosterone. J Clin Endocrinol Metab. 2022;107:e500–14.

Ball J, Carrington MJ, et al. Mild cognitive impairment in high-risk patients with chronic atrial fibrillation: a forgotten component of clinical management? Heart. 2013;99:542–7.

Ball J, Lochen ML, Carrington MJ, Wiley JF, Stewart S. Mild cognitive impairment impacts health outcomes of patients with atrial fibrillation undergoing a disease management intervention. Open Heart. 2018;5:e000755.

Banwell C, Dixon J, Bambrick H, Edwards F, Kjellstrom T. Socio-cultural reflections on heat in Australia with implications for health and climate change adaptation. Glob Health Action. 2012;5:1135.

Baranchuk A. The cardiovascular risk of snowfall and snow shovelling in Canada. CMAJ. 2017;189:E545.

Beale AL, Meyer P, Marwick TH, Lam CSP, Kaye DM. Sex differences in cardiovascular pathophysiology: why women are overrepresented in heart failure with preserved ejection fraction. Circulation. 2018;138:198–205.

Bhatt A, Flink L, Lu DY, Fang Q, Bibby D, Schiller NB. Exercise physiology of the left atrium: quantity and timing of contribution to cardiac output. Am J Physiol Heart Circ Physiol. 2021;320:H575–83.

Bianconi V, Mannarino MR, Cosentini E, Figorilli F, Colangelo C, Cellini G, Braca M, Lombardini R, Paltriccia R, Sahebkar A, Pirro M. The impact of statin therapy on in-hospital prognosis and endothelial function of patients at high-to-very high cardiovascular risk admitted for COVID-19. J Med Virol. 2023;95:e28678.

Blinman E. 2000 years of cultural adaptation to climate change in the Southwestern United States. Ambio. 2008;14:489–97.

Bolam KA, Bojsen-Moller E, Wallin P, Paulsson S, Lindwall M, Rundqvist H, Ekblom-Bak E. Association between change in cardiorespiratory fitness and prostate cancer incidence and mortality in 57 652 Swedish men. Br J Sports Med. 2024;58:366–72.

Bouscaren N, Millet GY, Racinais S. Heat stress challenges in marathon vs. ultra-endurance running. Front Sports Act Living. 2019;1:59.

Bukhman G, Mocumbi AO, Horton R. Reframing NCDs and injuries for the poorest billion: a Lancet commission. Lancet. 2015;386:1221–2.

Bukhman G, Mocumbi AO, Atun R, Becker AE, Bhutta Z, Binagwaho A, Clinton C, Coates MM, Dain K, Ezzati M, et al. The Lancet NCDI poverty commission: bridging a gap in universal health coverage for the poorest billion. Lancet. 2020;396:991–1044.

Bukhman G, Mocumbi A, Wroe E, Gupta N, Pearson L, Bermejo R, Dangou JM, Moeti M. The PEN-plus partnership: addressing severe chronic non-communicable diseases among the poorest billion. Lancet Diabetes Endocrinol. 2023;11:384–6.

Butt M, Dwivedi G, Khair O, Lip GY. Assessment of endothelial function: implications for hypertension and cardiovascular disorders. Expert Rev Cardiovasc Ther. 2009;7:561–3.

Cadenas-Sanchez C, Fernandez-Rodriguez R, Martinez-Vizcaino V, et al. A systematic review and cluster analysis approach of 103 studies of high-intensity interval training on cardiorespiratory fitness. Eur J Prev Cardiol. 2024;31:400–11.

Cambride Dictionary. Resilient. https://dictionary.cambridge.org/dictionary/english/resilient. Accessed April 2024.

Cambridge Dictionary. Vulnerable. https://dictionary.cambridge.org/dictionary/english/vulnerable. Accessed April 2024.

Cameron J, Worrall-Carter L, Page K, Stewart S, Ski CF. Screening for mild cognitive impairment in patients with heart failure: Montreal cognitive assessment versus mini mental state exam. Eur J Cardiovasc Nurs. 2013;12:252–60.

Capewell S, MacIntyre K, Stewart S, Chalmers JW, Boyd J, Finlayson A, Redpath A, Pell JP, McMurray JJ. Age, sex, and social trends in out-of-hospital cardiac deaths in Scotland 1986–1995: a retrospective cohort study. Lancet. 2001;358:1213–7.

Carbonell R, Urgeles S, Rodriguez A, Bodi M, Martin-Loeches I, Sole-Violan J, Diaz E, Gomez J, Trefler S, Vallverdu M, Murcia J, Albaya A, et al. Mortality comparison between the first and second/third waves among 3795 critical COVID-19 patients with pneumonia admitted to the ICU: a multicentre retrospective cohort study. Lancet Reg Health Eur. 2021;11:100243.

Chen L, Chan YK, Busija L, Norekval TM, Riegel B, Stewart S. Malignant and benign phenotypes of multimorbidity in heart failure: implications for clinical practice. J Cardiovasc Nurs. 2019;34:258–66.

Chen A, Waite L, Mocumbi AO, Chan YK, Beilby J, Ojji DB, Stewart S. Elevated blood pressure among adolescents in sub-Saharan Africa: a systematic review and meta-analysis. Lancet Glob Health. 2023;11:e1238–48.

Cheuvront SN, Caldwell AR, Cheuvront PJ, Kenefick RW, Troyanos C. Earlier Boston marathon start time mitigates environmental heat stress. Med Sci Sports Exerc. 2021;53:1999–2005.

Coates MM, Ezzati M, Robles Aguilar G, Kwan GF, Vigo D, Mocumbi AO, Becker AE, Makani J, Hyder AA, Jain Y, Stefan DC, Gupta N, Marx A, Bukhman G. Burden of disease among the world's poorest billion people: an expert-informed secondary analysis of global burden of disease estimates. PLoS ONE. 2021;16:e0253073.

Contreras S, Priesemann V. Risking further COVID-19 waves despite vaccination. Lancet Infect Dis. 2021;21:745–6.

Cordero RR, Feron S, Damiani A, Carrasco J, Karas C, Wang C, Kraamwinkel CT, Beaulieu A. Extreme fire weather in Chile driven by climate change and El Nino-Southern Oscillation (ENSO). Sci Rep. 2024;14:1974.

DeFilippis EM, Beale A, Martyn T, Agarwal A, Elkayam U, Lam CSP, Hsich E. Heart failure subtypes and cardiomyopathies in women. Circ Res. 2022;130:436–54.

Deng J, Dai A. Arctic sea ice-air interactions weaken El Nino-Southern Oscillation. Sci Adv. 2024;10:eadk3990.

Deo SV, Al-Kindi S, Motairek I, Elgudin YE, Gorodeski E, Nasir K, Rajagopalan S, Petrie MC, Sattar N. Neighbourhood-level social deprivation and the risk of recurrent heart failure hospitalizations in type 2 diabetes. Diabetes Obes Metab. 2023;35:589.

Dewan P, Ferreira JP, Butt JH, Petrie MC, Abraham WT, Desai AS, Dickstein K, Kober L, Packer M, Rouleau JL, Stewart S, Swedberg K, Zile MR, Solomon SD, Jhund PS, McMurray JJV. Impact of multimorbidity on mortality in heart failure with reduced ejection fraction: which comorbidities matter most? An analysis of PARADIGM-HF and ATMOSPHERE. Eur J Heart Fail. 2023;25:687–97.

Ding Y, Zhou Y, Ling P, Feng X, Luo S, Zheng X, Little PJ, Xu S, Weng J. Metformin in cardiovascular diabetology: a focused review of its impact on endothelial function. Theranostics. 2021;11:9376–96.

Docherty KF, Jackson AM, Macartney M, Campbell RT, Petrie MC, Pfeffer MA, McMurray JJV, Jhund PS. Declining risk of heart failure hospitalization following first acute myocardial infarction in Scotland between 1991–2016. Eur J Heart Fail. 2023;34:5217.

Donaldson GC, Ermakov SP, Komarov YM, McDonald CP, Keatinge WR. Cold related mortalities and protection against cold in Yakutsk, eastern Siberia: observation and interview study. BMJ. 1998;317:978–82.

Duggan J, Peters A, Antevil J, Faselis C, Samuel I, Kokkinos P, Trachiotis G. Long-term mortality risk according to cardiorespiratory fitness in patients undergoing coronary artery bypass graft surgery. J Clin Med. 2024;13:192.

Ern Yeoh S, Osmanska J, Petrie MC, Brooksbank KJM, Clark AL, Docherty KF, Foley PWX, Guha K, Halliday CA, Jhund PS, Kalra PR, McKinley G, Lang NN, Lee MMY, McConnachie A, McDermott JJ, Platz E, Sartipy P, et al. Dapagliflozin versus metolazone in heart failure resistant to loop diuretics. Eur Heart J. 2023;54:6897.

Forman DE, Maurer MS, Boyd C, Brindis R, Salive ME, Horne FM, Bell SP, Fulmer T, Reuben DB, Zieman S, Rich MW. Multimorbidity in older adults with cardiovascular disease. J Am Coll Cardiol. 2018;71:2149–61.

Fosstveit SH, Lohne-Seiler H, Feron J, Lucas SJE, Ivarsson A, Berntsen S. The intensity paradox: a systematic review and meta-analysis of its impact on the cardiorespiratory fitness of older adults. Scand J Med Sci Sports. 2024;34:e14573.

Froldi G, Dorigo P. Endothelial dysfunction in Coronavirus disease 2019 (COVID-19): gender and age influences. Med Hypotheses. 2020;144:110015.

Gifford RM, Todisco T, Stacey M, Fujisawa T, Allerhand M, Woods DR, Reynolds RM. Risk of heat illness in men and women: a systematic review and meta-analysis. Environ Res. 2019;171:24–35.

Gill SK, Teixeira A, Rama L, Prestes J, Rosado F, Hankey J, Scheer V, Hemmings K, Ansley-Robson P, Costa RJ. Circulatory endotoxin concentration and cytokine profile in response to exertional-heat stress during a multi-stage ultra-marathon competition. Exerc Immunol Rev. 2015;21:114–28.

Gohar A, Chong JPC, Liew OW, den Ruijter H, de Kleijn DPV, Sim D, Yeo DPS, Ong HY, Jaufeerally F, Leong GKT, Ling LH, Lam CSP, Richards AM. The prognostic value of highly sensitive cardiac troponin assays for adverse events in men and women with stable heart failure and a preserved vs. reduced ejection fraction. Eur J Heart Fail. 2017;19:1638–47.

Gomez-Casillas A, Gomez MV. The effect of social network sites usage in climate change awareness in Latin America. Popul Environ. 2023;45:7.

Gonzalez MA, Selwyn AP. Endothelial function, inflammation, and prognosis in cardiovascular disease. Am J Med. 2003;115(Suppl 8A):99S-106S.

Gwatkin DR, Guillot M, Heuveline P. The burden of disease among the global poor. Lancet. 1999;354:586–9.

Hammoudeh AJ, Haft JI. Coronary-plaque rupture in acute coronary syndromes triggered by snow shoveling. N Engl J Med. 1996;335:2001.

Hinderliter AL, Caughey M. Assessing endothelial function as a risk factor for cardiovascular disease. Curr Atheroscler Rep. 2003;5:506–13.

Hoogenboom WS, Pham A, Anand H, Fleysher R, Buczek A, Soby S, Mirhaji P, Yee J, Duong TQ. Clinical characteristics of the first and second COVID-19 waves in the Bronx, New York: a retrospective cohort study. Lancet Reg Health Am. 2021;3:100041.

Intergovernmental Panel on Climate Change. Climate change 2023: synthesis report. In: Contribution of working groups I, II and III to the sixth assessment report of the intergovernmental panel on climate change; 2023a.

Intergovernmental Panel on Climate Change. Climate change 2023: synthesis report. In: 3.1.2 Impacts and related risks. Contribution of working groups I, II and III to the sixth assessment report of the intergovernmental panel on climate change; 2023b.

Janos T, Ballester J, Cupr P, Achebak H. Countrywide analysis of heat- and cold-related mortality trends in the Czech Republic: growing inequalities under recent climate warming. Int J Epidemiol. 2024;53:579.

Jarvis JD, Townsend B. Universal health coverage for the poorest billion: justice and equity considerations. Lancet. 2021;397:473–4.

Jassat W, Mudara C, Ozougwu L, Tempia S, Blumberg L, Davies MA, Pillay Y, Carter T, Morewane R, Wolmarans M, von Gottberg A, Bhiman JN, et al. Difference in mortality among individuals admitted to hospital with COVID-19 during the first and second waves in South Africa: a cohort study. Lancet Glob Health. 2021;9:e1216–25.

Jiang N, Zhu C, Hu ZZ, McPhaden MJ, Chen D, Liu B, Ma S, Yan Y, Zhou T, Qian W, Luo J, Yang X, Liu F, Zhu Y. Enhanced risk of record-breaking regional temperatures during the 2023–24 El Nino. Sci Rep. 2024;14:2521.

Jin X, Chandramouli C, Allocco B, Gong E, Lam CSP, Yan LL. Women's participation in cardiovascular clinical trials from 2010 to 2017. Circulation. 2020;141:540–8.

Jujic A, Kenneback C, Johansson M, Nilsson PM, Holm H. The impact of age on endothelial dysfunction measured by peripheral arterial tonometry in a healthy population-based cohort: the Malmo offspring study. Blood Press. 2023;32:2234059.

Keates AK, Mocumbi AO, Ntsekhe M, Sliwa K, Stewart S. Cardiovascular disease in Africa: epidemiological profile and challenges. Nat Rev Cardiol. 2017;14:273–93.

King J, Lowery DR. Physiology, cardiac output StatPearls treasure Island (FL) ineligible companies. Disclosure: David Lowery declares no relevant financial relationships with ineligible companies; 2024.

Klassen SL, Okello E, Ferrer JME, Alizadeh F, Barango P, Chillo P, Chimalizeni Y, Dagnaw WW, Eisele JL, et al. Decentralization and integration of advanced cardiac care for the world's poorest billion through the PEN-plus strategy for severe chronic non-communicable disease. Glob Heart. 2024;19:33.

Kodama S, Saito K, Tanaka S, Maki M, Yachi Y, Asumi M, Sugawara A, Totsuka K, Shimano H, Ohashi Y, Yamada N, Sone H. Cardiorespiratory fitness as a quantitative predictor of all-cause mortality and cardiovascular events in healthy men and women: a meta-analysis. JAMA. 2009;301:2024–35.

Konigstein K, Wagner J, Infanger D, Knaier R, Neve G, Klenk C, Carrard J, Hinrichs T, Schmidt-Trucksass A. Cardiorespiratory fitness and endothelial function in aging healthy subjects and patients with cardiovascular disease. Front Cardiovasc Med. 2022;9:870847.

Kure CE, Chan YK, Ski CF, Thompson DR, Carrington MJ, Stewart S. Gender-specific secondary prevention? Differential psychosocial risk factors for major cardiovascular events. Open Heart. 2016;3:e000356.

Kwan GF, Mayosi BM, Mocumbi AO, Miranda JJ, Ezzati M, Jain Y, Robles G, Benjamin EJ, Subramanian SV, Bukhman G. Endemic cardiovascular diseases of the poorest billion. Circulation. 2016;133:2561–75.

Lam CSP, Myhre PL. Left ventricular ejection fraction in women: when normal isn't normal. Heart. 2023;109:1584–5.

Lang JJ, Prince SA, Merucci K, Cadenas-Sanchez C, Chaput JP, Fraser BJ, Manyanga T, McGrath R, Ortega FB, Singh B, Tomkinson GR. Cardiorespiratory fitness is a strong and consistent predictor of morbidity and mortality among adults: an overview of meta-analyses representing over 20.9 million observations from 199 unique cohort studies. Br J Sports Med. 2024;58:556–66.

Linke A, Erbs S, Hambrecht R. Effects of exercise training upon endothelial function in patients with cardiovascular disease. Front Biosci. 2008;13:424–32.

Lloyd-Jones DM, Larson MG, Beiser A, Levy D. Lifetime risk of developing coronary heart disease. Lancet. 1999;353:89–92.

Loader J, Chan YK, Hawley JA, Moholdt T, McDonald CF, Jhund P, Petrie MC, McMurray JJ, Scuffham PA, Ramchand J, Burrell LM, Stewart S. Prevalence and profile of "seasonal frequent flyers" with chronic heart disease: analysis of 1598 patients and 4588 patient-years follow-up. Int J Cardiol. 2019a;279:126–32.

Loader J, Khouri C, Taylor F, Stewart S, Lorenzen C, Cracowski JL, Walther G, Roustit M. The continuums of impairment in vascular reactivity across the spectrum of cardiometabolic health: a systematic review and network meta-analysis. Obes Rev. 2019b;20:906–20.

Ma Y, Olendzki BC, Li W, Hafner AR, Chiriboga D, Hebert JR, Campbell M, Sarnie M, Ockene IS. Seasonal variation in food intake, physical activity, and body weight in a predominantly overweight population. Eur J Clin Nutr. 2006;60:519–28.

Ma Y, Liu K, Hu W, Song S, Zhang S, Shao Z. Epidemiological characteristics, seasonal dynamic patterns, and associations with meteorological factors of rubella in Shaanxi Province, China, 2005–2018. Am J Trop Med Hyg. 2021;104:166–74.

MacIntyre K, Stewart S, Capewell S, Chalmers JW, Pell JP, Boyd J, Finlayson A, Redpath A, Gilmour H, McMurray JJ. Gender and survival: a population-based study of 201,114 men and women following a first acute myocardial infarction. J Am Coll Cardiol. 2001;38:729–35.

Madrigano J, Lane K, Petrovic N, Ahmed M, Blum M, Matte T. Awareness, risk perception, and protective behaviors for extreme heat and climate change in New York City. Int J Environ Res Public Health. 2018;15:38–95.

Malik ME, Falkentoft AC, Jensen J, Zahir D, Parveen S, Alhakak A, Andersson C, Petrie MC, Sattar N, McMurray JJV, Kober L, Schou M. Discontinuation and reinitiation of SGLT-2 inhibitors and GLP-1R agonists in patients with type 2 diabetes: a nationwide study from 2013 to 2021. Lancet Reg Health Eur. 2023;29:100617.

Martinez-Majander N, Gordin D, Joutsi-Korhonen L, Salopuro T, Adeshara K, Sibolt G, Curtze S, Pirinen J, Liebkind R, Soinne L, Sairanen T, Sinisalo J, Lehto M, Groop PH, Tatlisumak T, Putaala J. Endothelial dysfunction is associated with early-onset cryptogenic ischemic stroke in men and with increasing age. J Am Heart Assoc. 2021;10:e020838.

Marti-Soler H, Gubelmann C, Aeschbacher S, Alves L, Bobak M, Bongard V, Clays E, et al. Seasonality of cardiovascular risk factors: an analysis including over 230,000 participants in 15 countries. Heart. 2014;100:1517–23.

Masato G, Bone A, Charlton-Perez A, Cavany S, Neal R, Dankers R, Dacre H, Carmichael K, Murray V. Improving the health forecasting alert system for cold weather and heat-waves in England: a proof-of-concept using temperature-mortality relationships. PLoS ONE. 2015;10:e0137804.

Mateo-Gallego R, Madinaveitia-Nisarre L, Gine-Gonzalez J, et al. The effects of high-intensity interval training on glucose metabolism, cardiorespiratory fitness and weight control in subjects with diabetes: systematic review a meta-analysis. Diabetes Res Clin Pract. 2022;190:109979.

McAlister FA, Murphy NF, Simpson CR, Stewart S, MacIntyre K, Kirkpatrick M, Chalmers J, Redpath A, Capewell S, McMurray JJ. Influence of socioeconomic deprivation on the primary care burden and treatment of patients with a diagnosis of heart failure in general practice in Scotland: population based study. BMJ. 2004;328:1110.

McAlister KL, Zhang D, Moore KN, Chapman TM, Zink J, Belcher BR. A systematic review of the associations of adiposity and cardiorespiratory fitness with arterial structure and function in nonclinical children and adolescents. Pediatr Exerc Sci. 2023;35:174–85.

McDonagh TA, Metra M, Adamo M, Gardner RS, Baumbach A, Bohm M, Burri H, Butler J, Celutkiene J, Chioncel O, Cleland JGF, Coats AJS, Crespo-Leiro MG, Farmakis D, Gilard M, Heymans S, Hoes AW, Jaarsma T, Jankowska EA, Lainscak M, Lam CSP, Lyon AR, McMurray JJV, Mebazaa A, Mindham R, Muneretto C, et al. 2021 ESC guidelines for the diagnosis and treatment of acute and chronic heart failure. Eur Heart J. 2021;42:3599–726.

McDonagh TA, Metra M, Adamo M, Gardner RS, Baumbach A, Bohm M, Burri H, Butler J, Celutkiene J, Chioncel O, Cleland JGF, Crespo-Leiro MG, Farmakis D, Gilard M, Heymans S, Hoes AW, Jaarsma T, et al. 2023 focused update of the 2021 ESC guidelines for the diagnosis and treatment of acute and chronic heart failure. Eur Heart J. 2023;44:3627–39.

McLennan SN, Pearson SA, Cameron J, Stewart S. Prognostic importance of cognitive impairment in chronic heart failure patients: does specialist management make a difference? Eur J Heart Fail. 2006;8:494–501.

McLennan SN, Mathias JL, Brennan LC, Stewart S. Validity of the montreal cognitive assessment (MoCA) as a screening test for mild cognitive impairment (MCI) in a cardiovascular population. J Geriatr Psychiatry Neurol. 2011;24:33–8.

McMurray J, Berry C. Ongoing clinical trials with angiotensin II receptor antagonists in chronic heart failure and myocardial infarction. J Renin Angiotensin Aldosterone Syst. 2000;1:131–6.

McMurray JJ, Califf RM, Bethel AM, Haffner SM, Holman RR. Comparative effectiveness of angiotensin-converting enzyme inhibitors and angiotensin receptor blockers for hypertension on clinical end points: a cohort study. J Clin Hypertens. 2012;14:731.

McMurray JJV, Jackson AM, Lam CSP, Redfield MM, Anand IS, Ge J, Lefkowitz MP, et al. Effects of sacubitril-valsartan versus valsartan in women compared with men with heart failure and preserved ejection fraction: insights from PARAGON-HF. Circulation. 2020;141:338–51.

Mehran R, Steg PG, Pfeffer MA, Jering K, Claggett B, Lewis EF, Granger C, Kober L, Maggioni A, Mann DL, et al. The effects of angiotensin receptor-neprilysin inhibition on major coronary events in patients with acute myocardial infarction: insights from the PARADISE-MI trial. Circulation. 2022;146:1749–57.

Melaku YA, Gill TK, Taylor AW, Appleton SL, Gonzalez-Chica D, Adams R, Achoki T, Shi Z, Renzaho A. Trends of mortality attributable to child and maternal undernutrition, overweight/obesity and dietary risk factors of non-communicable diseases in sub-Saharan Africa, 1990–2015: findings from the global burden of disease study 2015. Public Health Nutr. 2019;22:827–40.

Men J, Zou S, Ma J, Xiang C, Li S, Wang J. Effects of high-intensity interval training on physical morphology, cardiorespiratory fitness and metabolic risk factors of cardiovascular disease in children and adolescents: a systematic review and meta-analysis. PLoS ONE. 2023;18:e0271845.

MIT Climate Portal. Will climate change drive humans extinct or destroy civilization? https://climate.mit.edu/. Accessed June 2024.

National Centers for Environmental Information. What's the difference between weather and climate? https://www.ncei.noaa.gov/news/weather-vs-climate. Accessed May 2024.

Nicholls N, Skinner C, Loughnan M, Tapper N. A simple heat alert system for Melbourne, Australia. Int J Biometeorol. 2008;52:375–84.

Nichols RB, McIntyre WF, Chan S, Scogstad-Stubbs D, Hopman WM, Baranchuk A. Snow-shoveling and the risk of acute coronary syndromes. Clin Res Cardiol. 2012;101:11–5.

Nogareda F, Regan AK, Couto P, Fowlkes AL, Gharpure R, Loayza S, Leite JA, Rodriguez A, Vicari A, Azziz-Baumgartner E, et al. Effectiveness of COVID-19 vaccines against hospitalisation in Latin America during three pandemic waves, 2021–2022: a test-negative case-control design. Lancet Reg Health Am. 2023;27:100626.

Ogah OS, Umuerri EM, Adebiyi A, Orimolade OA, Sani MU, Ojji DB, Mbakwem AC, Stewart S, Sliwa K. SARS-CoV 2 infection (Covid-19) and cardiovascular disease in Africa: health care and socio-economic implications. Glob Heart. 2021;16:18.

Okada S, Hiuge A, Makino H, Nagumo A, Takaki H, Konishi H, Goto Y, Yoshimasa Y, Miyamoto Y. Effect of exercise intervention on endothelial function and incidence of cardiovascular disease in patients with type 2 diabetes. J Atheroscler Thromb. 2010;17:828–33.

Packer M, McMurray JJ, Desai AS, Gong J, Lefkowitz MP, Rizkala AR, Rouleau JL, Shi VC, Solomon SD, Swedberg K, Zile M, Andersen K, Arango JL, Arnold JM, Belohlavek J, Bohm M, Boytsov S, Burgess LJ, Cabrera W, Calvo C, et al. Angiotensin receptor neprilysin inhibition compared with enalapril on the risk of clinical progression in surviving patients with heart failure. Circulation. 2015;131:54–61.

Paterick TE, Fletcher GF. Endothelial function and cardiovascular prevention: role of blood lipids, exercise, and other risk factors. Cardiol Rev. 2001;9:282–6.

Pennington M, Gomes M, Chrysanthaki T, Hendriks J, Wittenberg R, Knapp M, Black N, Smith S. The cost of diagnosis and early support in patients with cognitive decline. Int J Geriatr Psychiatry. 2018;33:5–13.

Pfeffer MA, Claggett B, Lewis EF, Granger CB, Kober L, Maggioni AP, Mann DL, McMurray JJV, Rouleau JL, Solomon SD, Steg PG, Berwanger O, Cikes M, et al. Impact of sacubitril/valsartan versus ramipril on total heart failure events in the PARADISE-MI trial. Circulation. 2022;145:87–9.

Playford D, Strange G, Celermajer DS, Evans G, Scalia GM, Stewart S, et al. Diastolic dysfunction and mortality in 436,360 men and women: the National Echo Database Australia (NEDA). Eur Heart J Cardiovasc Imaging. 2021;22:505–15.

Playford D, Stewart S, Harris SA, Chan YK, Strange G. Pattern and prognostic impact of regional wall motion abnormalities in 255,697 men and 236,641 women investigated with echocardiography. J Am Heart Assoc. 2023;12:e031243.

Prabhakaran D, Singh K, Kondal D, Raspail L, Mohan B, Kato T, Sarrafzadegan N, Talukder SH, Akter S, Amin MR, Goma F, et al. Cardiovascular risk factors and clinical outcomes among patients hospitalized with COVID-19: findings from the world heart federation COVID-19 study. Glob Heart. 2022;17:40.

Ravera A, Santema BT, de Boer RA, Anker SD, Samani NJ, Lang CC, Ng L, Cleland JGF, Dickstein K, Lam CSP, Van Spall HGC, Filippatos G, van Veldhuisen DJ, Metra M, Voors AA, Sama IE. Distinct pathophysiological pathways in women and men with heart failure. Eur J Heart Fail. 2022;24:1532–44.

Reriani MK, Lerman LO, Lerman A. Endothelial function as a functional expression of cardiovascular risk factors. Biomark Med. 2010;4:351–60.

Ruane L, Parsonage W, Hawkins T, Hammett C, Lam CS, Knowlman T, Doig S, Cullen L. Differences in presentation, management and outcomes in women and men presenting to an emergency department with possible cardiac chest pain. Heart Lung Circ. 2017;26:1282–90.

Sauter T, Haider DG, Ricklin ME, Exadaktylos AK. The snow, the men, the shovel, the risk? ER admissions after snow shovelling: 13 winters in Bern. Swiss Med Wkly. 2015;145:w14104.

Sparks JR, Wang X, Lavie CJ, Zhang J, Sui X. Cardiorespiratory fitness as a predictor of non-cardiovascular disease and non-cancer mortality in men. Mayo Clin Proc. 2024;5:987.

Srinivas G, Vialard J, Liu F, Voldoire A, Izumo T, Guilyardi E, Lengaigne M. Dominant contribution of atmospheric nonlinearities to ENSO asymmetry and extreme El Nino events. Sci Rep. 2024;14:8122.

Stewart ST, Zelinski EM, Wallace RB. Age, medical conditions, and gender as interactive predictors of cognitive performance: the effects of selective survival. J Gerontol B Psychol Sci Soc Sci. 2000;55:P381–3.

Stewart S, Riegel B, Boyd C, Ahamed Y, Thompson DR, Burrell LM, Carrington MJ, Coats A, Granger BB, Hides J, Weintraub WS, Moser DK, Dickson VV, McDermott CJ, Keates AK, Rich MW. Establishing a pragmatic framework to optimise health outcomes in heart failure and multimorbidity (ARISE-HF): a multidisciplinary position statement. Int J Cardiol. 2016;212:1–10.

Stewart S, Keates AK, Redfern A, McMurray JJV. Seasonal variations in cardiovascular disease. Nat Rev Cardiol. 2017;14:654–64.

Stewart S, Moholdt TT, Burrell LM, Sliwa K, Mocumbi AO, McMurray JJ, Keates AK, Hawley JA. Winter peaks in heart failure: an inevitable or preventable consequence of seasonal vulnerability? Card Fail Rev. 2019;5:83–5.

Stewart S, Playford D, Scalia GM, Currie P, Celermajer DS, Prior D, Codde J, et al. Ejection fraction and mortality: a nationwide register-based cohort study of 499,153 women and men. Eur J Heart Fail. 2021;23:406–16.

Stewart S, Patel SK, Lancefield TF, Rodrigues TS, Doumtsis N, Jess A, Vaughan-Fowler ER, Chan YK, Ramchand J, Yates PA, Kwong JC, McDonald CF, Burrell LM. Vulnerability to environmental and climatic health provocations among women and men hospitalized with chronic heart disease: insights from the RESILIENCE TRIAL cohort. Eur J Cardiovasc Nurs. 2024;23:278–86.

Strange G, Stewart S, Celermajer D, Prior D, Scalia GM, Marwick T, Ilton M, Joseph M, Codde J, et al. Poor long-term survival in patients with moderate aortic stenosis. J Am Coll Cardiol. 2019a;74:1851–63.

Strange G, Stewart S, Celermajer DS, Prior D, Scalia GM, Marwick TH, Gabbay E, Ilton M, Joseph M, Codde J, et al. Threshold of pulmonary hypertension associated with increased mortality. J Am Coll Cardiol. 2019b;73:2660–72.

Strange G, Playford D, Scalia GM, Celermajer DS, Prior D, Codde J, Chan YK, Bulsara MK, Stewart S, et al. Change in ejection fraction and long-term mortality in adults referred for echocardiography. Eur J Heart Fail. 2021;23:555–63.

Suman S, Pravalika J, Manjula P, Farooq U. Gender and CVD-does it really matters? Curr Probl Cardiol. 2023;48:101604.

Sydney Morning Herald; 2017. https://www.smh.com.au/healthcare/cardiologists-call-for-overhaul-of-malecentric-heart-guidelines-20161011-grzq7x.html. Accessed May 2024.

Tadic M, Cuspidi C, Hering D. Hypertension and cognitive dysfunction in elderly: blood pressure management for this global burden. BMC Cardiovasc Disord. 2016;16:208.

Tanaka A, Shimabukuro M, Machii N, Teragawa H, Okada Y, Shima KR, Takamura T, Taguchi I, Hisauchi I, Toyoda S, Matsuzawa Y, Tomiyama H, Yamaoka-Tojo M, Ueda S, Higashi Y, Node K. Secondary analyses to assess the profound effects of empagliflozin on endothelial function in patients with type 2 diabetes and established cardiovascular diseases: the placebo-controlled double-blind randomized effect of empagliflozin on endothelial function in cardiovascular high risk diabetes mellitus: multi-center placebo-controlled double-blind randomized trial. J Diabetes Investig. 2020;11:1551–63.

Terblanche-Greeff AC, Dokken JV, van Niekerk D, Loubser RA. Cultural beliefs of time orientation and social self-construal: Influences on climate change adaptation. Jamba. 2018;10:510.

Thorsson S, Rayner D, Palm G, Lindberg F, Carlstrom E, Borjesson M, Nilson F, Khorram-Manesh A, Holmer B. Is physiological equivalent temperature (PET) a superior screening tool for heat stress risk than wet-bulb globe temperature (WBGT) index? Eight years of data from the Gothenburg half marathon. Br J Sports Med. 2021;55:825–30.

Trubee NW, Vanderburgh PM, Diestelkamp WS, Jackson KJ. Effects of heat stress and sex on pacing in marathon runners. J Strength Cond Res. 2014;28:1673–8.

Ung GA, Nguyen KH, Hui A, Wong ND, Dineen EH. Impact of cardiorespiratory fitness and diabetes status on cardiovascular disease and all-cause mortality: an NHANES retrospective cohort study. Am Heart J plus. 2024;42:100395.

Vaduganathan M, Docherty KF, Claggett BL, Jhund PS, de Boer RA, Hernandez AF, Inzucchi SE, Kosiborod MN, Lam CSP, Martinez F, Shah SJ, Desai AS, McMurray JJV, Solomon SD. SGLT-2 inhibitors in patients with heart failure: a comprehensive meta-analysis of five randomised controlled trials. Lancet. 2022;400:757–67.

Van De Maele K, Bruyndonckx L. Quantifying cardiovascular risk: will measuring endothelial function suffice? Eur J Prev Cardiol. 2024;98:567.

Vikstrom L, Fjallstrom P, Gwon YD, Sheward DJ, Wigren-Bystrom J, Evander M, Bladh O, Widerstrom M, Molnar C, et al. Vaccine-induced correlate of protection against fatal COVID-19 in older and frail adults during waves of neutralization-resistant variants of concern: an observational study. Lancet Reg Health Eur. 2023;30:100646.

Vita JA, Keaney JF. Endothelial function: a barometer for cardiovascular risk? Circulation. 2002;106:640–2.

Vogel B, Acevedo M, Appelman Y, Bairey Merz CN, Chieffo A, Figtree GA, Guerrero M, Kunadian V, Lam CSP, Maas A, Mihailidou AS, Olszanecka A, Poole JE, Saldarriaga C, Saw J, Zuhlke L, Mehran R. The Lancet women and cardiovascular disease commission: reducing the global burden by 2030. Lancet. 2021;397:2385–438.

Wang Z, Hoy WE. Lifetime risk of developing coronary heart disease in Aboriginal Australians: a cohort study. BMJ Open. 2013;3:89.

Weintraub S, Randolph C, Bain L, Hendrix JA, Carrillo MC. Is cognitive decline measurable in preclinical Alzheimer's disease? Alzheimers Dement. 2017;13:322–3.

Whittington RM. Snow-shovelling and coronary deaths. Br Med J. 1977;1:577.

Yang B, Lin Y, Xiong W, Liu C, Gao H, Ho F, Zhou J, Zhang R, Wong JY, Cheung JK, Lau EHY, Tsang TK, Xiao J, Wong IOL, Martin-Sanchez M, Leung GM, Cowling BJ, Wu P. Comparison of control and transmission of COVID-19 across epidemic waves in Hong Kong: an observational study. Lancet Reg Health West Pac. 2024;43:100969.

Yeoh SE, Docherty KF, Jhund PS, Petrie MC, Inzucchi SE, Kober L, Kosiborod MN, Martinez FA, Ponikowski P, Sabatine MS, Bengtsson O, Boulton DW, Greasley PJ, Langkilde AM, Sjostrand M, Solomon SD, McMurray JJV. Relationship of dapagliflozin with serum sodium: findings from the DAPA-HF trial. JACC Heart Fail. 2022;10:306–18.

Zaman S, Chow C, Lam CSP, Saw J, Nicholls SJ, Figtree GA. Heart disease in women: where are we now and what is the future? Heart Lung Circ. 2021;30:1–2.

# Climate-Driven Variations in Cardiovascular Events

**Abstract**

Previous chapters have focussed on the broader picture of global health in the context of climate change, before examining why climatic conditions/acute weather events can influence an individual's heart and broader cardiovascular health through the lens of 'vulnerability to resilience'. In doing so, an argument for why and how we should alter our collective thinking around the role of climate and health (*essentially embracing a new paradigm in providing clinical car*e) is urgently needed. However, such a radical change would be pointless, or at least a low priority, if there was little evidence that health outcomes are indeed—(1) Shaped and influenced by the weather/climatic conditions, thereby resulting in clinically significant variations in event rates and, (2) Climate change is likely to exacerbate the problem in terms of provoking more events that might be preventable. Thus, in the context of a growing body of research and published data (much of which is gravitating towards a more simplistic "*heat is bad*" mindset), this chapter provides hard evidence that the timing and frequency of concrete events such as hospital admissions and deaths linked to cardiovascular disease and the main subtypes of heart disease are not random. Instead, they ebb and flow according to both predictable climatic transitions (seasons) and unpredictable weather conditions (heatwaves and cold spells) in different ways.

**Keywords**

Climatic variations · Climate influences · Heart health · Seasonal transitions · Cardiovascular events · Cardiovascular subtypes (acute coronary syndromes/acute myocardial infarction · atrial fibrillation · heart failure and cerebrovascular disease/strokes) · Non-climatic influences · Heatwaves · Cold periods · Excess deaths

## 5.1 Back to the Future—A Time for All Seasons

While it may seem odd to introduce a Chapter in a book on the understanding and management of heart disease written in the twenty-first century with ancient wisdom, but it is entirely appropriate to begin with the observation that the notion (or doctrine) that our health is "*greatly influenced*" by seasonal transitions in climatic conditions (and therefore the ambient weather experienced over a few months) was not only held by the Ancient Greek philosophers, but those of Ancient India and China (Dong 2007).

S. Stewart, *Heart Disease and Climate Change*, Sustainable Development Goals Series, https://doi.org/10.1007/978-3-031-73106-8_5

If the reader remembers only one quote or idea (attributed to the Greek philosopher Herodotus) from this book, then the following is most deserving of that honour *"we generally get ill when things changes—and by 'things' here I mean especially, but not exclusively, the seasons"* (Dong 2007). This long-standing wisdom appears more apt than ever before given the modern context of largely stable seasonal patterns of weather (expected transitions in climatic conditions) now being altered by climate change. Other Greek philosophers such as Hippocrates recognised the need for prevention and a holistic perspective on health—one might argue that a more reductionist approach to medicine and other health disciplines (especially when conducted in the environmentally controlled rooms of tertiary hospitals and primary care clinics) has meant this wisdom has been truly lost in translation.

Nevertheless, variations in the pattern of human illness (and specifically cardiac events) according to their geographic location, the time of the year and ambient weather conditions at that location make intuitive sense. As will be demonstrated, this concept is particularly relevant to those regions of the world that experience the four distinct seasons of winter, spring, summer, autumn as a result of the rotational tilt on the earth's axis as it completes one orbit around the sun. However, they do also occur in regions closer to the equator (where annual variations due to changes in the duration and angle of sunlight are far less when compared to those located closer to the poles). As discussed throughout this book, while a large body of evidence suggests that cardiovascular (and therefore cardiac-specific) events occur in peaks and troughs in response to the seasons (with winter, or just thereafter, still appearing to be the deadliest of seasons), they don't precisely correlate with the ambient climate, temperature variation, and latitude of affected populations. Indeed, as already discussed in the previous chapter, the available data suggest a more complex interaction between the climate and an individual's characteristics/behaviours that precludes a simple 'dose–response' to daylight exposure and

ambient temperature change. They also reflect broader environmental conditions such as the confounding influence of pollution (discussed in Chap. 6) and other health conditions such as respiratory disease (discussed in Chap. 7). To my knowledge, no one has been able to generate a precise model to precisely predict 'when' a person will suffer a cardiac event. Instead there is a growing body of predictive algorithms based on a person's risk profile as to "if" they will suffer a cardiovascular event over a certain period of time (Chybowska et al. 2024; Wang et al. 2023a; Temtem et al. 2024; Shah et al. 2024; Ghosh et al. 2024; Pennells et al. 2024). Clearly, this represents the holy grail in terms of sorting wheat from chaff for cost-effective primary to secondary prevention. However, it strikes me that nearly every screening tool/algorithm ignores, unlike the Ancients (Dong 2007), the additional need to consider when that might happen, by taking into account the seasons, acute weather events and, now, the complication of climate change.

## 5.2 A Historical Perspective on Climatic Influences on Heart Health

As highlighted in a high-level review of the importance of seasonal changes in the weather on cardiovascular disease (Stewart et al. 2017), according to the Köppen-Geiger climate classification system, there are five main climates radiating from the equator to the poles—(A) tropical, (B) dry/arid, (C) temperate, (D) continental and (E) polar (Beck et al. 2018; Kottek et al. 2006). As described in Chap. 3, humans have gravitated towards more temperate regions of the world given the discomfort levels and the more extreme adaptations (to maintain homeostasis) needed to live in the more extreme climates on the planet. Overall, there are almost 30 distinct climatic variations of these categories (based on seasonal variations in temperature and rainfall) and, of course, these are being challenged by climate change (Beck et al. 2018; Kottek et al. 2006). Many of these major and sub-climates are highlighted in this chapter, but it's important to note that many are not—simply because, like most of the global, cardiologic research portfolio, there are many low-resource countries and regions of the world who remain unstudied.

Within broadly recognised climatic zones, complex factors will shape local weather conditions—one only must sit for a few hours watching the rapid formation and flow of 'rivers' of cloud, rain, wind and markedly changing temperatures that swirl around Table Mountain, the Lions Head and the Twelve Apostles in Cape Town, South Africa (see Fig. 5.1) to appreciate the dynamic zone of microclimates as the Atlantic Ocean meets the rapidly rising African continent at that point. As such, urban areas and distinct geographical features (from ocean currents to adjacent mountainous regions) modulate both overall climatic conditions and mean temperatures. As evidenced by the increasing focus on the healthy 'built environment' concept, from multiple perspectives (including its contribution to activity levels and patterns of cardiovascular disease) (Jabeen et al. 2023; Patel et al. 2024;

Lan et al. 2023; Li et al. 2024; Nabaweesi et al. 2023; Turner et al. 2023; Xu et al. 2023), major urban areas have the potential to create their own microclimates.

The broader phenomenon of climatic influences on cardiac and other events such ischaemic and haemorrhagic strokes has been long recognised in the medical literature—especially in relation to acute myocardial infarction, which has been studied for more than 100 years. For example, in a UK paper published in 1926, the researchers reported that deaths linked to the circulatory system was well-known to be higher in winter versus summer months (Bundesen 1926). Moreover, a range of behaviours and physiological responses that might contribute to differential levels of cardiovascular activity according to the time of day and year have been recognised. Throughout the twentieth century, therefore, a range of research reports [but mostly confined to acute myocardial infarction and including a 1938 review of the available literature (Bean 1938)] documented both winter and summer peaks in cardiovascular-related morbidity and mortality (Bean 1938).

**Fig. 5.1**   The twelve apostles (table mountain) in Cape Town, South Africa

These historical perspectives are extremely important when seeking to understand how and why climate change will influence future cardiovascular events. For example, as an early indication of the complex mechanisms underlying the observed pattern of acute cardiovascular presentations, a 1955 study from Texas, USA, reported one of the earliest, potentially "paradoxical" findings of an overall summer peak in hospital admissions for acute myocardial infarction, but with a provocation of events after the onset of relatively cold weather/cold spell (Teng 1955). Why is this important? For many years it was accepted (or at least hypothesised) that there would be a gradient of increasing risk of experiencing an acute myocardial infarction or stroke (the two main manifestations of CVD at the time) from the equator upwards (or downwards), from hot/tropical to cold climates. In other words, absolute temperature and temperature changes modulated cardiovascular events. This is extremely important in the context of increasing temperatures as part of climate change—surely, if this is true, then a warming climate should mean less cold, less cold variations and therefore less cardiovascular events.

However, as always, it's not so simple as *"cold is bad/heat is good"*! The EuroWinter Group were not only instrumental in demonstrating this, but in presenting their findings at the 1997 European Society of Cardiology meeting in Stockholm, Sweden, inspired me to explore their findings further—something I've now done for over 3 decades through a variety of projects (Stewart et al. 2002, 2019, 2023).

Specifically, as originally reported, more than 50 years after the Texas study, the EuroWinter Group sought to—*"to assess whether increases in mortality per 1 °C fall in temperature differ in various European regions and to relate any differences to usual winter climate and measures to protect against cold"* (The Eurowinter Group 1997). To do this, they gathered cause-specific death data over a 5-year period (1988–1992) for men and women aged 50–59 and 65–74 years living in 24 different locations (comprising northern Finland/Arctic Circle), southern Finland, Baden-Württemburg, the Netherlands,

London, UK and northern Italy) with equivalent comparisons made to 1992 data derived from Athens in Greece and Palermo in southern Italy (The Eurowinter Group 1997). Thus, the study spanned from cold to temperate climates (including true Mediterranean climates of hot dry summers and relatively cool wet winters) (The Koppen Climatic Classification 2024). Surprisingly, the highest incidence of cold-related mortality occurred in countries with 'warmer' winters. As succinctly reported at the time—*"the percentage increases in all-cause mortality per 1 °C fall in temperature below 18 °C were greater in warmer regions than in colder regions (e.g., Athens 2.15% [95% CI 1.20–3.10] versus south Finland 0.27% [0.15–0.40]). At an outdoor temperature of 7 °C, the mean living-room temperature was 19.2 °C in Athens and 21.7 °C in south Finland; 13 and 72% of people in these regions, respectively, wore hats when outdoors at 7 °C"* (The Eurowinter Group 1997). Critically, they also highlighted a range of environmental and behavioural factors (limited bedroom heating and wearing warmer clothes respectively) that modulated what they were observing in terms of the interactions between the weather/climate and outcomes. It is these findings that prompted my initial thinking around bio-behavioural factors and eventually the vulnerable-to-resilient phenotypes presented in Chap. 4. A later report from the group, reinforces the concept of both "non-modifiable" and "modifiable" areas of climatic vulnerability. Specifically, among people aged 65–74 years (in whom the link between cold and mortality was strongest), they found higher levels of protection (lower than expected mortality per temperature reduction) against indoor and outdoor cold temperatures in those countries who experienced cold winters/temperatures. Alternatively, in places like London, UK (milder/temperate climate), in the same age-group, there were excess deaths linked to each change/fall in temperature during cold periods/winter (Keatinge et al. 2000).

The observations made by EuroWinter Group were subsequently supported by the highly influential WHO MONICA Cohort Study (Evans et al. 2001; Tunstall-Pedoe et al. 2000; Tunstall-Pedoe et al. 1999; Wolf et al. 1997; Thorvaldsen et al. 1997). For the uninitiated, the MONICA Study, led by public health/epidemiology luminaries such as Professors Hugh Tunstall-Pedoe and Carolyn Morrison (both of whom were fearsome and intimidating intellects) can be counted as one of the most influential cardiovascular studies to date. Specifically, in an analysis of 87,410 documented events (44% fatal) within a subset of the entire/larger WHO MONICA cohort (in this case 24 populations across 21 countries studied between 1980 and 1995), coronary event rates during cold episodes were more prominent among people aged 35–64 years (i.e. relatively young people) in relatively warmer climates. Alternatively, they found only slight increases found in people living in cold climates. Moreover, it was found that women were 1.07-fold more susceptible to this phenomenon than men (Barnett et al. 2005). Also consistent with other reports (published before and then after this report), the MONICA Investigators found a highly important 'lag effect' of up to 11 days between cold episodes and subsequent spikes in cardiac events (Barnett et al. 2005). This finding is extremely important when one considers the physiological differences in cold versus heat responses (as described in Chap. 3), the potential modulating impact of infectious diseases (see section below and Chap. 7) and then any attempts to modulate exposure to cold spells/winters versus heatwaves/summers in vulnerable people to reduce and prevent cardiovascular events (Chaps. 8 and 9). Consistent, with the potential to delineate between the two, a subsequent meta-analysis of outcomes (including 3.9 million deaths and 12.2 million hospitalisations among people aged 65 years or more) according to temperature change reported by Bunker et al. (2016), concluded that cardiovascular-related mortality increased by 3.4% per 1 °C rise in temperature. Such findings have immediate clinical to public health implications in the context of climate change! This 'heat effect' contrasted to an equivalent 1.7% increase in mortality associated with a 1 °C decline in temperature among individuals aged $\geq$65 years (once again noting the older age group being studied). Overall, rather than saying heat is worse than cold, these data suggested a greater and more immediate cardiovascular risk posed by heatwaves than cold spell episodes, with colder temperatures more closely linked to respiratory events and the potential for the hidden lag effect revealed by the MONICA Investigators (Barnett et al. 2005). Consistent with the pattern of increased mortality linked to relative cold conditions versus absolute cold conditions, an analysis of 74 million deaths in people of all ages living in 384 different locations worldwide during the period of 1995–2012, revealed that while cold temperatures contributed to the greatest number of deaths overall, it was 'moderately cold' rather than 'very cold' temperatures that contributed most to this phenomenon. Specifically, Gasparrini et al. reported that while "*7.71% (95% CI 7.43–7.91) of mortality was attributable to non-optimum temperature*", this overall finding was qualified by the fact that "*more temperature-attributable deaths were caused by cold (7.29%, 7.02–7.49) than by heat (0.42%, 0.39–0.44)*". Moreover, only 0.86% (95% CI 0.84–0.87) of deaths could be attributed to hot and cold temperature extremes. Thus, this extremely large study found a much higher impact of cold temperatures than hot temperatures (more than 7:1-fold higher) (Gasparrini et al. 2015).

## 5.3 Choose Your Poison—Death by Cold or by Heat?

Of course, it could be argued that human-induced climate change has radically altered the balance between the risk of death in cold versus hot temperatures—noting once again the radical idea that if the associations are correctly reported AND we can overcome the paradox of people living in milder climates dying when exposed to relatively colder weather, climate change might actually reduce our risk of premature mortality and thereby extend our collective longevity! As succinctly argued by Dimitriadou et al. (2022)—"*A point often overlooked is that climate change may be beneficial for some countries characterized by temperate climate, since milder winters may lead to a decrease in human mortality, while in low-latitude countries upcoming hotter and drier conditions may result in mosquito elimination".* For anyone following the logic and arguments made in this book (founded on the original findings/reports from the EuroWinter and MONICA groups), this argument is most probably flawed—if one accepts that shifting colder climates into relatively milder climates will invoke the paradoxical loss of cold adaptation and a resultant higher level of cardiovascular-related events. Nevertheless, they argue that contemporary models seeking to explain the future impact of climate change predict a neutral effect on mortality given that cold- (less) versus heat-events (more) will cancel each other out (Martinez-Solanas et al. 2021; Zhang et al. 2018)—something that this author cannot completely agree with when considering cardiovascular-related mortality, an increasing number of people at risk and more granular findings from places/climates like Australia (described later in this book). Based on a study of temperature and mortality revealing that colder/northern European cities have a lower temperature threshold for mortality than warmer/southern cities (see below), we can at least agree on the following statement that— *"Different climatic conditions affect different populations and since climate change will affect mortality globally but not uniformly; it is critical to know how different regions are affected"* (Dimitriadou et al. 2022).

It is worth revisiting at this point, the current (known/understood) ratio of cold versus heat-related impact on mortality (and by extension, cardiovascular-related deaths as a major component of death worldwide). Are they likely to cancel each other out with climate change?

In a recent report in *Lancet Planetary Health*, Masselot et al. reported on a study focussing on temperature associated deaths in people aged 20 years or more living in 854 urban European areas during a 20-year period of 2000–2019. Data were derived from the *Urban Audit dataset of Eurostat*, the *Multi-country Multi-city*

*Collaborative Research Network, Moderate Resolution Imaging Spectroradiometer* and the satellite-based Copernicus initiative (European Union 2024). As specifically reported, the study revealed *"an annual excess of 203,620 (95% CI 180,882–224,613) deaths attributed to cold".* This compared to a *tenfold lesser* figure of 20,173 (95% CI 17,261–22,934) deaths attributed to heat. Overall there was an age-standardised rate of 129 deaths (95% CI 114–142)/100,000 person-years versus 13 (95% CI 11–14) deaths/100,000 person-years associated with cold versus hot climatic conditions across Europe as a whole (Masselot et al. 2023). Observed mortality differentials were most evident in Eastern Europe (home to some of the poorest communities in the region), with a number of gradients and 'hot' and 'cold' spots for differential findings evident. For example, in relation to heat-related deaths, there was a gradient from northwest to southeast gradient, with a small amount of excess deaths in the north that increased in more southern countries—something that is both logical and likely to change if climate change brings [as it has already done as far as the Arctic Circle (Grigorieva et al. 2023)] more heatwaves to countries nearer the poles in both hemispheres. Consistent with pleas for a more nuanced approach to estimating and responding to the effects of climate change (something discussed later in this book) is the fact there was evidence that the pattern of deaths was modulated by a sea/land interaction in coastal locations cities. Specifically, there were higher cold-related and lower heat-related effects in these locations when compared to their inland neighbours. This is significant when one remembers the *paradoxical phenomenon* of a greater impact of cold on people living in milder climates and that water/the sea retains heat much more than land (where temperatures oscillate much more rapidly with the ambient weather conditions), thereby modulating extreme temperatures overall. Figures 5.2, 5.3, 5.4, and 5.5 compares the main findings of this study, with a particular focus on the ratio of excess cold-identified versus excess heat-identified deaths across the different regions and countries comprising

|                | Excess Cold Deaths | Attributed % | Excess Heat Deaths | Attributed % | Excess Ratio |
|----------------|--------------------|--------------|--------------------|--------------|--------------|
| Denmark        | 1,319              | 8.64         | 44                 | 0.29         | 30.0         |
| Estonia        | 727                | 8.49         | 29                 | 0.34         | 25.1         |
| Finland        | 2,231              | 9.28         | 69                 | 0.29         | 32.3         |
| Ireland        | 1,664              | 12.7         | 10                 | 0.08         | 166.0        |
| Latvia         | 2,095              | 9.72         | 126                | 0.59         | 16.6         |
| Lithuania      | 2,301              | 9.39         | 141                | 0.58         | 16.3         |
| Norway         | 1,085              | 9.89         | 28                 | 0.25         | 38.8         |
| Sweden         | 4,168              | 8.92         | 134                | 0.29         | 31.1         |
| United Kingdom | 42,915             | 9.87         | 762                | 0.18         | 56.3         |

Standardised Excess Death Rate (per 100,000 person-years cold)

243

153

**Fig. 5.2** Excess deaths from cold and hot temperatures—Northern Europe. These data are derived from data published by Masselot et al. (2023). Gradient colours are based on the standardised excess death rates per 100,000 person-years (cold)—in rank order the highest-to-lowest countries in this regard were Latvia (240 deaths/100,000 person-years), Ireland, the UK/Sweden, Estonia, Lithuania, Finland, Denmark and Norway (131 deaths/100,000 person-years). The attributable % refers to the proportion of deaths relative to total mortality and the excess ratio to the absolute number of excess cold-versus heat-related deaths for the country

Europe. In Northern Europe (predominantly "cold continental"), the climate typically ranges from extremely cold subarctic climate (according to the Köppen Climate Classification system 2024) to subarctic. Overall, the number of excess deaths attributable to cold versus heat was 58,505 versus 1343 (a ratio of 43.6 to 1 excess deaths) in Northern Europe—see Fig. 5.2. In Western Europe (predominantly 'temperate'), the climate typically ranges from subpolar oceanic climate (The Koppen Climatic Classification 2024), temperate oceanic and humid subtropical. Overall, the number of excess deaths attributable to cold versus heat was 46,141 versus 5433 (a markedly lower ratio of 8.49 excess deaths to 1) in Western Europe—see Fig. 5.3. Further east (Eastern Europe which has a predominant cold continental climate), the climate typically ranges from warm-to-hot summer humid continental (The Koppen Climatic Classification 2024).

| | Excess Cold Deaths | Attributed % | Excess Heat Deaths | Attributed % | Excess Ratio |
|---|---|---|---|---|---|
| Austria | 1,586 | 4.97 | 215 | 0.67 | 7.3 |
| Belgium | 2,093 | 8.49 | 247 | 0.53 | 8.5 |
| France | 17,730 | 9.28 | 1,388 | 0.47 | 12.8 |
| Germany | 19,728 | 12.7 | 2,886 | 0.64 | 6.4 |
| Luxembourg | 46 | 9.72 | 6 | 0.59 | 7.7 |
| Netherlands | 3,419 | 9.39 | 479 | 0.52 | 7.1 |
| Switzerland | 1,538 | 9.87 | 213 | 0.68 | 7.2 |

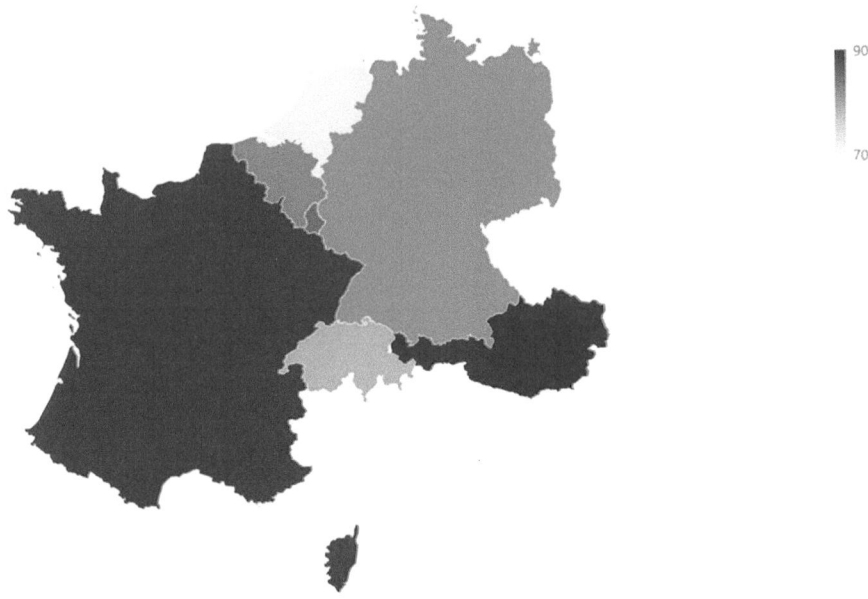

Standardised Excess Death Rate (per 100,000 person-years cold)

**Fig. 5.3** Excess deaths from cold and hot temperatures—Western Europe. These data are derived from data published by Masselot et al. (2023). Gradient colours are based on the standardised excess death rates per 100,000 person-years (cold)—in rank order the highest-to-lowest countries in this regard were Austria (90 deaths/100,000 person-years), France, Luxembourg, Belgium, Germany, Switzerland and the Netherlands (70 deaths/100,000 person-years). The attributable % refers to the proportion of deaths relative to total mortality and the excess ratio to the absolute number of excess cold- versus heat-related deaths for the country

Overall, the number of excess deaths attributable to cold versus heat was 45,408 versus 3659 (a similar/slightly higher [to Western Europe] ratio of 12.41 excess deaths to 1) in Eastern Europe—see Fig. 5.4. Lastly, in Southern Europe (predominant temperate climate), the climate typically ranges from cool-summer, warm-summer to a hot-summer Mediterranean climate (The Koppen Climatic Classification 2024). Overall, the number of excess deaths attributable to cold versus heat was 53,566 versus 9738 (the lowest ratio for Europe at 5.50 excess deaths to 1) in Southern Europe—see Fig. 5.5.

## 5.4   Climatic Influences on Heart Health

As described above, there is strong evidence to suggest that cardiovascular events fluctuate according to ambient climatic conditions. The predominant evidence suggests that there is a more marked (but often delayed), aggravated response to cold events/weather. This appears to be particularly true among those living in milder climates although, as seen above, the absolute ratio of reported excess cold-to-hot deaths is strongest in cold climates (~44 to 1) and least in

| | Excess Cold Deaths | Attributed % | Excess Heat Deaths | Attributed % | Excess Ratio |
|---|---|---|---|---|---|
| Bulgaria | 5,745 | 9.25 | 522 | 0.84 | 11.0 |
| Czechia | 3,716 | 8.00 | 235 | 0.51 | 15.8 |
| Hungary | 5,295 | 8.78 | 402 | 0.67 | 13.2 |
| Poland | 16,510 | 8.44 | 1263 | 0.65 | 13.1 |
| Romania | 12,939 | 10.01 | 1137 | 0.88 | 11.4 |
| Slovakia | 1,203 | 8.51 | 99 | 0.70 | 12.2 |

Standardised Excess Death Rate (per 100,000 person-years cold)

**Fig. 5.4** Excess deaths from cold and hot temperatures—Eastern Europe. These data are derived from data published by Masselot et al. (2023). Gradient colours are based on the standardised excess death rates per 100,000 person-years (cold)—in rank order the highest-to-lowest countries in this regard were Bulgaria (266 deaths/100,000 person-years), Romania, Hungary, Slovakia, Poland and Czechia (180 deaths/100,000 person-years). The attributable % refers to the proportion of deaths relative to total mortality and the excess ratio to the absolute number of excess cold- versus heat-related deaths for the country

warm-to-hot climates (~6 to 1) (Masselot et al. 2023). As has been emphasised repeatedly, there is undoubtedly a complex interaction between climatic conditions and individual responses that modulates health outcomes. For example, a highly insightful study conducted in (what is now) Czechia during the period of 1994–2009 focussed on spatial patterns of heat-related cardiovascular mortality (Urban et al. 2016). The investigators demonstrated that, when multiple districts were compared, the characteristics of physical environment (altitude and mean summer temperature) and urbanisation level (percentage of impervious surface and population density) had a significant influence on heat-related CVD mortality. Notably, in comparison to a previously discussed study conducted in the US (Hoogenboom et al. 2021), socio-economic status did not appear to be a significant modulator of outcome. Alternatively, a close comparison of districts with low or high index of socio-economic status revealed that, below a certain threshold, socio-economic status influenced excess mortality. Perhaps not surprisingly the highest heat-related mortality was observed in districts at a lower altitude. These areas had both hotter local climates and higher levels of urbanisation (Urban et al. 2016). Once again, these data suggest that living in an area that regularly challenges an individual's capacity to maintain optimal thermoregulation [i.e. when living in an area that exceeds a 'Goldilocks

| | Excess Cold Deaths | Attributed % | Excess Heat Deaths | Attributed % | Excess Ratio |
|---|---|---|---|---|---|
| Croatia | 1,714 | 7.52 | 412 | 1.81 | 4.2 |
| Cyprus | 507 | 8.29 | 74 | 1.20 | 6.9 |
| Greece | 4,693 | 6.98 | 738 | 1.10 | 6.4 |
| Italy | 23,283 | 6.75 | 5,034 | 1.46 | 4.6 |
| Malta | 178 | 7.73 | 35 | 1.51 | 5.1 |
| Portugal | 4,573 | 6.96 | 529 | 0.80 | 8.7 |
| Slovenia | 278 | 5.41 | 70 | 1.37 | 4.0 |
| Spain | 18,342 | 5.55 | 2,848 | 0.86 | 6.4 |

Standardised Excess Death Rate (per 100,000 person-years cold)

178

88

**Fig. 5.5** Excess deaths from cold and hot temperatures—Southern Europe. These data are derived from data published by Masselot et al. (2023). Gradient colours are based on the standardised excess death rates per 100,000 person-years (cold)—in rank order the highest-to-lowest countries in this regard were Croatia (178 deaths/100,000 person-years), Cyprus, Malta, Portugal, Greece, Italy, Slovenia and Spain (88 deaths/100,000 person-years). The attributable % refers to the proportion of deaths relative to total mortality and the excess ratio to the absolute number of excess cold- versus heat-related deaths for the country

equivalent', 'thermoneutral' air temperature range of 25–31 °C (Lim 2020)] will provoke a potentially adverse cardiovascular response.

As shown in Fig. 5.6 [a newly reproduced plot of seasonal findings from major epidemiological studies published prior to 2017 (Stewart et al. 2017)], many studies worldwide confirm both winter (blue shaded) and, to a lesser extent spring (green shaded) and summer (pink shaded) peaks in major events (hospitalisations and deaths) for nearly every subtype of CVD. This includes atrial fibrillation (AF), heart failure (HF), acute myocardial infarction (AMI) and other forms of acute coronary syndrome (ACS), sudden cardiac death (SCD), and both

major forms of stroke. Consistent with those studies highlighted earlier in the chapter, most studies included in this analysis were derived from the temperate climates of Europe and, to a lesser extent, North America. By contrast, there remains a paucity of large-scale reports from the more impoverished, yet substantial populations living in Africa, South America and the Indian subcontinent. This unequal distribution of global data (something that persists today) may well explain why there are more reports (even in the context of climate change) of 'winter peaks' followed by parallel 'summer troughs' in cardiovascular events. Nevertheless, as reported at the time (Stewart et al. 2017), a J-shaped pattern of

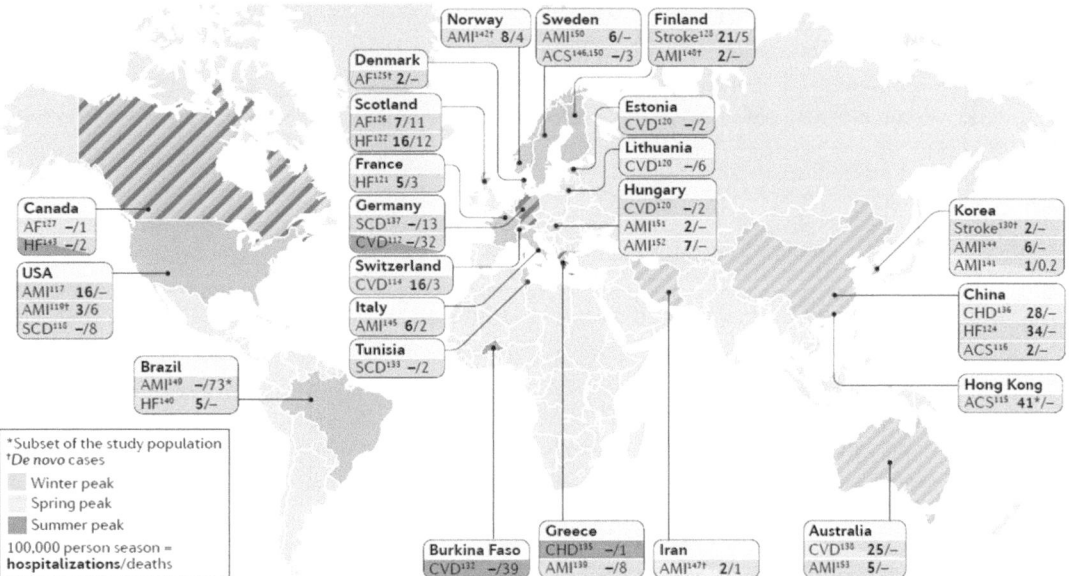

**Fig. 5.6** Pattern of peaks and troughs in cardiovascular events across the world. This figure plots the absolute difference in peak versus trough rates of cardiovascular events across the globe on a seasonal basis derived from large epidemiological studies examining seasonal variations in CVD-related hospitalisations and deaths, providing like-for-like, global comparisons in terms of variance of event rates (from peak to trough periods per 100,000 population at risk of event per annum) across a range of CVD subtypes—adapted from the original figure/data (Stewart et al. 2017)

cardiovascular-related events is evident at both ends of the cold-to-hot spectrum of temperature. This overall finding matches those revealed by more contemporary studies (Masselot et al. 2023), As also commented at the time (Stewart et al. 2017)—"*without standardized methodology for identifying cases (the majority of large-scale reports rely on retrospective analyses of pre-existing patient registries linked to mortality records) and standardized follow-up, meaningful comparisons between study findings is problematic*". This remains problematic when assessing a growing body of research reports focusing on population-based to granular studies of the association between climatic conditions and cardiovascular event rates. It is hard to see how this problem will change or be addressed given so many groups now focussed on climate change with their own ideas, definitions and methods.

Nevertheless, large-scale studies (>10,000 patients) examining the rate of cardiovascular events according to seasonal transitions in the weather/climatic conditions provide important insights into what might happen if climate change (as it has already purported to do) radically changes them. For example, in a German study of three Bavarian cities with a combined population >2.5 million, increased cardiovascular mortality at both the lower and upper scales of temperature were documented. At the warmer end of the spectrum, a 2-day increase in temperatures within the upper centile (20.0–24.8 °C) was associated with a 10% increase in cardiovascular mortality (95% CI 5–15%). At the lower end of the spectrum, a decline in temperatures (to the lower centile of −1.0 to 7.5 °C) over a longer period of 15 days was associated with an equivalent 8% increase in cardiovascular mortality (95% CI 2–14%)—noting once again the differential exposure rates (Breitner et al. 2014). In a comparison of two distinct populations (both ~1 million people) living in the markedly different climates of Dublin, Ireland, and Oslo, Norway, there were more marked variations in cardiovascular mortality linked to seasonal transitions in in the milder climes of Ireland (Eng and Mercer 1998)—noting these finding are consistent with higher ratio of excess cold-to-heat-related deaths in Ireland versus Norway revealed by Masselot et al. (2023). As in previous reports, in both countries, a negative correlation was observed between mortality and both air temperature and wind-chill factor (Eng and Mercer 1998). Of course, it would be disingenuous to suggest that peak event rates have only been found in relation to colder climatic conditions (compared to warmer ones). However, these are hard to find. For example, in a study on more than 11 million hospital episodes in Switzerland, Reavey et al. (2013) reported an overall pattern of fewer cardiovascular admissions and related deaths in those hospitalised during the summer months. As will be discussed in more detail below, the pattern of fewer events in the summer months was more pronounced in relation to AMI versus stroke (ischaemic and haemorrhagic combined) (Reavey et al. 2013).

## 5.4.1  Climatic Variations in Cardiovascular Subtypes

As will be explored in the wider context of cardiovascular specific conditions versus non-cardiovascular conditions (such as malignancy—which is probably the best *"control"* for revealing climatically-linked, temporal patterns of morbidity and mortality), any critical examination of the evidence linking climate provocations to heart health, needs to consider whether there is a generic or specific cardiovascular response/provocation in play. To be clear from my perspective, when first researching this subject, I needed to be reassured that there was evidence (based on the known pathophysiological pathways/natural history of different cardiovascular conditions) of a differential pattern evident within all cardiovascular events. It is up to the reader to determine if I've presented enough evidence to answer that question/issue.

Unsurprisingly, most historical data focuses on the acute manifestations of coronary artery disease. This includes AMI and, with more

relevance to the evolving pattern of presentations (at least in high-income countries), less severe forms of ACS. These studies span multiple populations and climate types (including those with a more dichotomous weather pattern). For example, a study of close to 55,000 presentations with an ACS in Hong Kong (a particularly dense, populous city exposed to a humid subtropical climate characterised by hot, humid summers and mild, drier winters) revealed a winter peak in hospitalisation with falls in daily temperatures (Chau et al. 2014). Similarly, a study of ACS presentations in Beijing, China (humid continental climate, influenced by a monsoonal weather pattern and short, cold and dry winters), also revealed a winter peak in events that was negatively correlated with mean daily temperatures and barometric pressures (Li et al. 2011). From a broader perspective, a study of >250,000 presentations with AMI across a wide-range of climates in the USA (including Hawaii in the Pacific and northerly Alaska) also revealed a consistent pattern of winter peaks (with ~50% more case presentations compared to summer). Observed trends persisted on a sex-specific and age-specific basis (Spencer et al. 1998). Hospital case-fatality was also more pronounced in winter (9% more deaths compared to summer) (Spencer et al. 1998). A more geographically focussed study of >62,000 sudden cardiac deaths in individuals aged ≥55 years in King County, Washington, USA, during 1980–2001 demonstrated that a 5 °C increase in temperature was associated with a decline in mortality in both men and women by a factor of 0.97 (95% CI 0.96–0.98) (Cagle and Hubbard 2005). Although these data were collected close to 25 years ago, in the context of climate change, the investigators reported that the lowest rate of death occurred on days hotter than 30 °C (Cagle and Hubbard 2005). As has been revealed by previous studies comparing the pattern out-of-hospital versus in-hospital deaths linked to AMI according to a person's age, sex and socio-economic profile (Capewell et al. 2001; MacIntyre et al. 2001), there are reasons why there might be differential responses to climatic provocations in this regard. Accordingly, an analysis of 2676 AMIs and 2066 SCDs captured by the *Olmsted County Cohort Study* between 1979 and 2002 revealed a seasonal pattern in SCD, but not AMI. Specifically, this study reported that SCD was increased 1.17-fold (95% CI 1.03–1.32) and 1.20-fold (95% CI 1.07–1.35) in low temperatures (0 °C versus 18–30 °C). This phenomenon strengthened in relation to unexpected SCD (Gerber et al. 2006). Once again, study findings have relevance to an era of climate change given that after adjustment for all climatic variables, a low temperature was associated with a significant increase in the risk of unexpected SCD (1.38-fold increase, 95% CI 1.10–1.73) whilst the association with winter (as a more indiscriminate variable) was equivocal (1.06-fold increase, 95% CI 0.83–1.35) (Gerber et al. 2006).

In respect to HF [noting the difficulty in determining the exact criteria used to identify cases and historical changes in diagnostic classification over time (McDonagh et al. 2021)], large population studies have consistently reported seasonally influenced patterns of related hospitalisation and related mortality with a strong winter influence. This includes climatically diverse reports from the warm temperate climate of France (Boulay et al. 1999), the much milder/colder, marine climate of Scotland (Stewart et al. 2002), the slightly warmer Mediterranean climate of Spain (San Roman Teran et al. 2008) and the subtropical climate of Hong Kong (Qiu 2013); all revealing similar winter peaks in morbidity and mortality.

This predominance of a cold/winter influence on morbidity and mortality extends to AF. Accordingly, climatically diverse reports from Denmark (Frost et al. 2006), Scotland (Murphy et al. 2004) and Canada (Upshur et al. 2004) [as well as smaller clinical studies from hotter climates (Kiu et al. 2004)] consistently demonstrate winter peaks in morbidity and subsequent mortality in those hospitalised with AF; as in other forms of CVD, the strongest variations occurring in older individuals.

The same climatically driven pattern of events (as a reminder this is often referred to as "*seasonality*") has also been well documented in most forms of stroke. For example, a key substudy of ~400,000 patients with stroke in three Finnish cities captured by the FINMONICA Study (Jakovljevic et al. 1996) found winter peaks in both sexes and for both ischaemic and haemorrhagic forms of stroke. Specifically, the rate of occurrence of ischaemic strokes presentations was 12% (95% CI 5–20%) and 11% (95% CI 4–19%) higher in men and women for winter compared to summer. Critically, for intracerebral haemorrhage/strokes, the equivalent differential rates were higher—being 28% (95% CI 3–58%) higher in men and 33% (95% CI 6–66%) higher in women. The greater incidence of ischaemic strokes in winter was particularly prominent among men aged 25–64 years, whilst associated 28-day case-fatality rate in ischaemic stroke cases showed significant seasonal variation (lowest mortality in summer) in women only (Jakovljevic et al. 1996). Once again, from a climate change perspective, the investigators found no variations in presentations for subarachnoid haemorrhage, although this is not a consistent finding (see below). From a pathophysiologic perspective, this was a logical observation given that this form of stroke is typically caused by a long-standing cerebral aneurysm or arteriovenous malformation that might spontaneously rupture at any time; thereby reinforcing the hypothesis that the mechanisms underlying climatically provoked cardiovascular events are more acute than chronic in nature (not withstanding previously observed lag-times between events and acute weather events). A more contemporary study of 172,000 hospital admissions for ischaemic stroke during the period 2009–2011 found that both lower absolute temperature and increased diurnal variations in temperature outside of winter were associated with higher event rates on an adjusted basis (Lichtman et al. 2016). These findings broadly support the finding in relation to SCD (Gerber et al. 2006), where it's the temperature/climatic conditions on any given day, rather than season (i.e. winter when more provocative conditions are more frequent in many part of the world) that provoke cardiovascular events. Once again this is important when considering the likely impact of climate change. In contrast to the previous study highlighting little or no associations with climatic conditions, in a study of 1477 consecutive haemorrhagic stroke events occurring in Seoul, South Korea (continental climate with monsoonal-affected summers), distinctive patterns of intracerebral and subarachnoid haemorrhage were observed with more events as temperatures declined (Han et al. 2016). Notably, while presentations with intracerebral haemorrhage was associated with increased particulate matter air pollution (1.09-fold, 95% CI 1.02–1.15), a presentation with subarachnoid haemorrhage correlated with elevated ozone levels (1.32-fold, 95% CI 1.10–1.58) (Han et al. 2016). This likely interaction between climatic conditions and pollution levels is something that can't be ignored from a climate change perspective and will be briefly discussed in the section below and then more comprehensively in Chap. 7.

## 5.5　Quantifying the Impact of Seasonal Transitions in Climatic Conditions

Figure 5.6 summarised the results of a systematic review and meta-analysis of 48 large (>10,000 events) population-based studies reporting the influence of seasonal transitions in various forms of CVD across 26 countries (Stewart 2016). Comprising 2.9 million cardiovascular hospitalisations and 1.6 million deaths it examines the relative difference in peak versus trough event rates adjusting for both mean temperatures and the type of climate [according to the Köppen–Geiger climate classification system (The Koppen Climatic Classification 2024)]. As indicated previously, mostly of these studies were conducted in the 'moist mid-latitude' or 'mild' climates of Europe. Overall, there were 1.20-fold (95% CI 1.14–1.25) more hospitalisations and 1.23-fold (95% CI 1.16–1.31) more fatal events (mostly documented in those previously hospitalised with CVD) during the peak compared with the trough season (predominantly winter versus summer) (Stewart

2016). With a few exceptions [such as a study conducted near the equator in Western Africa (Kynast-Wolf et al. 2010)], there was a predominance of winter peaks, along with some spring peaks that were likely early rather than later in the climatic transition period, for cardiovascular-related hospitalisations and deaths (Stewart et al. 2017). Consistent with the specific studies described in the sections above, along with the body of 48 studies examined in the meta-analysis, there was a positive correlation between the age of the cohort studied and the variation between observed peak versus trough event rates. The absolute difference in event rates across seasons (typically cold versus warm/hot climatic conditions) peaked as high as 25–39 events per 100,000 per annum compared to baseline (i.e. assuming every event is random and there is a basal rate of event with only minor/chance-induced findings). To place this in context, at minimum, for a city of 3 million people, this difference would equate to 750 more hospital admissions, occupying a minimum of 2000 hospital-bed days in the climatic season that provokes the most cardiovascular events. As always, there are caveats to such reports (with plans to update these same analyses with more contemporary studies). However, it was noticeable that the reported difference between peak and trough periods of AMI-related hospitalisation was consistently 2–8 events per 100,000 per annum. Moreover, the same general pattern was reported in relation to SCDs (Stewart 2016). Overall, therefore, in both relative and absolute terms, seasonally associated variations in cardiovascular events, including the major cardiac subtypes, has proven to be a major contributor to a consistent pattern of peak and trough health-care demands throughout the calendar year in many countries. As the major theme of this book, these observations can't be divorced from the rising problem of climate change; with arguments both for and against more 'volatility' in event rates and subsequent demands for health services because of warmer temperatures and more dynamic weather events.

## 5.6   Important Non-climatic Influences on Heart Health

One of the key problems in predicting what will happen next, especially in relation to cardio-vascular, and specifically cardiac, events in the setting of climate change, is the issue of confounding factors such as pollution levels and the interaction with common comorbid conditions such as infections respiratory illnesses. This obviously includes the recent emergence of COVID-19.

### 5.6.1   Air Pollution

As will be discussed in more Chap. 6, we cannot divorce climate change and its influence on the cardiovascular system from pollution—especially when local to global atmospheric conditions determine the location and levels of any freely transported (airborne) pollution we produce. As long recognised by the *American Heart Association* and other major cardiac societies (Brook et al. 2010; Brauer et al. 2021; Rajagopalan et al. 2020; Kaluzna-Oleksy et al. 2018), elevated atmospheric levels of a various pollutants, including particulate matter, nitrogen dioxide, ozone, elemental carbon, ammonia and soot, have been linked to increased rates of cardiovascular events. Studies have demonstrated that exposure to air pollutants can invoke a pro-inflammatory state (Viehmann et al. 2015; Hajat et al. 2015; Alexeeff et al. 2011; Bind et al. 2012; Fandino-Del-Rio et al. 2021), endothelial dysfunction (Brook et al. 2002; Wang et al. 2024; Li et al. 2023; Parsanathan and Palanichamy 2022; Munzel et al. 2018), a prothrombotic state (Jacobs et al. 2010; Becerra et al. 2016; Poursafa and Kelishadi 2010), and provoke elevated blood pressure/hypertension (Salerno et al. 2023; Wang et al. 2023b; Cao et al. 2023; Zhang et al. 2023; Ulusoy et al. 2023; Brook et al. 2023; Wen et al. 2023; Roswall et al. 2023; Tandon et al. 2023). In considering the mechanisms that compromise the cardiovascular system, it's important to not forget the direct respiratory 'insult' provoked by exposure to high levels of air pollutants increases the likelihood of developing a respiratory illness, particularly in vulnerable individuals (see below)—increasing the likelihood of respiratory-related pulmonary hypertension (Mocumbi et al. 2024). As will be further explored in Chap. 6 (discussing a broader range of pollutants that are likely to be 'cardiac toxic'), specific climatic conditions (including colder, wintry conditions) have been linked to an increase in cardiovascular events specifically associated with greater concentrations of air pollutants across a broad range of countries and climates (Stojic et al. 2016).

### 5.6.2   Infectious Disease

Prior to the emergence of the COVID-19 pandemic, with immunity levels and other factors influencing the pattern of cases (Takahashi et al. 2024; Xie et al. 2024; Pan et al. 2024), a large body of evidence showed a close association between influenza and other forms of infectious disease to an increased risk of cardiovascular morbidity and mortality in winter months (Stewart et al. 2017). As a seasonal condition, influenza and other respiratory conditions provoked by cold weather and our habit of huddling close together in warm spaces in winter (noting once again a biological to behavioural component!) undoubtedly play an important role in what has been reported pre-pandemic—seasonal ebbs and flows in cardiac/cardiovascular event rates. This association includes a potentially critical role for influenza in both the formation and progression of atherosclerosis, and provocation of an acute event as a result of a pro-inflammatory and prothrombotic state after infection (Auer et al. 2002a, b; Madjid et al. 2005; Peretz et al. 2019; Pleskov et al. 2003). However, as indicated, the COVID-19 epidemic has complicated the picture given the dynamics of COVID waves and variants has proven, thus far, to be largely independent from climatic conditions (Jalili et al. 2022). Moreover, as discussed in more detail in Chap. 7, the prospect of climate change (particularly in respect to warmer temperatures altering the patterns of infectious zones) complicates the picture further given the prospect of more cardiac provocations from infectious diseases on a global basis.

## 5.7    A Generic or Cardio-Specific Phenomenon? Lessons from the HUNT Study

Much of what is presented above is based on an original examination of the literature by my research group, with the results subsequently published in a high-level report in *Nature Reviews Cardiology* in 2017 (Stewart et al. 2017). This was followed-up with specific updates focussed on HF published in 2019 (Stewart et al. 2017, 2019) and, most recently, by a health services trial (to be described in more detail in Chap. 8) that sought to promote resilience in older people presenting to hospital with heart disease and multimorbidity (Stewart et al. 2023). ALL of these data (and indeed this book) are predicated on one fundamental argument—*that heart disease and most other major forms of CVD are particularly sensitive to climatic (and other environmental conditions)*. However, what if this is not the case? What if everything I've described (noting that much of the evidence has included all-cause deaths) is applicable to every major condition—thereby explaining the phenomenon of 'bed-block' at major hospitals during certain times of the year worldwide? That hypothesis, when presented, would state that heart health in relation to climatic influences is no different than any other common cause of morbidity and mortality (e.g. neoplasms and respiratory diseases).

To truly understand if this is the case and test the hypothesis (in the null form), you need a longitudinal, population-based study with minimal biases (in capturing and coding major events) that is truly 'agnostic' to the proposition. Fortunately, through the auspices of Dr Trine Moholdt, a senior cardiometabolic researcher at the University of Norwegian Science and Technology, we were able to test this hypothesis by examining data and outcomes derived

from the (world-renowned) Trøndelag Health (HUNT) Study (Vie et al. 2023; Breidablik et al. 2023; Asvold et al. 2023; Bhatta et al. 2018; Stenehjem et al. 2018; Jorgensen et al. 2017; Holmen et al. 2016; Rangul et al. 2012; Stensvold et al. 2011; Munkhaugen et al. 2009; Vatten et al. 2007; Droyvold et al. 2006; Romundstad et al. 2002). Context and validity are important points here. With a population of ~5.5 million people, Norway has a long tradition of undertaking insightful, longitudinal population cohort studies like the HUNT Study. This includes the Tromsø Study in Central/Northern Norway—something that prompted my visit and experience of an arctic storm/conditions described in Chap. 2 (Tiwari et al. 2017). From a climatic perspective, although the warm currents of the Gulf Stream act to moderate Norway's weather, given its northerly latitude (far beyond the more temperate latitudes where humans tend to live and congregate), the country and its relatively small population Norway still experiences extreme weather conditions [Central Norway's Köppen Climate Classification subtype is Continental Subarctic Climate (The Koppen Climatic Classification 2024)]. The coldest month of the year is January (with mean temperature of $-3\,°C$) and the warmest month is July (with a mean temperature of ~$13\,°C$) with a mean annual temperature of $4.8\,°C$ overall. Although Norway enjoys relatively clean air, the winter solstice and the darkest days of the year coincide with Christmas (something that became more significant as we examined health outcomes within the cohort).

The original wave of population screening (that formed the basis for HUNT1) was undertaken during the period of 1984–1986. Remarkably, 88% of eligible inhabitants within the region aged $\geq 20$ years in Nord-Trøndelag County were recruited and profiled at baseline and periodically thereafter. This included a total of 79,626 men and women who attended a clinical examination and filled out detailed questionnaires about their health and lifestyle. This profiling yielded data on each participant's socio-economic status, perceived levels of health and life satisfaction and lifestyle behaviours—all components of interest to the hypothesis being tested. Overall, the analysis we conducted included 40,353 women (50.1%) aged $50 \pm 8$ years and 39,273 men aged $48 \pm 18$ years. Two-thirds were married and just over half had <10 years of formal education. Many of those studied had high levels of antecedent cardiovascular risk and other forms of chronic disease. This included elevated levels of blood pressure, smoking and relatively high levels of sedentary behaviours and overweight status typical of the era in which these baseline measures were documented.

### 5.7.1 Causes of Death During Long-Term Follow of the HUNT Cohort

In 2024, it is estimated that heart diseases (including strokes and other cardiovascular conditions) cause 33% of global deaths, while cancers and respiratory disease cause 18 and 7% of deaths, respectively (Saloni-Dattani et al. 2024). Broadly consistent with this pattern, during 33.5 (IQR 17.1–34.4) years follow-up of the HUNT cohort, 19,879 (50.7%) men and 19,316 (49.3%) women died. This equated to an age adjusted rate of 5.3 and 4.6 deaths per 1000/ annum, respectively. The most common (documented) causes of death in men and women were cardiovascular related (8355 [43.6%] and 7969 deaths [43.0%], respectively) followed by cancer (5051 [26.4%] and 4150 deaths [22.4%]) in men and women, respectively) and then respiratory disease/illness (1599 [8.3%] and 1606 [8.7%] deaths in men and women, respectively). Thus, the historical hierarchy of contributing conditions to cause of death, both globally and in Norway even today (Norwegian Institute of Public Health 2024) was observed within this cohort—notwithstanding expected variances over time. Overall, these conditions accounted for 78% of all deaths in men and 74% of all deaths in women. Other causes of death included endocrine disorders/diabetes (1343 [3.4%] of all deaths in men and women combined), psychiatric disorders (1118 [2.9%])

and external factors including motor vehicle accidents and violence (951 [2.5%]). For, all analyses, the focus was on the 'big 3'—cardiovascular-related, cancer-related and respiratory-related deaths over time.

### 5.7.2 Climatic Influences on Deaths

As shown in Fig. 5.7, while there was a clear seasonal pattern/fluctuation in cardiovascular-related deaths (extending somewhat to respiratory-related deaths), it didn't extend to those caused by cancer/malignancies. Critically, the differential between cardiovascular- and respiratory-related deaths during the winter months (4446 [27.4%] and 1037 [32.4%] deaths) versus summer months (3832 [23.5%] and 661 [20.6%] deaths) explained 59% of the variance between observed 'peak' and 'trough' periods of the year. In absolute terms the difference was 1010 deaths. On adjusted basis, each winter there were 44 (95% CI 43–45) more deaths when compared to the equivalent 3 months of summer. As expected, the main contributor to the observed excess in deaths occurring in winter months was CVD (+21, 95% CI 20–22 deaths/ annum). The other main causes were respiratory disease (+13, 95% CI 13–14 deaths/annum) and a range of other miscellaneous conditions (+14, 95% CI 13–14 deaths/annum). Critical to the hypothesis being tested, over the 35-year period, involving thousands of deaths, the pattern of cancer-related deaths was deaths occurred at a far more stable (one might say random rate) relative to seasons and markedly different climatic conditions. Indeed, the absolute difference in cancer-related deaths occurring in winter versus summer months was only 10 deaths in total.

As will be discussed in Chap. 9, observed correlations between the bio-behavioural profiles of those participating in the HUNT cohort study (especially when framed by the model presented in Chap. 4) provides important clues as to how we might attenuate cardiovascular events being provoked by predictable to unpredictable climatic conditions, whilst acknowledging that factors such as short days/long nights

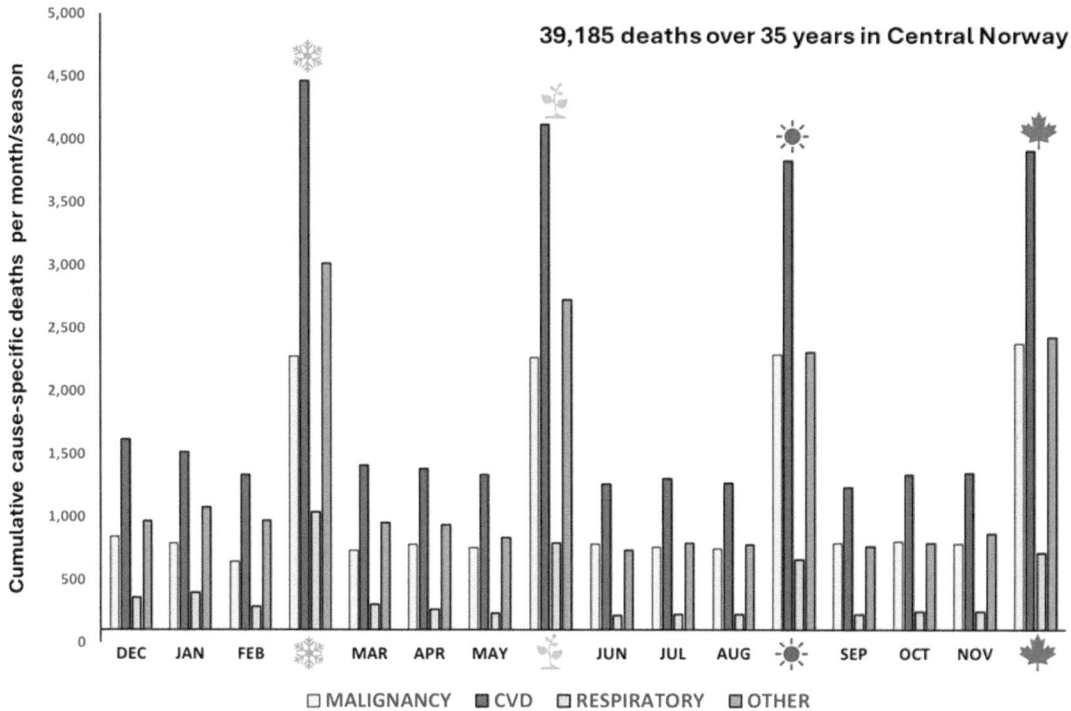

**Fig. 5.7** Hunting for climatically driven variations in cardiovascular and other conditions. This figure plots the cause of death by month and season, attributed to cancer, CVD, respiratory disease and other causes over a 35-year period within the HUNT cohort living in Central Norway—this figure is derived from original published data (Stewart et al. 2019)

and even provocative events such as Christmas (which proved to be the deadliest time of the year in Central Norway!) cannot be avoided. These are discussed further in Chap. 9, after considering additional aspects around climate change from several different perspectives in the next few chapters.

## 5.8   Fluctuating Fortunes When Dealing with Heart Disease

Overall, the evidence derived from a myriad of sources, including those where the same population, were exposed to the same climatic conditions, indicate a clear signal of 'non-random' cardiovascular events. Instead, both the timing of hospital admissions and deaths (many of which occur during the same cardiac crisis),

fluctuate with predictable (at least up until climate change) transitions in the weather and climate conditions, as well as less predictable climatic events. This doesn't appear to be the case for most people dying from cancer, although the rising importance of a cardio-oncology perspective may change the reporting and classification of related deaths. Thus, while clinicians are often concerned about 'who' is at highest risk, the argument for adding the 'when' and 'why' that person might suddenly deteriorate and require acute hospital treatment is a cogent one in terms of proffering 'climatic vulnerability' as the culprit. Before the 'how' one might go about preventing certain individuals from succumbing to such vulnerability is discussed, it is important to re-emphasise the complexity of interactions we have with the environment—especially in the context of climate change. In the next few chapters, the importance of pollution and a range of infectious disease in the context of climate change affecting vulnerable populations worldwide will be discussed.

# References

Alexeeff SE, Coull BA, Gryparis A, Suh H, Sparrow D, Vokonas PS, Schwartz J. Medium-term exposure to traffic-related air pollution and markers of inflammation and endothelial function. Environ Health Perspect. 2011;119:481–6.

Asvold BO, Langhammer A, Rehn TA, Kjelvik G, Grontvedt TV, Sorgjerd EP, Fenstad JS, Heggland J, Holmen O, Stuifbergen MC, Vikjord SAA, Brumpton BM, Skjellegrind HK, Thingstad P, Sund ER, Selbaek G, Mork PJ, Rangul V, Hveem K, Naess M, Krokstad S. Cohort profile update: the HUNT study, Norway. Int J Epidemiol. 2023;52:e80–91.

Auer J, Berent R, Weber T, Eber B. Influenza virus infection, infectious burden, and atherosclerosis. Stroke. 2002a;33:1454–5.

Auer J, Leitinger M, Berent R, Prammer W, Weber T, Lassnig E, Eber B. Influenza A and B IgG seropositivity and coronary atherosclerosis assessed by angiography. Heart Dis. 2002b;4:349–54.

Barnett AG, Dobson AJ, McElduff P, Salomaa V, Kuulasmaa K, et al. Cold periods and coronary events: an analysis of populations worldwide. J Epidemiol Community Health. 2005;59:551–7.

Bean WBMCA. Coronary occlusion, heart failure, and environmental temperatures. Am Heart J. 1938;16:701–13.

Becerra AZ, Georas S, Brenna JT, Hopke PK, Kane C, Chalupa D, Frampton MW, Block R, Rich DQ. Increases in ambient particulate matter air pollution, acute changes in platelet function, and effect modification by aspirin and omega-3 fatty acids: a panel study. J Toxicol Environ Health A. 2016;79:287–98.

Beck HE, Zimmermann NE, McVicar TR, Vergopolan N, Berg A, Wood EF. Present and future Koppen-Geiger climate classification maps at 1-km resolution. Sci Data. 2018;5:180214.

Bhatta L, Leivseth L, Mai XM, Chen Y, Henriksen AH, Langhammer A, Brumpton BM. Prevalence and trend of COPD from 1995–1997 to 2006–2008: THE HUNT study, Norway. Respir Med. 2018;138:50–6.

Bind MA, Baccarelli A, Zanobetti A, Tarantini L, Suh H, Vokonas P, Schwartz J. Air pollution and markers of coagulation, inflammation, and endothelial function: associations and epigene-environment interactions in an elderly cohort. Epidemiology. 2012;23:332–40.

Boulay F, Berthier F, Sisteron O, Gendreike Y, Gibelin P. Seasonal variation in chronic heart failure hospitalizations and mortality in France. Circulation. 1999;100:280–6.

Brauer M, Casadei B, Harrington RA, Kovacs R, et al. Taking a stand against air pollution—the impact on cardiovascular disease: a joint opinion from the world heart federation, American College of Cardiology, American Heart Association, and the European Society of Cardiology. Glob Heart. 2021;16:8.

Breidablik HJ, Hufthammer KO, Rangul V, Andersen JR, Meland E, Hetlevik O, Vie TL. Lower levels of physical activity volume are beneficial, and it's never too late to start: results from the HUNT study, Norway. Scand J Public Health. 2023;54:14034948231162728.

Breitner S, Wolf K, Peters A, Schneider A. Short-term effects of air temperature on cause-specific cardiovascular mortality in Bavaria, Germany. Heart. 2014;100:1272–80.

Brook RD, Brook JR, Urch B, Vincent R, Rajagopalan S, Silverman F. Inhalation of fine particulate air pollution and ozone causes acute arterial vasoconstriction in healthy adults. Circulation. 2002;105:1534–6.

Brook RD, Rajagopalan S, Pope CA, Brook JR, Bhatnagar A, Diez-Roux AV, Holguin F, et al. Particulate matter air pollution and cardiovascular disease: an update to the scientific statement from the American Heart Association. Circulation. 2010;121:2331–78.

Brook RD, Motairek I, Rajagopalan S, Al-Kindi S. Excess global blood pressure associated with fine particulate matter air pollution levels exceeding world health organization guidelines. J Am Heart Assoc. 2023;12:e029206.

Bundesen HNFIS. Low temperature, high barometer and sudden death. JAMA. 1926;87:1987.

Bunker A, Wildenhain J, Vandenbergh A, Henschke N, Rocklov J, Hajat S, Sauerborn R. Effects of air temperature on climate-sensitive mortality and morbidity outcomes in the elderly; a systematic review and meta-analysis of epidemiological evidence. EBioMedicine. 2016;6:258–68.

Cagle A, Hubbard R. Cold-related cardiac mortality in King County, Washington, USA 1980–2001. Ann Hum Biol. 2005;32:525–37.

Cao M, Zheng C, Zhou H, Wang X, Chen Z, Zhang L, Cao X, Tian Y, Han X, Liu H, Liu Y, Xue T, Wang Z, Guan T. Air pollution attenuated the benefits of physical activity on blood pressure: evidence from a nationwide cross-sectional study. Ecotoxicol Environ Saf. 2023;262:115345.

Capewell S, MacIntyre K, Stewart S, Chalmers JW, Boyd J, Finlayson A, Redpath A, Pell JP, McMurray JJ. Age, sex, and social trends in out-of-hospital cardiac deaths in Scotland 1986–1995: a retrospective cohort study. Lancet. 2001;358:1213–7.

Chau PH, Wong M, Woo J. Ischemic heart disease hospitalization among older people in a subtropical city—Hong Kong: does winter have a greater impact than summer? Int J Environ Res Public Health. 2014;11:3845–58.

Chybowska AD, Gadd DA, Cheng Y, Bernabeu E, Campbell A, Walker RM, McIntosh AM, Wrobel N, Murphy L, Welsh P, Sattar N, Price JF, McCartney

DL, Evans KL, Marioni RE. Epigenetic contributions to clinical risk prediction of cardiovascular disease. Circ Genom Precis Med. 2024;17:e004265.

Dimitriadou L, Nastos P, Eleftheratos K, Kapsomenakis J, Zerefos C. Mortality related to air temperature in European cities, based on threshold regression models. Int J Environ Res Public Health. 2022;19:1542.

Dong Q. Seasonal changes and seasonal regimen in hippocrates. J Camb Stud. 2007;6:128–43.

Droyvold WB, Nilsen TI, Kruger O, Holmen TL, Krokstad S, Midthjell K, Holmen J. Change in height, weight and body mass index: longitudinal data from the HUNT study in Norway. Int J Obes. 2006;30:935–9.

Eng H, Mercer JB. Seasonal variations in mortality caused by cardiovascular diseases in Norway and Ireland. J Cardiovasc Risk. 1998;5:89–95.

European Union. COPERNICUS Project. Europe's eyes on the world; 2024. https://defence-industry-space. ec.europa.eu/eu-space/copernicus-earth-observation_ en. Accessed May 2024.

Evans A, Tolonen H, Hense HW, Ferrario M, Sans S, et al. Trends in coronary risk factors in the WHO MONICA project. Int J Epidemiol. 2001;30(Suppl 1):S35–40.

Fandino-Del-Rio M, Kephart JL, Williams KN, Malpartida G, et al. Household air pollution and blood markers of inflammation: a cross-sectional analysis. Indoor Air. 2021;31:1509–21.

Frost L, Vukelic-Andersen L, Mortensen LS, Dethlefsen C. Seasonal variation in stroke and stroke-associated mortality in patients with a hospital diagnosis of nonvalvular atrial fibrillation or flutter. A population-based study in Denmark. Neuroepidemiology. 2006;26:220–5.

Gasparrini A, Guo Y, Hashizume M, Lavigne E, Zanobetti A, Schwartz J, Tobias A, Tong S, Rocklov J, Forsberg B, et al. Mortality risk attributable to high and low ambient temperature: a multicountry observational study. Lancet. 2015;386:369–75.

Gerber Y, Jacobsen SJ, Killian JM, Weston SA, Roger VL. Seasonality and daily weather conditions in relation to myocardial infarction and sudden cardiac death in Olmsted County, Minnesota, 1979 to 2002. J Am Coll Cardiol. 2006;48:287–92.

Ghosh AK, Venkatraman S, Nanna MG, Safford MM, Colantonio LD, Brown TM, Pinheiro LC, Peterson ED, Navar AM, Sterling MR, Soroka O, Nahid M, Banerjee S, Goyal P. Risk prediction for atherosclerotic cardiovascular disease with and without race stratification. JAMA Cardiol. 2024;9:55–62.

Grigorieva EA, Alexeev VA, Walsh JE. Universal thermal climate index in the Arctic in an era of climate change: Alaska and Chukotka as a case study. Int J Biometeorol. 2023;67:1703–21.

Hajat A, Allison M, Diez-Roux AV, Jenny NS, Jorgensen NW, Szpiro AA, Vedal S, Kaufman JD. Long-term exposure to air pollution and markers of inflammation, coagulation, and endothelial activation: a repeat-measures analysis in the multi-ethnic study of atherosclerosis (MESA). Epidemiology. 2015;26:310–20.

Han MH, Yi HJ, Ko Y, Kim YS, Lee YJ. Association between hemorrhagic stroke occurrence and meteorological factors and pollutants. BMC Neurol. 2016;16:59.

Holmen J, Holmen TL, Tverdal A, Holmen OL, Sund ER, Midthjell K. Blood pressure changes during 22-year of follow-up in large general population—the HUNT study Norway. BMC Cardiovasc Disord. 2016;16:94.

Hoogenboom WS, Pham A, Anand H, Fleysher R, Buczek A, Soby S, Mirhaji P, Yee J, Duong TQ. Clinical characteristics of the first and second COVID-19 waves in the Bronx, New York: a retrospective cohort study. Lancet Reg Health Am. 2021;3:100041.

Jabeen A, Afzal MS, Pathan SA. A review of the role of built environment and temperature in the development of childhood obesity. Cureus. 2023;15:e49657.

Jacobs L, Emmerechts J, Mathieu C, Hoylaerts MF, Fierens F, Hoet PH, Nemery B, Nawrot TS. Air pollution related prothrombotic changes in persons with diabetes. Environ Health Perspect. 2010;118:191–6.

Jakovljevic D, Salomaa V, Sivenius J, Tamminen M, Sarti C, Salmi K, Kaarsalo E, Narva V, Immonen-Raiha P, Torppa J, Tuomilehto J. Seasonal variation in the occurrence of stroke in a finnish adult population. The FINMONICA stroke register. Finnish monitoring trends and determinants in cardiovascular disease. Stroke. 1996;27:1774–9.

Jalili M, Sayehmiri K, Ansari N, Pourhossein B, Fazeli M, Jalilian AF. Association between influenza and COVID-19 viruses and the risk of atherosclerosis: meta-analysis study and systematic review. Adv Respir Med. 2022;90:338–48.

Jorgensen P, Langhammer A, Krokstad S, Forsmo S. Mortality in persons with undetected and diagnosed hypertension, type 2 diabetes, and hypothyroidism, compared with persons without corresponding disease: a prospective cohort study; the HUNT study Norway. BMC Fam Pract. 2017;18:98.

Kaluzna-Oleksy M, Aunan K, Rao-Skirbekk S, Kjellstrom T, Ezekowitz JA, Agewall S, Atar D. Impact of climate and air pollution on acute coronary syndromes: an update from the European society of cardiology congress 2017. Scand Cardiovasc J. 2018;52:1–3.

Keatinge WR, Donaldson GC, Bucher K, Jendritzky G, Cordioli E, Martinelli M, Katsouyanni K, Kunst AE, McDonald C, Nayha S, Vuori I, Eurowinter G. Winter mortality in relation to climate. Int J Circumpolar Health. 2000;59:154–9.

Kiu A, Horowitz JD, Stewart S. Seasonal variation in AF-related admissions to a coronary care unit in a "hot" climate: fact or fiction? J Cardiovasc Nurs. 2004;19:138–41.

Kottek M, Grieser J, Beck C, Rudolf B, Rubel F. World map of the Köppen-Geiger climate classification updated. Meteorol Z. 2006;15:259.

Kynast-Wolf G, Preuss M, Sie A, Kouyate B, Becher H. Seasonal patterns of cardiovascular disease mortality of adults in Burkina Faso, West Africa. Trop Med Int Health. 2010;15:1082–9.

Lan F, Pan J, Zhou Y, Huang X. Impact of the built environment on residents' health: evidence from the China labor dynamics survey in 2016. J Environ Public Health. 2023;2023:3414849.

Li Y, Du T, Lewin MR, Wang H, Ji X, Zhang Y, Xu T, Xu L, Wu JS. The seasonality of acute coronary syndrome and its relations with climatic parameters. Am J Emerg Med. 2011;29:768–74.

Li J, Liu F, Liang F, Yang Y, Lu X, Gu D. Air pollution exposure and vascular endothelial function: a systematic review and meta-analysis. Environ Sci Pollut Res Int. 2023;30:28525–49.

Li S, Jia X, Peng C, Zhu Y, Cao B. Effects of temperature cycles on human thermal comfort in built environment under summer conditions. Sci Total Environ. 2024;912:168756.

Lichtman JH, Leifheit-Limson EC, Jones SB, Wang Y, Goldstein LB. Average temperature, diurnal temperature variation, and stroke hospitalizations. J Stroke Cerebrovasc Dis. 2016;25:1489–94.

Lim CL. Fundamental concepts of human thermoregulation and adaptation to heat: a review in the context of global warming. Int J Environ Res Public Health. 2020;17:546.

MacIntyre K, Stewart S, Capewell S, Chalmers JW, Pell JP, Boyd J, Finlayson A, Redpath A, Gilmour H, McMurray JJ. Gender and survival: a population-based study of 201,114 men and women following a first acute myocardial infarction. J Am Coll Cardiol. 2001;38:729–35.

Madjid M, Awan I, Ali M, Frazier L, Casscells W. Influenza and atherosclerosis: vaccination for cardiovascular disease prevention. Expert Opin Biol Ther. 2005;5:91–6.

Martinez-Solanas E, Quijal-Zamorano M, Achebak H, Petrova D, Robine JM, Herrmann FR, Rodo X, Ballester J. Projections of temperature-attributable mortality in Europe: a time series analysis of 147 contiguous regions in 16 countries. Lancet Planet Health. 2021;5:e446–54.

Masselot P, Mistry M, Vanoli J, Schneider R, Iungman T, Garcia-Leon D, Ciscar JC, Feyen L, Orru H, Urban A, Breitner S, et al. Excess mortality attributed to heat and cold: a health impact assessment study in 854 cities in Europe. Lancet Planet Health. 2023;7:e271–81.

McDonagh TA, Metra M, Adamo M, Gardner RS, Baumbach A, Bohm M, Burri H, Butler J, Celutkiene J, Chioncel O, Cleland JGF, Coats AJS, Crespo-Leiro MG, Farmakis D, Gilard M, Heymans S, Hoes AW, Jaarsma T, et al. 2021 ESC guidelines for the diagnosis and treatment of acute and chronic heart failure. Eur Heart J. 2021;42:3599–726.

Mocumbi A, Humbert M, Saxena A, Jing ZC, Sliwa K, Thienemann F, Archer SL, Stewart S. Pulmonary hypertension. Nat Rev Dis Primers. 2024;10:1.

Munkhaugen J, Lydersen S, Wideroe TE, Hallan S. Prehypertension, obesity, and risk of kidney disease: 20-year follow-up of the HUNT I study in Norway. Am J Kidney Dis. 2009;54:638–46.

Munzel T, Gori T, Al-Kindi S, Deanfield J, Lelieveld J, Daiber A, Rajagopalan S. Effects of gaseous and solid constituents of air pollution on endothelial function. Eur Heart J. 2018;39:3543–50.

Murphy NF, Stewart S, MacIntyre K, Capewell S, McMurray JJ. Seasonal variation in morbidity and mortality related to atrial fibrillation. Int J Cardiol. 2004;97:283–8.

Nabaweesi R, Hanna M, Muthuka JK, Samuels AD, Brown V, Schwartz D, Ekadi G. The built environment as a social determinant of health. Prim Care. 2023;50:591–9.

Norwegian Institute of Public Health. Causes of death and life-expectancy; 2024. https://www.fhi.no/en/ch/cause-of-death-and-life-expectancy/. Accessed May 2024.

Pan Q, Chen X, Yu Y, Zang G, Tang Z. The outbreak of seasonal influenza after the COVID-19 pandemic in China: unraveling the "immunity debt." Infect Dis Now. 2024;54:104834.

Parsanathan R, Palanichamy R. Air pollution impairs endothelial function and blood pressure. Hypertens Res. 2022;45:380–1.

Patel J, Katapally TR, Khadilkar A, Bhawra J. The interplay between air pollution, built environment, and physical activity: perceptions of children and youth in rural and urban India. Health Place. 2024;85:103167.

Pennells L, Kaptoge S, Di Angelantonio E. Adapting cardiovascular risk prediction models to different populations: the need for recalibration. Eur Heart J. 2024;45:129–31.

Peretz A, Azrad M, Blum A. Influenza virus and atherosclerosis. QJM. 2019;112:749–55.

Pleskov VM, Bannikov AI, Gurevich VS, Pleskova IV. Influenza viruses and atherosclerosis: the role of atherosclerotic plaques in prolonging the persistent form of influenza infection. Vestn Ross Akad Med Nauk. 2003;31:10–3.

Poursafa P, Kelishadi R. Air pollution, platelet activation and atherosclerosis. Inflamm Allergy Drug Targets. 2010;9:387–92.

Qiu H. Is greater temperature change within a day associated with increased emergency hospital admissions for heart failure? Circ Heart Fail. 2013;6:930–5.

Rajagopalan S, Brauer M, Bhatnagar A, Bhatt DL, Brook JR, Huang W, Munzel T, Newby D, Siegel J, Brook RD, et al. Personal-level protective actions against particulate matter air pollution exposure: a scientific statement from the American Heart Association. Circulation. 2020;142:e411–31.

Rangul V, Bauman A, Holmen TL, Midthjell K. Is physical activity maintenance from adolescence to young adulthood associated with reduced CVD risk factors, improved mental health and satisfaction with life: the HUNT study, Norway. Int J Behav Nutr Phys Act. 2012;9:144.

Reavey M, Saner H, Paccaud F, Marques-Vidal P. Exploring the periodicity of cardiovascular events in Switzerland: variation in deaths and hospitalizations across seasons, day of the week and hour of the day. Int J Cardiol. 2013;168:2195–200.

Romundstad S, Holmen J, Hallan H, Kvenild K, Kruger O, Midthjell K. Microalbuminuria, cardiovascular disease and risk factors in a nondiabetic/nonhypertensive population. The Nord-Trondelag health study (HUNT, 1995–1997), Norway. J Intern Med. 2002;252:164–72.

Roswall N, Poulsen AH, Hvidtfeldt UA, Hendriksen PF, Boll K, Halkjaer J, Ketzel M, Brandt J, Frohn LM, Christensen JH, Im U, Sorensen M, Raaschou-Nielsen O. Exposure to ambient air pollution and lipid levels and blood pressure in an adult, Danish Cohort. Environ Res. 2023;220:115179.

Salerno P, Motairek I, Dallan LAP, Bourges-Sevenier B, Rajagopalan S, Al-Kindi SG. Excess systolic blood pressure associated with fine particulate matter air pollution above the WHO guidelines in Brazil. Arq Bras Cardiol. 2023;120:e20230347.

Saloni-Dattani FS, Hannah R, Max R. Causes of death 2024. https://ourworldindata.org/causes-of-death. Accessed May 2024.

San Roman Teran CM, Guijarro Merino R, Guil Garcia M, Villar Jimenez J, Martin Perez M, Gomez Huelgas R, Efficiency Group of the Internal Medicine S, Andalusian Society of Internal M. Analysis of 27,248 hospital discharges for heart failure: a study of an administrative database 1998–2002. Rev Clin Esp. 2008;208:281–7.

Shah BR, Austin PC, Ivers NM, Katz A, Singer A, Sirski M, Thiruchelvam D, Tu K. Risk prediction scores for type 2 diabetes microvascular and cardiovascular complications derived and validated with real-world data from 2 provinces: the DIabeteS COmplications (DISCO) risk scores. Can J Diabetes. 2024;48:188-194e5.

Spencer FA, Goldberg RJ, Becker RC, Gore JM. Seasonal distribution of acute myocardial infarction in the second national registry of myocardial infarction. J Am Coll Cardiol. 1998;31:1226–33.

Stenehjem JS, Hjerkind KV, Nilsen TIL. Adiposity, physical activity, and risk of hypertension: prospective data from the population-based HUNT study Norway. J Hum Hypertens. 2018;32:278–86.

Stensvold D, Nauman J, Nilsen TI, Wisloff U, Slordahl SA, Vatten L. Even low level of physical activity is associated with reduced mortality among people with metabolic syndrome, a population based study (the HUNT 2 study, Norway). BMC Med. 2011;9:109.

Stewart S. Seasonal variations in cardiovascular-related mortality but not hospitalization are modulated by temperature and not climate type: a systematic review and meta-analysis of 4.5 million events in 26 countries. Circulation. 2016;134:A16759–A16759.

Stewart S, McIntyre K, Capewell S, McMurray JJ. Heart failure in a cold climate. Seasonal variation in heart failure-related morbidity and mortality. J Am Coll Cardiol. 2002;39:760–6.

Stewart S, Keates AK, Redfern A, McMurray JJV. Seasonal variations in cardiovascular disease. Nat Rev Cardiol. 2017;14:654–64.

Stewart S, Moholdt TT, Burrell LM, Sliwa K, Mocumbi AO, McMurray JJ, Keates AK, Hawley JA. Winter peaks in heart failure: an inevitable or preventable consequence of seasonal vulnerability? Card Fail Rev. 2019;5:83–5.

Stewart S, Patel SK, Lancefield TF, Rodrigues TS, Doumtsis N, Jess A, Vaughan-Fowler ER, Chan YK, Ramchand J, Yates PA, Kwong JC, McDonald CF, Burrell LM. Vulnerability to environmental and climatic health provocations among women and men hospitalised with chronic heart disease: insights from the RESILIENCE TRIAL cohort. Eur J Cardiovasc Nurs. 2023;23:278–86.

Stojic SS, Stanisic N, Stojic A, Sostaric A. Single and combined effects of air pollutants on circulatory and respiratory system-related mortality in Belgrade, Serbia. J Toxicol Environ Health A. 2016;79:17–27.

Takahashi H, Nagamatsu H, Yamada Y, Toba N, Toyama-Kousaka M, Ota S, et al. Surveillance of seasonal influenza viruses during the COVID-19 pandemic in Tokyo, Japan, 2018–2023, a single-center study. Influenza Other Respir Viruses. 2024;18:e13248.

Tandon S, Grande AJ, Karamanos A, Cruickshank JK, Roever L, Mudway IS, Kelly FJ, Ayis S, Harding S. Association of ambient air pollution with blood pressure in adolescence: a systematic-review and meta-analysis. Curr Probl Cardiol. 2023;48:101460.

Temtem M, Mendonca MI, Gomes-Serrao M, Santos M, Sa D, Sousa F, Soares C, Rodrigues R, Henriques E, Freitas S, Borges S, Rodrigues M, Guerra G, et al. Predictive improvement of adding coronary calcium score and a genetic risk score to a traditional risk model for cardiovascular event prediction. Eur J Prev Cardiol. 2024;31:709–15.

Teng HCHHE. The relationship between sudden changes in weather and the occurrence of acute myocardial infarction. Am Heart J. 1955;49:9–20.

The Eurowinter Group. Cold exposure and winter mortality from ischaemic heart disease, cerebrovascular disease, respiratory disease, and all causes in warm and cold regions of Europe. Lancet. 1997;349:1341–6.

The Köppen Climate Classification. https://www.mindat.org/climate.php. Accessed May 2024.

Thorvaldsen P, Kuulasmaa K, Rajakangas AM, Rastenyte D, Sarti C, Wilhelmsen L. Stroke trends in the WHO MONICA project. Stroke. 1997;28:500–6.

Tiwari S, Lochen ML, Jacobsen BK, Hopstock LA, Nyrnes A, Njolstad I, Mathiesen EB, Arntzen KA, Ball J, Stewart S, Wilsgaard T, Schirmer H. Atrial fibrillation is associated with cognitive decline in stroke-free subjects: the Tromso Study. Eur J Neurol. 2017;24:1485–92.

Tunstall-Pedoe H, Kuulasmaa K, Mahonen M, Tolonen H, Ruokokoski E, Amouyel P. Contribution of trends in survival and coronary-event rates to changes in coronary heart disease mortality: 10-year results from 37 WHO MONICA project populations. Monitoring trends and determinants in cardiovascular disease. Lancet. 1999;353:1547–57.

Tunstall-Pedoe H, Vanuzzo D, Hobbs M, Mahonen M, Cepaitis Z, Kuulasmaa K, Keil U. Estimation of contribution of changes in coronary care to improving survival, event rates, and coronary heart disease mortality across the WHO MONICA project populations. Lancet. 2000;355:688–700.

Turner MM, Ghayoomi M, Duderstadt K, Brewer J, Kholodov A. Climate change and seismic resilience: key considerations for Alaska's infrastructure and built environment. PLoS ONE. 2023;18:e0292320.

Ulusoy S, Ozkan G, Varol G, Erdem Y, Derici U, Yilmaz R, et al. The effect of ambient air pollution on office, home, and 24-hour ambulatory blood pressure measurements. Am J Hypertens. 2023;36:431–8.

Upshur RE, Moineddin R, Crighton EJ, Mamdani M. Is there a clinically significant seasonal component to hospital admissions for atrial fibrillation? BMC Health Serv Res. 2004;4:5.

Urban A, Burkart K, Kysely J, Schuster C, Plavcova E, Hanzlikova H, Stepanek P, Lakes T. Spatial patterns of heat-related cardiovascular mortality in the Czech Republic. Int J Environ Res Public Health. 2016;13:323.

Vatten LJ, Trichopoulos D, Holmen J, Nilsen TI. Blood pressure and renal cancer risk: the HUNT Study in Norway. Br J Cancer. 2007;97:112–4.

Vie TL, Hufthammer KO, Rangul V, Andersen JR, Meland E, Breidablik HJ. Patterns of physical activity over 34 years in a large sample of adults: the HUNT study, Norway. Scand J Public Health. 2023;12:14034948231174948.

Viehmann A, Hertel S, Fuks K, Eisele L, Moebus S, Mohlenkamp S, Nonnemacher M, Jakobs H, Erbel R, Jockel KH, et al. Long-term residential exposure to urban air pollution, and repeated measures of systemic blood markers of inflammation and coagulation. Occup Environ Med. 2015;72:656–63.

Wang Z, Sun Z, Yu L, Wang Z, Li L, Lu X. Machine learning-based prediction of composite risk of cardiovascular events in patients with stable angina pectoris combined with coronary heart disease: development and validation of a clinical prediction model for Chinese patients. Front Pharmacol. 2023a;14:1334439.

Wang T, Han Y, Chen X, Chen W, Li H, Wang Y, Qiu X, Gong J, Li W, Zhu T. Particulate air pollution and blood pressure: signaling by the arachidonate metabolism. Hypertension. 2023b;80:2687–96.

Wang K, Lei L, Li G, Lan Y, Wang W, Zhu J, Liu Q, Ren L, Wu S. Association between ambient particulate air pollution and soluble biomarkers of endothelial function: a meta-analysis. Toxics. 2024;12:879.

Wen T, Liao D, Wellenius GA, Whitsel EA, Margolis HG, Tinker LF, Stewart JD, Kong L, Yanosky JD. Short-term air pollution levels and blood pressure in older women. Epidemiology. 2023;34:271–81.

Wolf HK, Tuomilehto J, Kuulasmaa K, Domarkiene S, Cepaitis Z, Molarius A, Sans S, Dobson A, Keil U, Rywik S. Blood pressure levels in the 41 populations of the WHO MONICA Project. J Hum Hypertens. 1997;11:733–42.

Xie Y, Choi T, Al-Aly Z. Long-term outcomes following hospital admission for COVID-19 versus seasonal influenza: a cohort study. Lancet Infect Dis. 2024;24:239–55.

Xu J, Jing Y, Xu X, Zhang X, Liu Y, He H, Chen F, Liu Y. Spatial scale analysis for the relationships between the built environment and cardiovascular disease based on multi-source data. Health Place. 2023;83:103048.

Zhang B, Li G, Ma Y, Pan X. Projection of temperature-related mortality due to cardiovascular disease in beijing under different climate change, population, and adaptation scenarios. Environ Res. 2018;162:152–9.

Zhang J, Zhang F, Xin C, Duan Z, Wei J, Zhang X, Han S, Niu Z. Associations of long-term exposure to air pollution, physical activity with blood pressure and prevalence of hypertension: the China health and retirement longitudinal study. Front Public Health. 2023;11:1137118.

## Abstract

While it might be tempting to simplify our interactions with weather and longer-term climatic conditions as a simple byproduct of varying atmospheric conditions, this would be a mistake. As will be outlined in this chapter, air pollution (a nasty consequence of the mainly human activities and technology driving climate change!) by itself, is likely cardio-toxic. Moreover, concentrations of outdoor air pollution and its impact on the cardiovascular system varies according to the prevailing climatic conditions, topography and human structures/activity. Other forms of pollution, including indoor air pollution, metal pollutants, microplastics and noise pollution are also important factors in eroding the capacity of an individual's cardiovascular (and broader cardiopulmonary) system to maintain homeostasis when confronted with provocative climatic conditions. It is for this reason that 'pollution' has been elevated to the status of being a "*non-traditional, major risk factor*" for cardiovascular disease. But how much does climate and climate change influence it's impact on our heart health? To answer this question, in this chapter the synergistic threat of pollution as both a consequence and cause of climate change will be explored.

## Keywords

Pollution · Air quality · Particulate matter · Ozone · Atmospheric conditions · Metal pollutants · Microplastics · Noise pollution · Topography · Wildfires/bushfires

## 6.1 Choking on Our Own Success

Having never been to Beijing in China, the seemingly new 'poster city' (along with other highly populous cities like Jakarta in Indonesia) for warning the world of the growing dangers posed by air pollution, I'm always struck by memories of my first-ever plane landing into the City of Angels (Los Angeles), located on the Pacific Coast in California, USA. After a long 15-h journey across the hauntingly beautiful and empty azure of the Pacific Ocean, it was memorable for two things. Firstly, I couldn't see the famous Hollywood sign! Secondly, after seeing a yellow, murky haze captured by the early morning sunlight becoming more distinct as we approached continental North America, we were suddenly thrust into what seemed like a blanket of gloom—Los Angeles' infamous layer of smog. According to the Köppen classification system (The Koppen Climatic Classification 2024), Los Angeles enjoys a Mediterranean

S. Stewart, *Heart Disease and Climate Change*, Sustainable Development Goals Series,
https://doi.org/10.1007/978-3-031-73106-8_6

climate of warm-to-hot dry summers and relatively cold-wet winters. As explained by Ed Avol (Professor of Preventive Medicine at University of Southern California), the approximate 4 million people living in Los Angeles (with 10 million more living in adjacent communities) are trapped by a problem of their own making and the geography/topography of the land on which they live. This is especially true under certain climatic conditions. As he has succinctly explained—"*Late summer is a challenging time for air quality, and it's likely to get worse with a warming climate. We have ideal conditions here in L.A. for ozone due to long, hot, sunny stagnant days. We tend to see these multi-day weather events, where smog builds up during the day and doesn't completely blow away overnight. Some pollution carries over into the next day, sloshing back and forth across the basin—inland by day and back toward the coast at night—so it cooks more and more and builds up over the course of a few days*" (Today UoC 2019). A critical component of the problem, apart from the pollution generated by so many people driving their cars and California's enormous industrial activity producing harmful emissions (by itself California is the fourth biggest economy in the world), therefore, is the trapping of air pollutants. These trapped pollutants, from prevailing winds, a typical temperature inversion that prevents rising air and the typical absence of rain to wash away pollutants (a problem in of itself!) are converted into a visible layer of ozone when reacting to sunlight—noting the role of ambient temperature, direction of sunlight and length of day(s) in co-contributing to the creation of this 'toxic cocktail'. Despite all the efforts to clean-up and reduce emissions, Los Angeles represents a classical geographic laboratory for understanding the synergistic harm (to humans) of adverse climatic conditions when they interact with harmful pollution levels.

As illustrated in Fig. 6.1, in a '*tale of two cities*', one only must compare the visual appearance of Los Angeles with New York located on the North-Eastern Atlantic seaboard of continental North America, to understand the true importance of considering the physical context in which air pollution occurs and is perceived to be a problem. With a much larger population of ~8.5 million people producing similar amounts of pollution, New York is far less renowned for its air pollution. However, air pollution levels routinely reach harmful levels there (https://waqi.info/;        https://www.aqi.in/au/real-time-most-polluted-city-ranking). New York's climate is characterised by a humid subtropical to continental climate that delivers both cold, wintry winters and hot, humid summers. In comparison

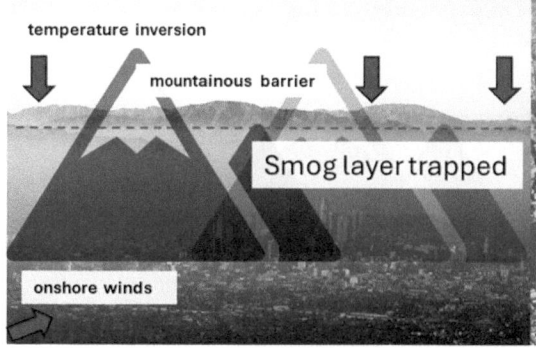

Los Angeles, South-Western Pacific Seaboard, USA (warm-hot dry summers/mild wet winters)

New York, North-Eastern Atlantic Seaboard, USA (warm humid summers/cold wet winters)

**Fig. 6.1** A tale of cities—Los Angeles and New York. Two US cities with very different topographies, that generate similar amounts of pollutants/per person (noting the famous freeways of California have long produced high-levels of harmful car emissions that are being phased out with the introduction of cleaner fuels, more efficient combustion engines and electric vehicles) but look very different

to Los Angeles, New York experiences far more rain, and along with its far more open and flat topography/geography lends itself to a far different dynamic in respect to the *visibility* and concentration of air pollution above the city. If this rather shallow comparison proves anything, it at least demonstrates the complexity of factors that shape—(a) the relative concentrations of harmful pollutants (in all their forms) we are routinely exposed to and, (b) our perceptions of these, given that we tend to ignore the things that we can't immediately see, touch or smell. The latter becomes obviously relevant as this chapter not only reviews the major forms of pollution and the mechanisms by which there are likely to adversely affect the heart and cardiovascular systems, but also the ones clinicians might not consider (i.e. from indoor air pollution to noise pollution) and the growing issue of 'microplastics' as a ubiquitous and growing form of pollution that remains hidden from sight in the environment and even our tissues.

## 6.2    Air Pollution

As convincingly argued, in a contemporary report by Ramos (2021), for an online European Society of Cardiology publication, we (clinicians, the public and health administrators alike) are seeming oblivious to the growing threat of air pollution (see text box below) to our cardiovascular, and specifically, heart health. Indeed, it is now considered to contribute to more than three million deaths per annum from a combination of ischaemic heart disease or stroke, and these figures will rise in the future (Hu et al. 2023). This requires a clinical to public health response (Brauer et al. 2021; Newby et al. 2015). Specifically, the Global Burden of Disease group estimated (from its 2019 Study) that air pollution contributes to 6.7 million global deaths (fourth leading cause) AND that consistent with the 3 million deaths figure, 45% of these are due to CVD, compared to a remarkably low 8% due to respiratory disease (Abbasi-Kangevari et al. 2023; Diabetes and Air 2019; Hu et al. 2019; Pan et al. 2023; Zhao et al.

2022). Whilst undoubtedly contributing to their development, this global contribution (of air pollution) now dwarfs that of other traditional risk factors such as diabetes, smoking and obesity.

As defined by the WHO—"*Air pollution is contamination of the indoor or outdoor environment by any chemical, physical or biological agent that modifies the natural characteristics of the atmosphere*" (https://www.who.int/health-topics/air-pollution#tab=tab_1). Of particular importance to climatic conditions and climate change, this contamination includes inhalable particulate matter (PM), carbon monoxide (CO), ozone ($O_3$), nitrogen dioxide ($NO_2$) and sulphur dioxide ($SO_2$). Particulate Matter (PM) varies in composition and size and is usually classified into 3 size groups: coarse particles ($PM_{10}$, diameter <10 and ≥2.5 μm), fine particles ($PM_{2.5}$, diameter <2.5 μm) and ultrafine particles (<0.1 μm) (Ramos 2021). The most concerning of these $PM_{2.5}$ (see below), originates from both combustible and non-combustible sources. This includes industrial emissions, brake and tire wear, resuspended soil and dust, bush/wildfires, and the burning of agricultural products, biomass and the fossil-fuel coal (with many coal-fired power plants still operating worldwide). As can be appreciated by the contrast pictures in Fig. 6.1, and some of the research reports detailed below, the concentration of harmful PM from a geographical perspective (i.e. localised air pollution) is very much dependent on prevailing winds, precipitation, and all-important temperature inversions. So, the climate plays an important modulating role in the levels of air pollution we are exposed to.

### 6.2.1    Outdoor Air Pollution

It is important to be cognisant, having described the predominantly heat-driven problems for Los Angeles (with future implications as temperatures rise), that the convergence of harmful climatic conditions and high pollution levels occurs at both end of the temperature spectrum—once again noting this depends on the specific conditions they generate in a location. As such, increased rates of cardiovascular events and other harmful effects associated with greater concentrations of air pollutants have been reported in a range of countries and climates during colder days/times of the year (Zuniga et al. 2016; Pintaric et al. 2016; Stojic et al. 2016; Su et al. 2016; Chang et al. 2015; Vinnikov et al. 2020; Wine et al. 2022; Yilmaz et al. 2021; Zhang et al. 2023). Concurrently, similar reports have noted peaks in cardiovascular events linked to warm/hot climatic conditions and high-levels of air pollution in a diversity (from temperate to Mediterranean) of climates (Tsangari et al. 2016; Samoli et al. 2016).

In a 2-year (2013–2014) study of 147,624 hospital admissions for stroke in Beijing, China (as noted a city now renowned for its high-levels of air pollution and radical efforts to combat it), Huang and colleagues reported significant associations between increasing levels of fine particulate matter (ranging in size from ≤2.5 to ≤10 μm) and elevated risk of admission for either ischaemic or (to a lesser extent) haemorrhagic stroke (Huang et al. 2016). Critically, in the context of climate change, these observed associations were modulated by ambient temperatures, with an enhanced risk of admission for both forms of stroke on warmer days (>13.5 °C) (Huang et al. 2016). Equivalent studies conducted in multiple cities across China as part of the *Pearl River Delta Study* (Lin et al. 2016) and a large study of 445,860 adults living in 100 U.S. metropolitan areas followed from 1982 to 2004, generated similar findings (Thurston et al. 2016). The specific role of increasing levels of sulphur dioxide in provoking non-ST and ST-elevated AMI in Turkey (Sen et al. 2016) and nitrogen dioxide provoking arrhythmic events across nine different regions of the Spain (2005–2010) have also been reported (Santurtun et al. 2017).

As emphasised in a recent "State-of-the-Art" review of pollution and cardiovascular disease by Rajagopalan and colleagues—"*Fine particulate matter (PM_{2.5}) air pollution is the most important environmental risk factor contributing to global cardiovascular (CV) mortality and disability. Short-term elevations in PM_{2.5}*

*increase the relative risk of acute CV events by 1–3% within a few days. Longer-term exposures over several years increase this risk by a larger magnitude (∼10%), which is partially attributable to the development of cardiometabolic conditions (e.g., hypertension and diabetes mellitus)*" (Rajagopalan et al. 2018). Such conclusions are based on large observational studies including a recent study of the impact of short-term increases in $PM_{2.5}$ and ozone levels on the 61 people who form the US Medicare population (Di et al. 2017). This study revealed that such increases were associated with a 1.011-fold (95% CI 1.009–1.012) and 1.005-fold (95% CI 1.004–1.006) rise in daily mortality rates (Di et al. 2017). From a longer-term exposure to air pollution perspective, exposure to an annual average $PM_{2.5}$ levels of 6–15.6 μg/m$^3$) within the same cohort was associated with a 7.3% (95% CI 7.1–7.5%) increase in all-cause mortality, while exposure to 10 parts of ozone per billion (annual exposure) was associated with a 1.1-fold (95% CI 1.0–1.2%) increase (Di et al. 2017).

Of course, these findings could reflect a 'generic' unhealthy response to air pollution exposure. As reported in the same 'State-of-the-Art' review of the relevant literature (Rajagopalan et al. 2018), a systematic review and meta-analysis of the 34 studies revealed that short-term exposure to $PM_{2.5}$ "*increased the relative risk for acute myocardial infarction (MI) by 2.5% per 10 mg/m$^3$ (relative risk: 1.025; 95% CI 1.015–1.036)*" (Mustafic et al. 2012). As commented upon (Rajagopalan et al. 2018), while these might seem modest, their effect is felt by millions of people worldwide and is a major contributor to the growing burden outlined above (Hu et al. 2023). Some of the postulated mechanisms by which air pollution are described below, but in framing these and the studies covered above, it is important to factor-in the repeated and, for some, near-continuous exposure to air pollution that many people experience—thereby proving a cogent mechanism for longer-term, adverse effects on the heart and wider cardiovascular system; particularly in the formation of atherosclerosis (Nawrot et al. 2011; Woo et al. 2021; Stachyra et al. 2017; Bai and Sun 2016; Brook and

Rajagopalan 2010). Consistent with this mechanistic association a meta-analysis of 59 studies, revealed that within the Chinese population, each $10 \text{ mg/m}^3$ increment in $PM_{2.5}$ was associated with an absolute 0.63% increase in cardiovascular mortality (Lu et al. 2015). Further afield, a system review and meta-analysis of 94 studies across 28 different countries and conducted up to 2014, revealed that a $10 \text{ mg/m}^3$ increase in $PM_{2.5}$ and $PM_{10}$ concentration was associated with a 1% increase in the risk (relative) for being hospitalised for and/or dying from a stroke (Shah et al. 2015). An updated systematic review and meta-analysis (Toubasi and Al-Sayegh 2023) of studies reporting on the association of ischaemic stroke and air pollution is worthy of particular mention here. This analysis comprised just over 18 million cases of ischaemic stroke derived from 110 observational studies (noting that 59% of studies were conducted in Asia, while 25% and 17% were from Europe and the Americas, respectively). With specific reference to an upcoming chapter on low-income countries—none were conducted in Africa. As specifically reported the incidence of stroke incidence was significantly associated with an increase in a range of PMs including increased concentrations of—"$NO_2$ $(RR=1.28;$ $95\%$ $CI$ $1.21–1.36)$, $O_3$ $(RR=1.05;$ $95\%$ $CI$ $1.03–1.07)$, $CO$ $(RR=1.26;$ $95\%$ $CI$ $1.21–1.32)$, $SO_2$ $(RR=1.15;$ $95\%$ $CI$ $1.11–1.19)$, $PM_1$ $(RR=1.09;$ $95\%$ $CI$ $1.06–1.12)$, $PM_{2.5}$ $(RR=1.15;$ $95\%$ $CI$ $1.13–1.17)$ and $PM_{10}$ $(RR=1.14;$ $95\%$ $CI$ $1.12–1.16)$". Furthermore, increased concentrations of—"$NO_2$ $(RR=1.33;$ $95\%$ $CI$ $1.07–1.65)$, $SO_2$ $(RR=1.60;$ $95\%$ $CI$ $1.05–2.44)$, $PM_{2.5}$ $(RR=1.09;$ $95\%$ $CI$ $1.04–1.15)$ and $PM_{10}$ $(RR=1.02;$ $95\%$ $CI$ $1.00–1.04)$" were found to be associated with an increased risk of dying from stroke (Toubasi and Al-Sayegh 2023).

Beyond considering the broad risk of associations outlined above, who is most at risk of being affected? It appears that living close to a roadway (Kulick et al. 2018) and/or living in poverty (Wing et al. 2017) appear to be positively associated with ischaemic stroke and stroke severity. Furthermore, an analysis of the ESCAPE cohort revealed potentially strong associations between $PM_{2.5}$ exposure levels and stroke among those aged 60 years or older (1.40-fold, 95% CI 1.05–1.87 more) those who didn't smoke (1.74-fold, 95% CI 1.06–2.88 more) and those with $PM_{2.5}$ exposure <25 mg/$m^3$ (1.33-fold, 95% CI 1.01–1.77) per 5 mg/$m^3$ increase in $PM_{2.5}$ (Cesaroni et al. 2014). There is little doubt that air pollution (in all its forms) is detrimental to the health of anyone with living a sub-type of cardiovascular disease. It's effects on eroding physiological resilience should be considered in context of the vulnerable phenotypes described in Chap. 4 and more broadly this book. But what are the specific effects/mechanisms, that drive increased morbidity and mortality—especially from cardiovascular to cardiac perspective?

It is, once again, beyond the specific expertise of this author, to describe the pathophysiological effects of air pollution, but several reviews are insightful in this regard. As noted in high-level review in *Nature Reviews Cardiology*—"*The health effects of air pollution might depend on chronic exposure, pre-existing medical conditions and sources or composition of the pollutants*" (Al-Kindi et al. 2020). In that review they nominate a number of pathological pathways/mechanisms that undoubtedly contribute to observed associations, but remain difficult to untangle given the nature and duration of the exposure (to air pollution), the role of the climate in modulating such exposure and the complicated research methods needed to reveal specific pathological pathways. Nevertheless, in 3 layers of cascading "cause and effect" the following appear to be important (Al-Kindi et al. 2020) [noting increasing interest in the role of epigenetic responses to air pollution (Micheu et al. 2020)]—see Table 6.1.

As depicted in Fig. 6.2, it comes as no surprise that the most immediate health impact of acute exposure to air pollution are respiratory-related (left-side), with those with pre-existing conditions most at risk. However, as depicted on the right-side of the figure, there are serious systemic effects that (as described below) compromise the cardiovascular system (Air ALASot 2024).

**Table 6.1**  Key stages of air pollution exposure leading to illness

| Pollutant inhalation → insult/exposure → | Physiologic response → | Pathophysiological mechanisms → illness |
|---|---|---|
| • Inflammation<br>• Oxidative stress<br>• Ion channel activation | • Biological intermediaries<br>• Autonomic imbalance<br>• HPA axis activation | • Epigenomic changes<br>• Tissue inflammation<br>• Thrombus formation<br>• Atherosclerotic plaque instability<br>• Endothelial dysfunction/damage<br>• Vasomotor/vascular dysfunction/damage |

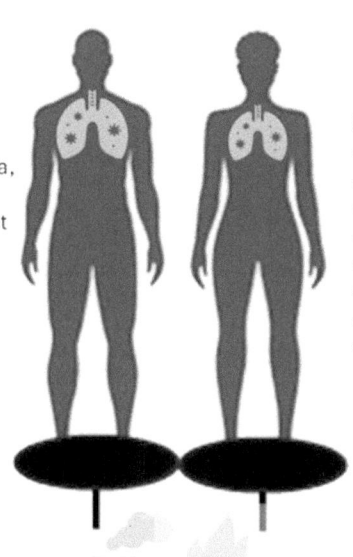

**Respiratory**
- Breathing difficulties – dyspnoea, wheezing, stridor & coughing
- Upper & Lower Respiratory Tract Infections
- Uncontrolled Asthma
- COPD exacerbation
- Pulmonary Hypertension
- Lung Cancer

**Systemic**
- Cor Pulmonale/Right Heart Failure
- Acute Myocardial Infarction
- Cerebrovascular Event/Stroke
- Immune System Compromise
- Metabolic Disorders
- Uncontrolled Asthma
- Cognitive Impairment
- Compromised Pregnancy – Pre-term Births & Low-Birth Weights

↑↑↑ hazardous air quality – acute-on-chronic exposures/exacerbations to cardiopulmonary health across the life-span

**Fig. 6.2**  Adverse health impact of air pollution. This figure summarises the pathological impact of/pathways triggered by inhaling air pollution—from a respiratory to systemic perspective

## 6.2.2  Indoor Air Pollution

In framing this section of the book, it is pertinent to make the general observation that there is a persistent (and troubling) focus on recognising and solving the world's problems through the 'lens' of wealth, and this most probably applies to considering the other half of air pollution—indoor air pollution (and specifically household air pollution). While rich and poor alike can be routinely exposed to outdoor air pollution (if they wish to go anywhere in a typically populous, urban environment), it is typically poorer people who must make do with cheaper and less environmentally friendly fuels to heat/cool their home and cook food (Mocumbi et al. 2019). Indeed, as will be explored more thoroughly in Chap. 9, a long list socio-economic indicators, such as low access to clean fuels, high unemployment, those living in extreme poverty and, a high Gini inequality coefficient correlate with greater exposure to higher levels of indoor air pollution (Mocumbi et al. 2019; Sliwa et al. 2016) and, consequently higher levels of often insidious, but deadly pulmonary hypertension (Mocumbi et al. 2019, 2024, 2015; Namuyonga and Mocumbi 2021; Stewart et al. 2019).

Although, outdoor air pollution is more directly linked to the climate, there is no doubt more extreme weather/climatic conditions will potentially force more people inside—and therefore expose them to greater levels of indoor air pollution. As with outdoor pollution, exposure to household air pollution is associated with poor cardiovascular health and outcomes (Li et al. 2024, 2022; Lu et al. 2021). In a derivative of the 2019 Global Burden of Disease Study, Lu and colleagues examined spatiotemporal trends in the burden of stroke linked to household air pollution from solid fuels across 204 countries (Lu et al. 2021). In the year 2019, they estimated that stroke attributable to household air pollution contributed to 14.7 million DALYs and 600,000 million deaths from stroke worldwide. As with climate associations, despite a declining rate of associated events, the role and trends of pollution contributing to ischaemic versus intracerebral stroke events differed. Moreover, some countries such as Zimbabwe and Philippines showed discouraging upward trends. Overall, household air pollution remains a problem in older people, low-income countries and in provoking intracerebral strokes (Lu et al. 2021).

Figure 6.3 presents a more cardiovascular-specific explanation of why air pollution provokes such a poor cardiovascular response (relative to respiratory conditions) during short- and long-term exposures—although many questions remain in relation to the duration of exposure needed to cause clinically significant harm (Ji 2022). Nevertheless a number of mechanisms, largely focussed on the development of atherosclerosis have been proposed. This includes modulating an individual's blood pressure levels (probably via endothelial dysfunction), provoking myocardial ischaemia, and triggering an acute coronary syndrome or

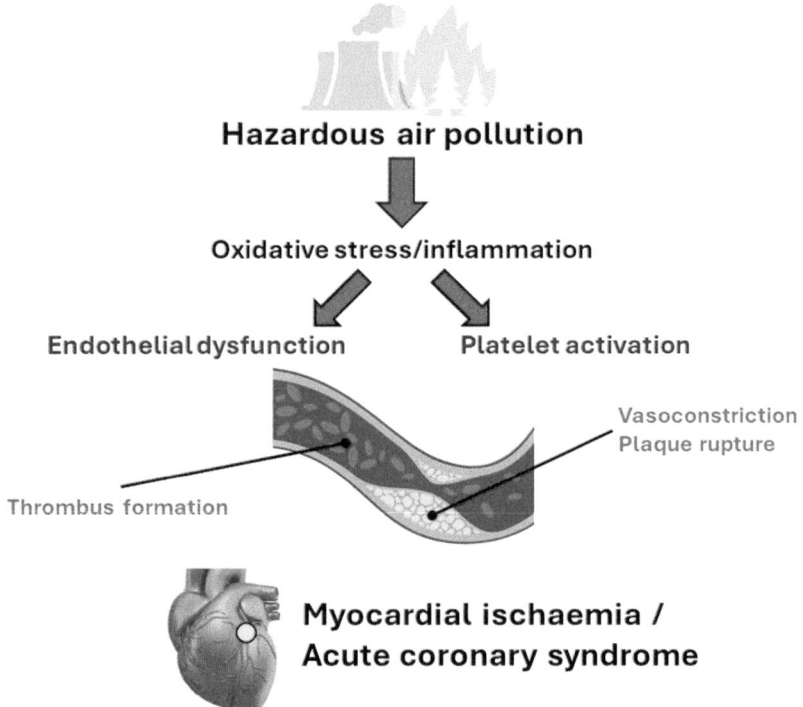

**Fig. 6.3**  From air pollution to myocardial ischaemia. This figure summarises the cardio-specific impact of air pollution in triggering myocardial ischaemia (Cosselman et al. 2015)

myocardial infarction. As the authors of the 2015 European Society of Cardiology's expert position paper on air pollution and cardiovascular disease, noted—"*Exposure to PM2.5 potentiates plaque burden and vascular dysfunction in models of atherosclerosis, and is also associated with plaque vulnerability, alterations in vasomotor tone, increased reactive oxygen species and pro-inflammatory mediators*" (Ji 2022).

## 6.3	Modulating Impact of Atmospheric Conditions on Air Pollution

A recent study of cold air pooling in the semi-arid, urban Lanzhou Basin of China, with many similarities to the problems faced by those living in the Los Angeles basis, revealed important interactions between this atmospheric condition and fine particulate matter (PM<2.5 μm) in that location. Specifically, Zhang and colleagues reported that during cold air pooling (in winter) a phenomenon of—"*aerosols accumulating in the lower basin heated the atmosphere during the daytime and facilitated boundary layer development* via *the "stove effect" (absorption aerosol heats lower atmosphere to promote boundary layer development)*" was observed (Zhang et al. 2024). Notably, they couldn't find evidence of a "*significant dome effect*" (whereby absorbed aerosol heats the upper boundary layer to suppress boundary layer development) that has been document in other locations (Zhang et al. 2024).

It is important to note that teasing out the complex relationship between atmospheric conditions and air pollution remains problematic—let alone its specific impact on the heart health of those breathing in the air under certain circumstances. Prompted by the notion that very few technologies (including those applied in environmentally 'cleaner cars') are tested in temperatures below 7 °C, Wine and colleagues examined the relevant literature (Wine et al. 2022). As they report, most papers studied—"*removed the possible direct effect of temperature on pollution and health by adjusting for*

*temperature*", with only 8 prospectively exploring the potential modifying effect of temperatures (on pollution then health). Nevertheless, 5 studies identified how "*extreme cold and warm temperatures aggravated mortality/morbidity associated with ozone, particles, and carbon-monoxide*" compared to the remainder (3 studies) that found no such associations (Wine et al. 2022). It makes physiological sense that the aggravating/harmful effects of air pollution is largely mediated by their entry into/irritation of lung tissues—leading to acute respiratory distress and chronic inflammation and/or aggravation of pre-existing lung disease (Nakao et al. 2016). It's worth clarifying at this point that deadly 'asthma storms' are caused by high pollen levels captured with a dynamic weather system (causing the pollen to be "*bathed in*" and absorb water) rather than air pollution However, such events are still more likely to occur in setting of climate change (Beggs 2024; Price et al. 2023; Thien 2018).

Beyond the seasonality seen in most respiratory infections, we are now confronted with a new viral infection (COVID-19) that will undoubtedly complicate the interactions between the climate, pollution, and cardiovascular disease. As has been well-established now, COVID-19 and its common variants can (Ogah et al. 2021) and does (Fischer et al. 2024; Kemerley et al. 2024; Karki et al. 2024; Yeow et al. 2024; Espiritu et al. 2024) adversely affect both people who are healthy as well as those living with cardiovascular diseases. Long-COVID is condition that is likely to contribute to the type of 'climatic vulnerability' described in previous chapters. However, how many people are affected remains uncertain (Szanyi et al. 2024). For example, a recent report from a patient cohort in Scotland suggested that 1.7% had long-COVID (Jeffrey et al. 2024). This compared to another report suggesting that almost one in five people in the US who had been infected with COVID-19 had symptoms indicative of this condition (with those not vaccinated most affected) (Nguyen et al. 2024). Nevertheless, an intriguing study has sought to investigate the confluence of COVID-19 and air pollution is worth mentioning. Conducted in Tehran, Iran of the spring and summer of 2020. As reported by Khorsandi et al. (2021)—"*short-term exposure to ambient $PM_{2.5}$, $PM_{10}$, $O_3$, and elevated temperatures was associated with higher rates of COVID-19-related hospital admissions/mortality throughout the summer*". While further studies are needed, these findings do point to a confluence of air pollution, atmospheric conditions and susceptibility to conditions that harm the cardiovascular system that will likely shape the pattern and impact of future endemic to pandemic diseases.

## 6.4 Metal Pollutants

Heavy metals, naturally occurring elements that are 5 times denser than water, are of increasing health concern because of their multiple uses (and therefore release into the environment) via multiple pathways of human activity. This includes a myriad of industrial, domestic, agricultural, medical and technological applications. As with air pollution, their potential toxicity depends on several factors including the dose and exposure levels and the health status/age of those affected (Wu et al. 2016). Unlike air pollution, there are multiple routes of exposure. From a cardiovascular perspective, metal pollution (in the form of more common heavy metals such as lead and cadmium—see below) have been implicated in the development and progression of cardiovascular disease (CVD) (Sliwa et al. 2024; Bandara and Weller 2017; Cosselman et al. 2015).

### 6.4.1 Lead

The main pathways of environmental exposure to lead is largely via air pollution, soil pollution and the contamination of drinking water. Studies have demonstrated that lead exposure promotes oxidative stress, endothelial dysfunction, proliferation of vascular smooth muscle cells and fibroblasts (Navas-Acien et al. 2007). Via the downregulation of nitric oxide, stimulation of the renin–angiotensin–aldosterone system and increased sympathetic nervous activity, lead toxicity has also been linked to elevated blood pressure/hypertension (Vaziri 2008). In respect to the latter, a meta-analysis of studies comprising more than 50,000 individuals, suggested that systolic blood pressure levels rose 0.8–1.2 mmHg for each twofold increase in the serum lead level (Navas-Acien et al. 2007). Consistent with the role of epigenetics in perpetuating the initial 'insult' of pollution, lead exposure has been shown to influence epigenetic modifications via DNA hypomethylation/expression of microRNAs linked to oxidative stress and inflammation (Cosselman et al. 2015, 2020).

### 6.4.2 Cadmium

During the twentieth century cadmium use (and therefore release into the environment) has increased exponentially. This includes nickel–cadmium batteries, plastic stabilisers, metal coating, industrial releases and fuel combustion. Human cadmium exposure can occur in multiple ways due to the use of phosphate fertilisers (that contaminate soil and then plants), with tobacco smoking and dietary ingestion of leafy and root vegetables, grains, as well as cocoa (Cosselman et al. 2015, 2020). Other dietary sources for cadmium, along with a range of other heavy metals (e.g. mercury, lead, chromium and nickel) includes the consumption of contaminated seafood and meat (Bosch et al. 2016). As a long-term 'legacy' effect, although less than 5% of ingested cadmium is absorbed, whatever quantity is absorbed into the human body has a biological half-life of up to 30 years (Cosselman et al. 2015, 2020).

As with lead and other heavy metals, observational studies have shown an association between elevated serum and urine levels of cadmium levels an increased incidence and then mortality from cardiovascular causes—including coronary artery disease, stroke and heart failure (Cosselman et al. 2015, 2020). While the exact mechanism is not completely elucidated, it is thought that cadmium damages vascular tissue, causes endothelial dysfunction and promotes atherosclerosis via oxidative stress mechanisms (Peters et al. 2010; Messner et al. 2009).

## 6.5   Microplastics

While toxic metals have long been a product of human activities (essentially since the start of the Industrial Revolution), plastics are relatively new phenomena and threat to our heart health. While most people would have trouble marrying the images of rivers and harbours chocked with plastic bottles, and marine-life snagged in plastic bags [it estimated that 11 million metric tonnes of *plastic* enters the world's oceans per annum (Trust 2022)] being a threat to our hearts, it's micro- and nanoplastics that are the problem in this regard (Persiani et al. 2023). As explained by Persiani and colleagues, the small plastic pollutants that enter the ecosystem sized less than 5 mm in diameter are called microplastics. If, as intended, they are produced via an industrial process they are referred to as 'primary' microplastics (Browne et al. 2011), while those resulting from the degradation of large pieces of plastic (typically by thermo-oxidation or mechanical degradation) are called 'secondary' microplastics (Ghanadi and Padhye 2024; Sababadichetty et al. 2024). More concerningly, from a pathological perspective, smaller particles of plastic less than 1 $\mu$m (nanoplastics) are becoming ubiquitous and are readily digested in contaminated food and water. Disturbingly, these nanoplastics can act as vectors for toxic heavy metals (see above) and even pathogens (Alimba and Faggio 2019; Catarci Carteny et al. 2023; Chen et al. 2024; Attaelmanan et al. 2023; Fu et al. 2021).

How does this new form of pollutant influence our heart health? A very recent study (March 2024) published in the *New England Journal of Medicine*, provides a clear warning why the cardiac community should be concerned. Specifically, Marfella and colleagues completed a prospective, multicentre study 304 patients undergoing a carotid endarterectomy for asymptomatic disease. They then followed

up 257 of them for $33.7 \pm 6.9$ months. Overall, clinically meaningful levels of the plastic polyethylene were detected in the carotid artery plaque of 150 patients (58.4%). A further 31 patients (12.1%) had measurable quantities of polyvinyl chloride detected. Consequently, those with microplastics detected in their excised plaque had a 4.5-fold (95% CI, 2.00–10.3) increased risk of experiencing the composite endpoint of myocardial infarction, stroke, or death from any cause compared to those with no microplastics detected (Marfella et al. 2024). Based on the review published by Persiani and colleagues and the recent discovery of microplastics in atherosclerotic plaques (Browne et al. 2011), Fig. 6.4 summarises the pathways microplastics are likely to drive more cardiac events in the future.

## 6.6   Considering an Interconnected Biosphere

As eloquently argued by Briffa and colleagues, we cannot divorce the interaction of our 'biosphere' from the other elements/domains that influence how and where we live—see Fig. 6.5 (Briffa and Blundell 2020). Once released, nearly all pollutants (including heavy metal pollutants) infiltrate all parts of our planet with soil pollution, water pollution and air pollution all areas of concern given they can exist in a diversity of states. This includes in solution/suspension of surface waters, liquid droplets and sedimentary, particulate matter that can, when disturbed/redistributed by climatic and environmental disturbances (the very definition of climate change!) enter the food chain or directly be ingested/absorbed by an individual (Briffa and Blundell 2020).

## 6.7   Noise Pollution

Although not directly related to climatic conditions, it is hard to ignore the influence of noise pollution in leaving at risk individuals vulnerable to other provocations to their

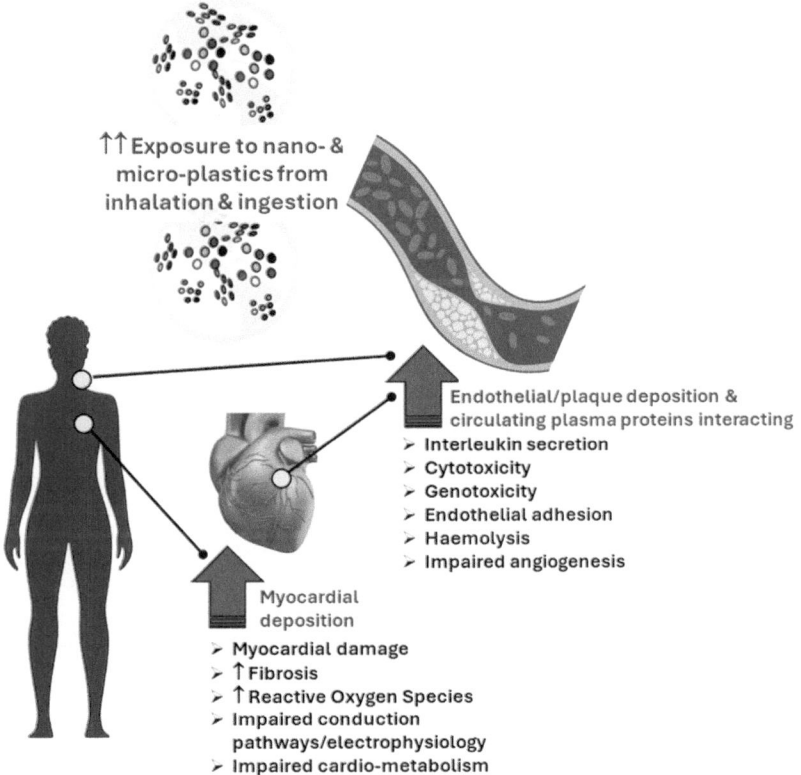

**Fig. 6.4** Cardiovascular injury caused by microplastics. This figure summarises the pathological impact of ingesting microplastics on the cardiovascular system and heart (Persiani et al. 2023)

cardiovascular health—including challenging climatic conditions. Although this remains an active areas of research (Dzhambov and Dimitrova 2016; Gan et al. 2012), it has been established that traffic-related noise levels, as well as chronic exposure to even lower decibels of noise, can prompt a low-level but chronic stressor response in turn (Sliwa et al. 2024). Chronic stress, in all its forms, is a known to trigger for elevated blood pressure, endothelial dysfunction, inflammatory cytokine surge and oxidative stress (Gathright et al. 2024; Li et al. 2015; Bergmann et al. 2017; Wolfram et al. 2003; Vitaliano et al. 2002). All are mediating factors for cardiovascular disease and it is reasonable to argue that prolonged exposure to noise pollution would be another reason why someone might become 'climatically vulnerable' on an acute to chronic basis.

## 6.8 Listening to the Signals of Pollution

From the sounds of respiratory distress to a strident cardiac murmur, there is little doubt the pathways for pollution to adversely impact the health of people in every corner of the planet (even if they didn't substantially contribute to pollution levels) is growing. Thus, pollution in all its forms, lies at the heart of climate change (from causality to consequence). Much of the mitigation of risk posed by pollution is focussed on avoidance, but for many poor people, ironically, avoidance of indoor air pollution might mean not eating and being able to stay warm—thereby placing them at increasing risk from climatic provocations to their health. As repeated multiple times in this book, therefore, the problems and issues being discussed are not simple

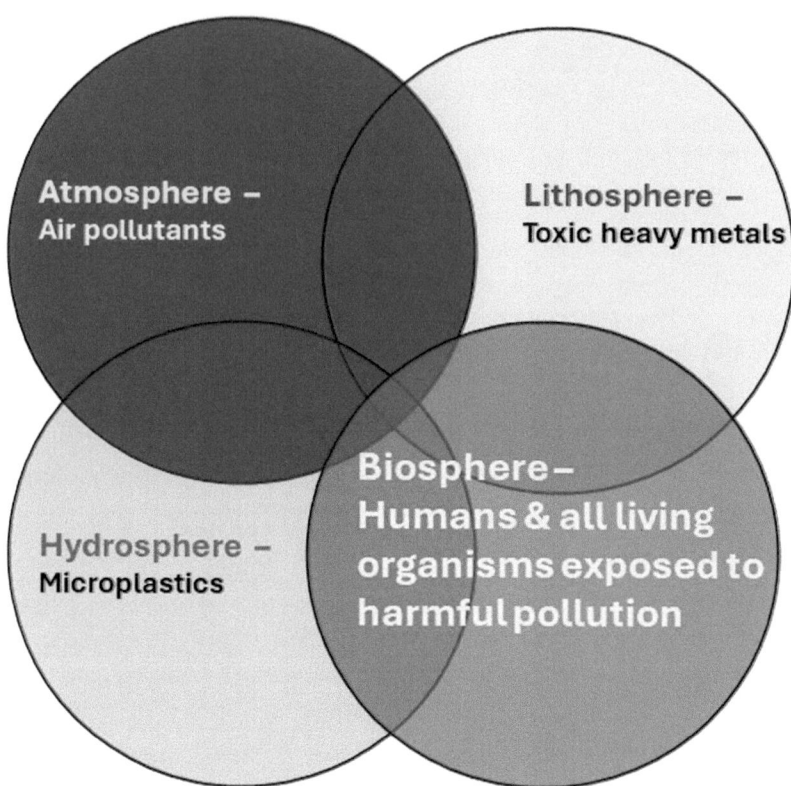

**Fig. 6.5** An interconnected planet. This figure is a reminder that all the planetary spheres are interconnected (including humans as part of the biosphere) and that pollution has infiltrated all parts of the earth (Briffa and Blundell 2020)

ones. Instead, they have multiple layers of complexity. In the next chapter, this theme will be extended further when considering the consequences of climate change on the pattern of air pollution from devastating wildfires to parasitic diseases.

## References

Abbasi-Kangevari M, Malekpour MR, Masinaei M, Moghaddam SS, Ghamari SH, Abbasi-Kangevari Z, Rezaei N, Rezaei N, Mokdad AH, Naghavi M, et al. Effect of air pollution on disease burden, mortality, and life expectancy in North Africa and the Middle East: a systematic analysis for the global burden of disease study 2019. Lancet Planet Health. 2023;7:e358–69. https://doi.org/10.1016/S2542-5196(23)00053-0.

Air ALASot. Health impact of air pollution; 2024. https://www.lung.org/research/sota/health-risks.

Alimba CG, Faggio C. Microplastics in the marine environment: current trends in environmental pollution and mechanisms of toxicological profile. Environ Toxicol Pharmacol. 2019;68:61–74. https://doi.org/10.1016/j.etap.2019.03.001.

Al-Kindi SG, Brook RD, Biswal S, Rajagopalan S. Environmental determinants of cardiovascular disease: lessons learned from air pollution. Nat Rev Cardiol. 2020;17:656–72. https://doi.org/10.1038/s41569-020-0371-2.

Attaelmanan AG, Aslam H, Ali T, Dronjak L. Mapping of heavy metal contamination associated with microplastics marine debris: a case study: Dubai, UAE. Sci Total Environ. 2023;891:164370. https://doi.org/10.1016/j.scitotenv.2023.164370.

Bai Y, Sun Q. Fine particulate matter air pollution and atherosclerosis: Mechanistic insights. Biochim Biophys Acta. 2016;1860:2863–8. https://doi.org/10.1016/j.bbagen.2016.04.030.

Bandara P, Weller S. Cardiovascular disease: time to identify emerging environmental risk factors. Eur J Prev Cardiol. 2017;24:1819–23. https://doi.org/10.1177/2047487317734898.

Beggs PJ. Thunderstorm asthma and climate change. JAMA. 2024;331:878–9. https://doi.org/10.1001/jama.2023.26649.

Bergmann N, Ballegaard S, Krogh J, Bech P, Hjalmarson A, Gyntelberg F, Faber J. Chronic psychological stress seems associated with elements of the metabolic syndrome in patients with ischaemic heart disease. Scand J Clin Lab Invest. 2017;77:513–9. https://doi.org/10.1080/00365513.2017.1354254.

Bosch AC, O'Neill B, Sigge GO, Kerwath SE, Hoffman LC. Heavy metals in marine fish meat and consumer health: a review. J Sci Food Agric. 2016;96:32–48. https://doi.org/10.1002/jsfa.7360.

Brauer M, Casadei B, Harrington RA, Kovacs R, et al. Taking a stand against air pollution—the impact on cardiovascular disease: a joint opinion from the world heart federation, American College of Cardiology, American Heart Association, and the European society of cardiology. Glob Heart. 2021;16:8. https://doi.org/10.5334/gh.948.

Briffa JSE, Blundell R. Heavy metal pollution in the environment and their toxicological effects on humans. Heylion. 2020;6:e04691.

Brook RD, Rajagopalan S. Particulate matter air pollution and atherosclerosis. Curr Atheroscler Rep. 2010;12:291–300. https://doi.org/10.1007/s11883-010-0122-7.

Browne MA, Crump P, Niven SJ, Teuten E, Tonkin A, Galloway T, Thompson R. Accumulation of microplastic on shorelines worldwide: sources and sinks. Environ Sci Technol. 2011;45:9175–9. https://doi.org/10.1021/es201811s.

Catarci Carteny C, Amato ED, Pfeiffer F, Christia C, Estoppey N, Poma G, Covaci A, Blust R. Accumulation and release of organic pollutants by conventional and biodegradable microplastics in the marine environment. Environ Sci Pollut Res Int. 2023;30:77819–29. https://doi.org/10.1007/s11356-023-27887-1.

Cesaroni G, Forastiere F, Stafoggia M, Andersen ZJ, Badaloni C, Beelen R, Caracciolo B, de Faire U, Erbel R, Eriksen KT, et al. Long term exposure to ambient air pollution and incidence of acute coronary events: prospective cohort study and meta-analysis in 11 European cohorts from the ESCAPE Project. BMJ. 2014;348:f7412. https://doi.org/10.1136/bmj.f7412.

Chang CC, Chen PS, Yang CY. Short-term effects of fine particulate air pollution on hospital admissions for cardiovascular diseases: a case-crossover study in a tropical city. J Toxicol Environ Health A. 2015;78:267–77. https://doi.org/10.1080/15287394.2014.960044.

Chen L, Chang N, Qiu T, Wang N, Cui Q, Zhao S, Huang F, Chen H, Zeng Y, Dong F, et al. Meta-analysis of impacts of microplastics on plant heavy metal(loid) accumulation. Environ Pollut. 2024;348:123787. https://doi.org/10.1016/j.envpol.2024.123787.

Cosselman KE, Navas-Acien A, Kaufman JD. Environmental factors in cardiovascular disease. Nat Rev Cardiol. 2015;12:627–42. https://doi.org/10.1038/nrcardio.2015.152.

Cosselman KE, Allen J, Jansen KL, Stapleton P, Trenga CA, Larson TV, Kaufman JD. Acute exposure to traffic-related air pollution alters antioxidant status in healthy adults. Environ Res. 2020;191:110027. https://doi.org/10.1016/j.envres.2020.110027.

Di Q, Dai L, Wang Y, Zanobetti A, Choirat C, Schwartz JD, Dominici F. Association of short-term exposure to air pollution with mortality in older adults. JAMA. 2017;318:2446–56. https://doi.org/10.1001/jama.2017.17923.

Diabetes GBD, Air Pollution C. Estimates, trends, and drivers of the global burden of type 2 diabetes attributable to PM(2.5) air pollution, 1990–2019: an analysis of data from the global burden of disease study 2019. Lancet Planet Health. 2022;6:e586–600. https://doi.org/10.1016/S2542-5196(22)00122-X

Dzhambov AM, Dimitrova DD. Association between noise pollution and prevalent ischemic heart disease. Folia Med. 2016;58:273–81. https://doi.org/10.1515/folmed-2016-0041.

Espiritu AI, Pilapil JCA, Aherrera JAM, Sy MCC, Anlacan VMM, Villanueva EQI, Jamora RDG. Outcomes of patients with COVID-19 and coronary artery disease and heart failure: findings from the Philippine CORONA study. BMC Res Notes. 2024;17:14. https://doi.org/10.1186/s13104-023-06677-5.

Fischer AJ, Hellmann AR, Diller GP, Maser M, Szardenings C, Marschall U, Bauer U, Baumgartner H, Lammers AE. Impact of COVID-19 infections among unvaccinated patients with congenital heart disease: results of a nationwide analysis in the first phase of the pandemic. J Clin Med. 2024;13:1282. https://doi.org/10.3390/jcm13051282.

Fu Q, Tan X, Ye S, Ma L, Gu Y, Zhang P, Chen Q, Yang Y, Tang Y. Mechanism analysis of heavy metal lead captured by natural-aged microplastics. Chemosphere. 2021;270:128624. https://doi.org/10.1016/j.chemosphere.2020.128624.

Gan WQ, Davies HW, Koehoorn M, Brauer M. Association of long-term exposure to community noise and traffic-related air pollution with coronary heart disease mortality. Am J Epidemiol. 2012;175:898–906. https://doi.org/10.1093/aje/kwr424.

Gathright EC, Hughes JW, Sun S, Storlazzi LE, DeCosta J, Balletto BL, Carey MP, Scott-Sheldon LAJ, Salmoirago-Blotcher E. Effects of stress management interventions on heart rate variability in adults with cardiovascular disease: a systematic review and meta-analysis. J Behav Med. 2024;47:374–88. https://doi.org/10.1007/s10865-024-00468-4.

Ghanadi M, Padhye LP. Revealing the long-term impact of photodegradation and fragmentation on HDPE in the marine environment: origins of microplastics and dissolved organics. J Hazard Mater. 2024;465:133509. https://doi.org/10.1016/j.jhazmat.2024.133509.

Hu J, Zhou R, Ding R, Ye DW, Su Y. Effect of PM(2.5) air pollution on the global burden of lower respiratory infections, 1990–2019: a systematic analysis from the global burden of disease study 2019. J Hazard Mater. 2023;459:132215. https://doi.org/10.1016/j.jhazmat.2023.132215.

Hu W, Fang L, Zhang H, Ni R, Pan G. Changing trends in the air pollution-related disease burden from 1990 to 2019 and its predicted level in 25 years. Environ Sci Pollut Res Int. 2023;30:1761–73. https://doi.org/10.1007/s11356-022-22318-z.

Huang F, Luo Y, Guo Y, Tao L, Xu Q, Wang C, Wang A, Li X, Guo J, Yan A, et al. Particulate matter and hospital admissions for stroke in Beijing, China: modification effects by ambient temperature. J Am Heart Assoc. 2016;5:437. https://doi.org/10.1161/JAHA.116.003437.

Jeffrey K, Woolford L, Maini R, Basetti S, Batchelor A, Weatherill D, White C, Hammersley V, Millington T, Macdonald C, et al. Prevalence and risk factors for long COVID among adults in Scotland using electronic health records: a national, retrospective, observational cohort study. EClinicalMedicine. 2024;71:102590. https://doi.org/10.1016/j.eclinm.2024.102590.

Ji JS. Air pollution and cardiovascular disease onset: hours, days, or years? Lancet Public Health. 2022;7:e890–1. https://doi.org/10.1016/S2468-2667(22)00257-2.

Karki M, Bhattarai P, Mohan R, Mushtaq F. COVID-19 unveiling heart failure in the realm of rheumatic heart disease. Cureus. 2024;16:e52903. https://doi.org/10.7759/cureus.52903.

Kemerley A, Gupta A, Thirunavukkarasu M, Maloney M, Burgwardt S, Maulik N. COVID-19 associated cardiovascular disease-risks, prevention and management: heart at risk due to COVID-19. Curr Issues Mol Biol. 2024;46:1904–20. https://doi.org/10.3390/cimb46030124.

Khorsandi B, Farzad K, Tahriri H, Maknoon R. Association between short-term exposure to air pollution and COVID-19 hospital admission/mortality during warm seasons. Environ Monit Assess. 2021;193:426. https://doi.org/10.1007/s10661-021-09210-y.

Kulick ER, Wellenius GA, Boehme AK, Sacco RL, Elkind MS. Residential proximity to major roadways and risk of incident ischemic stroke in NOMAS (The Northern Manhattan Study). Stroke. 2018;49:835–41. https://doi.org/10.1161/STROKEAHA.117.019580.

Li J, Zhang M, Loerbroks A, Angerer P, Siegrist J. Work stress and the risk of recurrent coronary heart disease events: a systematic review and meta-analysis. Int J Occup Med Environ Health. 2015;28:8–19. https://doi.org/10.2478/s13382-014-0303-7.

Li Z, Ma Y, Xu Y. Burden of lung cancer attributable to household air pollution in the Chinese female population: trend analysis from 1990 to 2019 and future predictions. Cad Saude Publica. 2022;38:e00050622. https://doi.org/10.1590/0102-311XEN050622.

Li C, Jing Z, Shen J, Liu S, Liu L, Yang X, Yu H. Global burden of vision loss attributable to household air pollution from 1990 to 2019: a trend analysis and 10-year prediction. Chin Med J. 2024;9:3051. https://doi.org/10.1097/CM9.0000000000003051.

Lin H, Liu T, Xiao J, Zeng W, Li X, Guo L, Zhang Y, Xu Y, Tao J, Xian H, et al. Mortality burden of ambient fine particulate air pollution in six Chinese cities: results from the pearl river delta study. Environ Int. 2016;96:91–7. https://doi.org/10.1016/j.envint.2016.09.007.

Lu F, Xu D, Cheng Y, Dong S, Guo C, Jiang X, Zheng X. Systematic review and meta-analysis of the adverse health effects of ambient PM2.5 and PM10 pollution in the Chinese population. Environ Res. 2015;136:196–204. https://doi.org/10.1016/j.envres.2014.06.029.

Lu H, Tan Z, Liu Z, Wang L, Wang Y, Suo C, Zhang T, Jin L, Dong Q, Cui M, et al. Spatiotemporal trends in stroke burden and mortality attributable to household air pollution from solid fuels in 204 countries and territories from 1990 to 2019. Sci Total Environ. 2021;775:145839. https://doi.org/10.1016/j.scitotenv.2021.145839.

Marfella R, Prattichizzo F, Sardu C, Fulgenzi G, Graciotti L, Spadoni T, D'Onofrio N, Scisciola L, La Grotta R, Frige C, et al. Microplastics and nanoplastics in atheromas and cardiovascular events. N Engl J Med. 2024;390:900–10. https://doi.org/10.1056/NEJMoa2309822.

Messner B, Knoflach M, Seubert A, Ritsch A, Pfaller K, Henderson B, Shen YH, Zeller I, Willeit J, Laufer G, et al. Cadmium is a novel and independent risk factor for early atherosclerosis mechanisms and in vivo relevance. Arterioscler Thromb Vasc Biol. 2009;29:1392–8. https://doi.org/10.1161/ATVBAHA.109.190082.

Micheu MM, Birsan MV, Szep R, Keresztesi A, Nita IA. From air pollution to cardiovascular diseases: the emerging role of epigenetics. Mol Biol Rep. 2020;47:5559–67. https://doi.org/10.1007/s11033-020-05570-9.

Mocumbi AO, Thienemann F, Sliwa K. A global perspective on the epidemiology of pulmonary hypertension. Can J Cardiol. 2015;31:375–81. https://doi.org/10.1016/j.cjca.2015.01.030.

Mocumbi AO, Stewart S, Patel S, Al-Delaimy WK. Cardiovascular effects of indoor air pollution from solid fuel: relevance to Sub-Saharan Africa. Curr Environ Health Rep. 2019;6:116–26. https://doi.org/10.1007/s40572-019-00234-8.

Mocumbi A, Humbert M, Saxena A, Jing ZC, Sliwa K, Thienemann F, Archer SL, Stewart S. Pulmonary hypertension. Nat Rev Dis Primers. 2024;10:1. https://doi.org/10.1038/s41572-023-00486-7.

Mustafic H, Jabre P, Caussin C, Murad MH, Escolano S, Tafflet M, Perier MC, Marijon E, Vernerey D,

Empana JP, et al. Main air pollutants and myocardial infarction: a systematic review and meta-analysis. JAMA. 2012;307:713–21. https://doi.org/10.1001/jama.2012.126.

Nakao M, Yamauchi K, Ishihara Y, Solongo B, Ichinnorov D. Effects of air pollution and seasonality on the respiratory symptoms and health-related quality of life (HR-QoL) of outpatients with chronic respiratory disease in Ulaanbaatar: pilot study for the comparison of the cold and warm seasons. Springerplus. 2016;5:1817. https://doi.org/10.1186/s40064-016-3481-x.

Namuyonga J, Mocumbi AO. Pulmonary hypertension in children across Africa: the silent threat. Int J Pediatr. 2021;2021:9998070. https://doi.org/10.1155/2021/9998070.

Navas-Acien A, Guallar E, Silbergeld EK, Rothenberg SJ. Lead exposure and cardiovascular disease: a systematic review. Environ Health Perspect. 2007;115:472–82. https://doi.org/10.1289/ehp.9785.

Nawrot TS, Perez L, Kunzli N, Munters E, Nemery B. Public health importance of triggers of myocardial infarction: a comparative risk assessment. Lancet. 2011;377:732–40. https://doi.org/10.1016/S0140-6736(10)62296-9.

Newby DE, Mannucci PM, Tell GS, Baccarelli AA, Brook RD, Donaldson K, Forastiere F, Franchini M, Franco OH, Graham I, et al. Expert position paper on air pollution and cardiovascular disease. Eur Heart J. 2015;36:83–93b. https://doi.org/10.1093/eurheartj/ehu458.

Nguyen KH, Bao Y, Mortazavi J, Allen JD, Chocano-Bedoya PO, Corlin L. Prevalence and factors associated with long COVID symptoms among US adults, 2022. Vaccines. 2024;12:99. https://doi.org/10.3390/vaccines12010099.

Ogah OS, Umuerri EM, Adebiyi A, Orimolade OA, Sani MU, Ojji DB, Mbakwem AC, Stewart S, Sliwa K. SARS-CoV 2 infection (Covid-19) and cardiovascular disease in Africa: health care and socio-economic implications. Glob Heart. 2021;16:18. https://doi.org/10.5334/gh.829.

Organisation WH. Air Pollution. https://www.who.int/health-topics/air-pollution#tab=tab_1

Pan G, Cheng J, Pan HF, Fan YG, Ye DQ. Global chronic obstructive pulmonary disease burden attributable to air pollution from 1990 to 2019. Int J Biometeorol. 2023;67:1543–53. https://doi.org/10.1007/s00484-023-02504-5.

Persiani E, Cecchettini A, Ceccherini E, Gisone I, Morales MA, Vozzi F. Microplastics: a matter of the heart (and vascular system). Biomedicines. 2023;11:264. https://doi.org/10.3390/biomedicines11020264.

Peters JL, Perlstein TS, Perry MJ, McNeely E, Weuve J. Cadmium exposure in association with history of stroke and heart failure. Environ Res. 2010;110:199–206. https://doi.org/10.1016/j.envres.2009.12.004.

Pintaric S, Zeljkovic I, Pehnec G, Nesek V, Vrsalovic M, Pintaric H. Impact of meteorological parameters and air pollution on emergency department visits for cardiovascular diseases in the city of Zagreb, Croatia. Arh Hig Rada Toksikol. 2016;67:240–6. https://doi.org/10.1515/aiht-2016-67-2770.

Price D, Hughes KM, Dona DW, Taylor PE, Morton DAV, Stevanovic S, Thien F, Choi J, Torre P, Suphioglu C. The perfect storm: temporal analysis of air during the world's most deadly epidemic thunderstorm asthma (ETSA) event in Melbourne. Ther Adv Respir Dis. 2023;17:17534666231186726. https://doi.org/10.1177/17534666231186726.

Rajagopalan S, Al-Kindi SG, Brook RD. Air pollution and cardiovascular disease: JACC state-of-the-art review. J Am Coll Cardiol. 2018;72:2054–70. https://doi.org/10.1016/j.jacc.2018.07.099.

Ramos P. Air pollution: a new risk factor for cardiovascular disease. e-J Cardiol Pract. 2021;22:546.

Real-time 100 Most Polluted Cities in the World. https://www.aqi.in/au/real-time-most-polluted-city-ranking. Accessed 12/4/2024.

Sababadichetty L, Miltgen G, Vincent B, Guilhaumon F, Lenoble V, Thibault M, Bureau S, Tortosa P, Bouvier T, Jourand P. Microplastics in the insular marine environment of the Southwest Indian Ocean carry a microbiome including antimicrobial resistant (AMR) bacteria: a case study from Reunion Island. Mar Pollut Bull. 2024;198:115911. https://doi.org/10.1016/j.marpolbul.2023.115911.

Samoli E, Atkinson RW, Analitis A, Fuller GW, Green DC, Mudway I, Anderson HR, Kelly FJ. Associations of short-term exposure to traffic-related air pollution with cardiovascular and respiratory hospital admissions in London, UK. Occup Environ Med. 2016;73:300–7. https://doi.org/10.1136/oemed-2015-103136.

Santurtun A, Sanchez-Lorenzo A, Villar A, Riancho JA, Zarrabeitia MT. The influence of nitrogen dioxide on arrhythmias in spain and its relationship with atmospheric circulation. Cardiovasc Toxicol. 2017;17:88–96. https://doi.org/10.1007/s12012-016-9359-x.

Sen T, Astarcioglu MA, Asarcikli LD, Kilit C, Kafes H, Parspur A, Yaymaci M, Pinar M, Tufekcioglu O, Amasyali B. The effects of air pollution and weather conditions on the incidence of acute myocardial infarction. Am J Emerg Med. 2016;34:449–54. https://doi.org/10.1016/j.ajem.2015.11.068.

Shah AS, Lee KK, McAllister DA, Hunter A, Nair H, Whiteley W, Langrish JP, Newby DE, Mills NL. Short term exposure to air pollution and stroke: systematic review and meta-analysis. BMJ. 2015;350:h1295. https://doi.org/10.1136/bmj.h1295.

Sliwa K, Acquah L, Gersh BJ, Mocumbi AO. Impact of socioeconomic status, ethnicity, and urbanization on risk factor profiles of cardiovascular disease in Africa. Circulation. 2016;133:1199–208. https://doi.org/10.1161/CIRCULATIONAHA.114.008730.

Sliwa K, Viljoen CA, Stewart S, Miller MR, Prabhakaran D, Kumar RK, Thienemann F, Piniero D, Prabhakaran P, Narula J, et al. Cardiovascular disease in low- and middle-income countries associated with environmental factors. Eur J Prev Cardiol. 2024;31:688–97. https://doi.org/10.1093/eurjpc/zwad388.

Stachyra K, Kiepura A, Olszanecki R. Air pollution and atherosclerosis: a brief review of mechanistic links between atherogenesis and biological actions of inorganic part of particulate matter. Folia Med Cracov. 2017;57:37–46.

Stewart S, Al-Delaimy W, Sliwa K, Yacoub M, Mocumbi A. Clinical algorithm to screen for cardiopulmonary disease in low-income settings. Nat Rev Cardiol. 2019;16:639–41. https://doi.org/10.1038/s41569-019-0268-0.

Stojic SS, Stanisic N, Stojic A, Sostaric A. Single and combined effects of air pollutants on circulatory and respiratory system-related mortality in Belgrade, Serbia. J Toxicol Environ Health A. 2016;79:17–27. https://doi.org/10.1080/15287394.2015.1101407.

Su C, Breitner S, Schneider A, Liu L, Franck U, Peters A, Pan X. Short-term effects of fine particulate air pollution on cardiovascular hospital emergency room visits: a time-series study in Beijing, China. Int Arch Occup Environ Health. 2016;89:641–57. https://doi.org/10.1007/s00420-015-1102-6.

Szanyi J, Howe S, Blakely T. The importance of reporting accurate estimates of long COVID prevalence. Lancet. 2024;403:1136–7. https://doi.org/10.1016/S0140-6736(23)01120-0.

The Köppen Climate Classification. https://www.mindat.org/climate.php. Accessed May, 2024.

Thien F. Melbourne epidemic thunderstorm asthma event 2016: lessons learnt from the perfect storm. Respirology. 2018;23:976–7. https://doi.org/10.1111/resp.13410.

Thurston GD, Burnett RT, Turner MC, Shi Y, Krewski D, Lall R, Ito K, Jerrett M, Gapstur SM, Diver WR, et al. Ischemic heart disease mortality and long-term exposure to source-related components of US fine particle air pollution. Environ Health Perspect. 2016;124:785–94. https://doi.org/10.1289/ehp.1509777.

Today UoC. L.A.'s summer smog siege concerns USC health scientists; 2019. https://today.usc.edu/l-a-s-summer-smog-siege-concerns-usc-health-scientists/. Accessed May, 2024.

Toubasi A, Al-Sayegh TN. Short-term exposure to air pollution and ischemic stroke: a systematic review and meta-analysis. Neurology. 2023;101:e1922–32. https://doi.org/10.1212/WNL.0000000000207856.

Trust P. Our Ocean is choking on plastic—but it's a problem we can solve; 2022. https://www.pewtrusts.org/en/trend/archive/winter-2022/our-ocean-is-choking-on-plastic-but-its-a-problem-we-can-solve. Accessed May, 2024.

Tsangari H, Paschalidou AK, Kassomenos AP, Vardoulakis S, Heaviside C, Georgiou KE, Yamasaki EN. Extreme weather and air pollution effects on cardiovascular and respiratory hospital admissions in Cyprus. Sci Total Environ. 2016;542:247–53. https://doi.org/10.1016/j.scitotenv.2015.10.106.

Vaziri ND. Mechanisms of lead-induced hypertension and cardiovascular disease. Am J Physiol Heart Circ Physiol. 2008;295:H454-465. https://doi.org/10.1152/ajpheart.00158.2008.

Vinnikov D, Tulekov Z, Raushanova A. Occupational exposure to particulate matter from air pollution in the outdoor workplaces in Almaty during the cold season. PLoS ONE. 2020;15:e0227447. https://doi.org/10.1371/journal.pone.0227447.

Vitaliano PP, Scanlan JM, Zhang J, Savage MV, Hirsch IB, Siegler IC. A path model of chronic stress, the metabolic syndrome, and coronary heart disease. Psychosom Med. 2002;64:418–35. https://doi.org/10.1097/00006842-200205000-00006.

Wine O, Osornio-Vargas A, Campbell SM, Hosseini V, Koch CR, Shahbakhti M. Cold climate impact on air-pollution-related health outcomes: a scoping review. Int J Environ Res Public Health. 2022;19:1473. https://doi.org/10.3390/ijerph19031473.

Wing JJ, Sanchez BN, Adar SD, Meurer WJ, Morgenstern LB, Smith MA, Lisabeth LD. Synergism of short-term air pollution exposures and neighborhood disadvantage on initial stroke severity. Stroke. 2017;48:3126–9. https://doi.org/10.1161/STROKEAHA.117.018816.

Wolfram R, Oguogho A, Palumbo B, Sinzinger H. Evidence for enhanced oxidative stress in coronary heart disease and chronic heart failure. Adv Exp Med Biol. 2003;525:197–200. https://doi.org/10.1007/978-1-4419-9194-2_42.

Woo KS, Chook P, Hu YJ, Lao XQ, Lin CQ, Lee P, Kwok C, Wei AN, Guo DS, Yin YH, et al. The impact of particulate matter air pollution (PM2.5) on atherosclerosis in modernizing China: a report from the CATHAY study. Int J Epidemiol. 2021;50:578–88. https://doi.org/10.1093/ije/dyaa235.

World Air Quality Index. https://waqi.info/. Accessed April, 2024.

Wu X, Cobbina SJ, Mao G, Xu H, Zhang Z, Yang L. A review of toxicity and mechanisms of individual and mixtures of heavy metals in the environment. Environ Sci Pollut Res Int. 2016;23:8244–59. https://doi.org/10.1007/s11356-016-6333-x.

Yeow RY, O'Leary MP, Reddy AR, Kamdar NS, Hayek SS, de Lemos JA, Sutton NR. Survival characteristics of older patients hospitalized with COVID-19: insights from the American heart association COVID-19 cardiovascular disease registry. J Am Med Dir Assoc. 2024;25:348–50. https://doi.org/10.1016/j.jamda.2023.11.027.

Yilmaz S, Sezen I, Irmak MA, Kulekci EA. Analysis of outdoor thermal comfort and air pollution under the influence of urban morphology in cold-climate cities: Erzurum/Turkey. Environ Sci Pollut Res Int. 2021;28:64068–83. https://doi.org/10.1007/s11356-021-14082-3.

Zhang H, Yin L, Zhang Y, Qiu Z, Peng S, Wang Z, Sun B, Ding J, Liu J, Du K, et al. Short-term effects of air pollution and weather changes on the occurrence of acute aortic dissection in a cold region. Front Public Health. 2023;11:1172532. https://doi.org/10.3389/fpubh.2023.1172532.

Zhang M, Tian P, Zhao Y, Song X, Liang J, Li J, Zhang Z, Guan X, Cao X, Ren Y, et al. Impact of aerosol-boundary layer interactions on PM(2.5) pollution during cold air pool events in a semi-arid urban basin. Sci Total Environ. 2024;922:171225. https://doi.org/10.1016/j.scitotenv.2024.171225.

Zhao S, Wang H, Chen H, Wang S, Ma J, Zhang D, Shen L, Yang X, Chen Y. Global magnitude and long-term trend of ischemic heart disease burden attributed to household air pollution from solid fuels in 204 countries and territories, 1990–2019. Indoor Air. 2022;32:e12981. https://doi.org/10.1111/ina.12981.

Zuniga J, Tarajia M, Herrera V, Urriola W, Gomez B, Motta J. Assessment of the possible association of air pollutants PM10, $O_3$, $NO_2$ with an increase in cardiovascular, respiratory, and diabetes mortality in panama city: a 2003 to 2013 data analysis. Medicine. 2016;95:e2464. https://doi.org/10.1097/MD.0000000000002464.

## Abstract

This chapter brings together a range of themes covered within the book thus far. It reflects on the diversity of problems different parts of the world are facing as the impact of climate change gathers pace. These problems encompass more cardiac complications arising from the air pollution generated from climate-triggered fires to the proliferation of vector-borne infectious diseases. In doing so, it discusses the complexity of pathways provoked by climatic change that might increase and even extend the burden of heart disease globally. Concurrently, the prospect of some unexpected benefits (in terms of disease reduction due to climate change) are identified. From that global perspective, some important changes in the 'infectious threats' posed to people living in the major continents are presented, noting, once again, how the poorest people in the world will likely bear the brunt of any changes in the pattern of disease provoked by climate change.

### Keywords

Infectious diseases · Parasites · Vector-borne diseases · Rheumatic heart disease · Tuberculosis · Chagas disease

## 7.1 A Complexity of Interrelated Factors

In the scope of one book, it is nigh impossible to convey all the different factors that come together and erode the cardiovascular health of whole communities to individuals—thereby leaving them vulnerable to both the current and future impact of climate change in all its variances. Consider the case of pulmonary hypertension (PHT), an insidious condition that is currently estimated to affect around 1% of the world's population (although there is scant information to inform such a figure) (Humbert et al. 2023; Mocumbi et al. 2024). The two most common pathways to this condition worldwide are left heart disease (Group 2 PHT—typically associated with left ventricular systolic/diastolic dysfunction, valvular heart disease, congenital/acquired inflow/outflow tract obstruction, cardiomyopathy) and respiratory conditions (Group 3 PHT—typically associated with chronic obstructive pulmonary disease, interstitial lung disease, chronic exposure to high altitude, hypoventilation syndromes, pulmonary diseases with restrictive and obstructive pattern) (Humbert et al. 2023; Mocumbi et al. 2024). While this book has mainly focussed on the commonly described forms of heart disease seen in high-income countries, PHT is a fascinating starting

S. Stewart, *Heart Disease and Climate Change*, Sustainable Development Goals Series,
https://doi.org/10.1007/978-3-031-73106-8_7

point for considering the wider and less obvious impact of climate change on cardiovascular health and related outcomes worldwide.

As discussed in Chap. 2, one of the major effects of the intensification of the *El Nino-Southern Oscillation* phenomenon has been a marked increase in the incidence and intensity of wildfires/bushfires (depending on the local vernacular) on either side of Pacific Ocean. This major climatic event affects high- and low-income communities alike (Cordero et al. 2024). If one, considers the likely burden of PHT (in all its forms) within the entire North American population for example, it is probable that major changes in climatic conditions affecting air quality (such as major, uncontrolled wildfires fuelled by drought conditions, more intense winds and scorching temperature) will provoke more cardiopulmonary complications in vulnerable people living in those regions where air quality is most compromised (Hasnain et al. 2024; Heaney et al. 2022; Jegasothy et al. 2023; Masri et al. 2022). Things become even more complicated when one considers additional extraneous factors such

as the COVID-19 pandemic (Naqvi et al. 2023). As identified by a recent '*State of the Air*' report, key cities located along the western seaboard of the US are already likely  hot-spots for cardiopulmonary events driven by periodic wildfire-induced degradations in their air-quality on top of already high levels of vehicular and industrial pollution (American Lung Association 2024)—see Fig. 7.1, With high rates of obesity (Hu 2023) and a significant proportion of the population not adequately vaccinated against serious COVID-19 infection (Johns Hopkins University of Medicine 2023), there is a strong confluence of factors (both controllable and uncontrollable in the short-to-medium term) that will likely increase climatic-provoked cardiopulmonary crises in this particular region of US/world. Thus, the burden of disease on the South-West coast of North America imposed by conditions such as PHT and other conditions sensitive to climatic conditions/respiratory disease, will likely differ from that of the East Coast of North America (where the drivers of cardiovascular disorders, including climatic conditions) are very different.

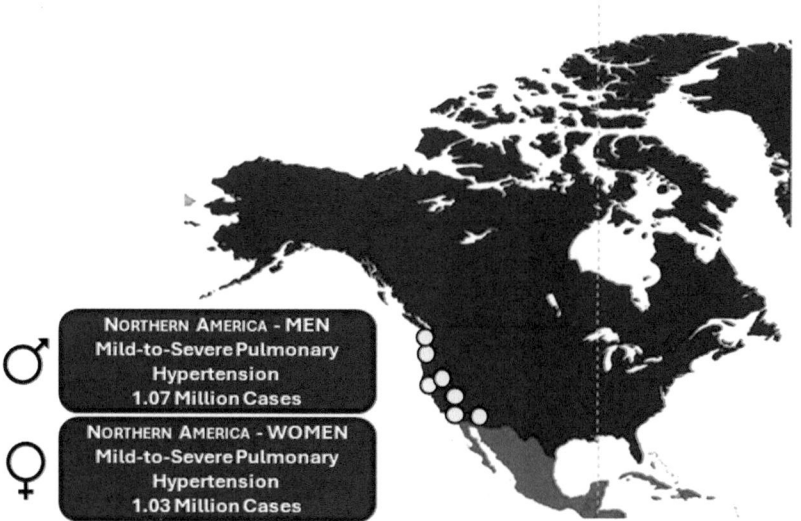

**Fig. 7.1** A confluence of wildfire pollution and cardiopulmonary disease. This figure shows the identified cities (yellow dots) on the West Coast of North America/US that are most vulnerable to a combination of wildfire-induced air pollution and that derived from more direct pollution (https://www.lung.org/research/sota/healthrisks). This 'high risk' region (for respiratory-related conditions that might induce/provoke pre-existing pulmonary hypertension [PHT]) is placed in the context of the likely number of male and female cases of PHT overall in the US—the critical point being that we are (at this stage) unable to document the specific pattern of PHT in different regions of the US and the world. Moreover, climate change is likely to change those patterns

As revealed by the *Heart of Soweto Study* in a large urban enclave in South Africa (Sliwa et al. 2010, 2008; Stewart et al. 2011a, b), in a vastly different part of the world, these complexity of factors are 'magnified' in low-to-middle income countries in whom a combination of historical and new threats to the cardiovascular health of communities undergoing socio-economic change is occurring at an accelerated rate within the setting of climate change. This study, confirmed by a more PHT focussed, multicentre study conducted in different parts of sub-Saharan Africa (Thienemann et al. 2016), raised a broad range of issues relevant to considering the importance of infectious pathways to heart disease in poverty-stricken people worldwide. This included a high burden of PHT provoked by respiratory diseases (including tuberculosis and exposure to pollutants), HIV-infection and multi-factorial contributors too complex to classify. It also raised the issue of endemic diseases linked to poverty—including rheumatic heart disease as a latent 'timebomb' in adults living in sub-Saharan Africa (Sliwa et al. 2010).

While the two narratives described above may appear disparate, they are proffered as exemplars of how the same phenomenon affecting everyone on the planet (i.e. climate change) produces very different consequences depending on—(a) where you live and (b) how empowered you are (as an individual and nation) to address them. In this context, it worth re-emphasising that the major health threats described below have none of the artificial boundaries (and border controls) we as humans apply to ourselves! Accordingly, the following sections consider how some of the main forms of 'infectious' triggers for heart disease might contribute to more cardiovascular events due to climate change—with a major focus on the poorer and least empowered regions of the world.

## 7.2 Infectious Causes of Heart Disease and Climate Change

According to a recent report from the World Economic Forum (World Economic Forum 2024), "*climate change is likely to cause an additional 14.5 million deaths and $12.5 trillion in economic losses worldwide*". In breaking down these dramatic figures, heatwaves alone, are estimated to contribute to 7.1 million deaths (World Economic Forum 2024). However, without offering any sophisticated modelling, as argued repeatedly throughout this book, it is likely that more 'mundane' variations in climatic conditions will provoke the majority of (preventable) cardiovascular events. This will include neglected pathways to heart conditions linked to infectious diseases. Indeed, in a recent 'call to action' to apply a new paradigm of modern medicine that considers the growing impact of climate change (Conrad 2023), the likely growth in infectious disease was specifically highlighted— "*Infectious diseases are also evolving as a result of climate change. The 2022 Intergovernmental Panel on Climate Change reports that the prevalence of vector-borne diseases has increased in recent decades. Warmer summers and milder temperatures have allowed pathogens to gain footholds in regions where populations have little immunity and warning systems are poorly developed*" (Conrad 2023). So, what are the type of infectious diseases commonly associated with cardiac complications that will likely thrive under climatic change and which populations are most at risk? Without providing an exhaustive list of diseases, the following sections highlight some of the more important/endemic infectious diseases (many of which are modulated by exposure to parasites) that have serious cardiac consequences for many people on a global basis.

### 7.2.1 Rheumatic Heart Disease

Rheumatic heart disease (RHD) is one of the most singularly distinctive conditions linked to poverty (Sliwa et al. 2010; Karthikeyan et al. 2023; Marijon et al. 2021; Zuhlke et al. 2015). Initially triggered by acute rheumatic fever following a *group A streptococcal* infection of the tonsillo-pharynx, subsequently, this infection can trigger an inflammatory reaction that involves fibrosis of the cardiac valves, valvular cardiomyopathy and then the syndrome of heart failure (HF) with all its deadly consequences. A recent study estimated that there were ~2.8 million new cases in 2019. This represented a close to 50% increase since 1990. Thus, although almost eradicated in many parts of the world with the application of prophylactic penicillin, this condition persists because of persistent poverty, and a lack of access to preventative health care for younger people (Karthikeyan et al. 2023). Critically, as shown by the *Heart of Soweto Study*, this condition (if not immediately fatal) leaves a growing legacy of younger adults presenting with later-stage valvular heart disease in sub-Saharan Africa (Sliwa et al. 2010). Global trends in the burden of RHD reveal why this is concerning for all lower-income regions of the world given that while the incidence decreased in higher-income regions, in poorer regions it increased. Specifically, as reported by Ou and colleagues—"*In terms of regions, the largest percent increase in incident number was found in western Sub-Saharan Africa (156%), while the largest decrease was in Central Europe (−39.8%)*" (Ou et al. 2022). This region of the world is not alone in this respect, noting similarly large increases in de novo new RHD cases and deaths have been predicted in China over the next 25 years (Zheng et al. 2022). The question of course, is will climate change (with expected warmer climates) increase the number of new RHD cases, or simply provoke a response from those with latent forms of the condition? If one tracks as far back as 1937, there has been long agreement that the primary cause of RHD is closely related to where children live—"*Despite the lack of accurate data, however, there is general agreement that the disease is common and severe in temperate zones, that it is less common in warmer and subtropical climates, and that it is rare in the tropics*" (Paul and Dixon 1937). Despite these early observations, there is little research or consideration [when estimating the future burden of disease (Hu et al. 2022)] of changing climatic conditions. It is possible that climate change will reduce the probability of primary infection with *group A streptococcus*, but we can't be sure (Ordunez et al. 2019).

## 7.2.2  Tuberculosis

Tuberculosis (TB) is primarily a respiratory infection caused by *Mycobacterium tuberculosis* (Adefuye et al. 2022). Transmission of TB is spread via the inhalation of small particle aerosols (1–5 μm in size) when a person with pulmonary TB expels air (Vyklyuk et al. 2024; Liu et al. 2024; Dinkele et al. 2024; Castro-Rodriguez et al. 2024). According to the *World Health Organization Global Tuberculosis Report 2021*, 9.9 million people were infected with TB in 2020, with 1.3 million people dying (representing 13% of those infected) (World Health Organisation WH 2022). Of these, a higher prevalence was seen in South-East Asia, Africa, and the Western Pacific regions, with a ratio of 1.7 males to 1 females affected, evident (Adefuye et al. 2022). One of the major complications of TB is cardiac involvement. For example, a contemporary meta-analysis based on 83,500 TB cases from four cohort studies, revealed that compared to people without TB, TB positive cases were 1.76-fold (95% CI 1.05–2.95) more likely to develop coronary heart disease (Wongtrakul et al. 2020). TB is also likely to increase the probability of ischaemic stroke. A case–control study reported by Sheu and colleagues involving 2283 patients treated for TB and 6849 randomly matched controls, showed that the former were 1.52-fold (95% CI 1.21–1.91) more likely to suffer an ischaemic stroke during 3-year follow-up (Sheu et al. 2010).

Overall, there are many factors influencing survival rates worldwide for those infected with TB, with drug-resistance to current therapies and co-infection, as well as the age of those affected important (Muflihah et al. 2024; Medrano et al. 2024; Taniguchi et al. 2024; Hu et al. 2024). A systematic review and meta-analysis of co-infection with TB and HIV, revealed the price of such co-infection being 16% (95% CI 13–2%) all-cause mortality during following-up compared to 4 and 6% all-cause mortality associated with lone HIV or TB infection, respectively (Hu et al. 2024).

As explicitly highlighted by Kharwadkar and colleagues (Kharwadkar et al. 2022)—"*There is a need to acknowledge TB as a climate-sensitive disease to facilitate its eradication*". This observation is based on predictive models suggesting a rise in the risk factors that lead to TB (Denholm 2020; Sergi et al. 2019). This includes those identified by the *Global Tuberculosis Report* 2021—diabetes mellitus, undernutrition, overcrowding, poverty and indoor air pollution (most of which will be exacerbated by climate change). As indicated above, it also encompasses a confluence of—(1) Co-infection with HIV [a not uncommon pathway to PHT (Thienemann et al. 2016)] in sub-Saharan Africa (home to most cases worldwide); and (2) Climate change (Abayomi and Cowan 2014). Accordingly, in their systematic review and meta-analysis of 53 studies focussing on the major risk for TB, Kharwadkar and colleagues reported a preponderance of evident suggesting that climate change would increase susceptibility to infection (Kharwadkar et al. 2022). Consequently, the number of pulmonary TB cases presenting with cardiac complications will almost inevitably rise with climate change and this rise will most likely occur within the most vulnerable populations worldwide.

## 7.3 Vector-Borne and Parasitic Heart Disease and Climate Change

In a comprehensive review of the subject, Hidron and colleagues provide a clear warning to much of the world's population that climate change (Khraishah et al. 2022) will bring about increasing exposure to a range of parasitic disease with serious cardiac complications (Hidron et al. 2010). As noted in this review, the heart and the lungs are the organs most affected by parasitic infections. Accordingly, any increased exposure to parasitic-mediated infections that invoke a pan-cardiac/myocardial, pathological response (most notably Trypanosoma cruzi [*Chagas' disease*], and other forms of vector-borne parasites that provoke pericarditis/

pericardial effusion, will likely increase the associated burden of disease (Hidron et al. 2010).

### 7.3.1 Trypanosoma Cruzi (Chagas' Disease)

Caused by a flagellate protozoan parasite (Trypanosoma cruzi) that is mainly vector-borne via the triatomine insect, Chagas' disease remains endemic to South America, Central America and parts of Southern United States and Mexico with similar climates (Hidron et al. 2010). Once infected, the accumulation of large numbers of amastigotes in the cardiac myocytes of the affected individual triggers immune-mediated damage that results in degeneration of muscle fibres, coagulation necrosis of myocytes and surrounding tissues. This is commonly followed by a sequalae of epicardial and pericardial involvement. The main cardiac manifestations of Chagas' infection, therefore, are *Chagas Cardiomyopath*y characterised by consequent heart failure, thromboembolism and arrhythmias (Torres et al. 2022). Unfortunately, although treatable if detected early, like other insidious forms of heart disease, many people don't present until late with myocardial damage and late-stage HF (Yacoub et al. 2008). A systematic review and meta-analysis of 25 studies involving 10,638 patients with 53,346 patient-years of follow-up in which 2739 fatal events occurred, suggested that those who develop Chagas' disease have a significantly higher risk (1.74-fold [95% CI 1.49–2.03]) of dying from any cause compared to those without (although a large degree of heterogeneity in study findings on all-cause mortality provides an important caveat) (Yacoub et al. 2008).

Long considered a disease of poverty in rural areas, it has been suggested that increasingly more cases of Chagas' disease are now appearing among people living in peri-urban and even urban areas (Moncayo 2003). Reflecting these potentially significant changes in the pattern of disease, a recent Global Burden of Disease report suggested that the worldwide prevalence of cases decreased by 11.3% (from 7.29

to 6.47 million) during the period 1990 to 2019 (Gomez-Ochoa et al. 2022). Critically, the highest burden of disease was found to be in older individuals (particularly men) and, unfortunately still among the poorest regions with the lowest sociodemographic index. Despite being home to the most cases in the world (with the top 5 countries in this regard—Bolivia [4994 cases/100,000 population], Venezuela [1655 cases/100,000], Argentina [1524 cases/100,000], Chile [1114 cases/100,000] and Mexico [1030 cases/100,000]), Latin America has seen impressive declines (from 41 to 72% less cases since 1990) in the incidence of Chagas' disease. In contrast, both North America (including the US) and Europe (notably Spain) have observed increasing cases (Gomez-Ochoa et al. 2022).

Within this context, there is good reason to believe that climate change will influence the vectors of parasites that spread and cause diseases affecting the cardiopulmonary system, thereby exposing new populations (and those visiting them) to potential infection (Flores-Lopez et al. 2022; Garrido et al. 2019). However, just as the jury is out on the net effect of climate change based on cold- versus heat-initiated events in vulnerable individuals, a recent analysis of the geographic distribution of five species of triatomines that act as vectors for Chagas' disease in Venezuela (second highest concentration of cases in the world) concluded that—"*possible future effects of global climate change on the Venezuelan population vulnerability show a slightly decreasing trend*" (Ceccarelli and Rabinovich 2015).

## 7.4 Other Possible 'Vectors' to Infectious-Triggered Heart Disease

Although other endemic conditions like *Schistosomiasis*, a designated Neglected Tropical Diseases, rarely provoke cardiac pathology, the number of cases worldwide are hard to ignore. Caused by exposure (and ingestion) of trematodes (parasitic worms), this condition affects over 230 million people globally,

with nearly 700 million at risk in more than 74 countries. Mainly provoking a hepatic pathology, because of the sheer numbers, *Schistosomiasis* provoked cardiac complications (whilst 'rare') are not uncommon (Villamizar-Monsalve et al. 2024). Similarly, although *Malaria*, caused by the plasmodium falciparum parasite transmitted by mosquitos, can often result in life-threatening complications, they are rarely cardiac related. Nevertheless, the sheer number of infected cases means that malaria-provoked cardiac complications are still not uncommon in those countries most affected by this devastating condition. In the latest report from the WHO, it was reported that "*of the 249 million cases noted in 2022, 233 million (around 94%) were in the WHO African Region, with Nigeria (27%), the Democratic Republic of the Congo (12%), Uganda (5%), and Mozambique (4%), accounting for nearly 50% of all cases*" (The 2023).

Other diseases of poverty (the *neglected diseases*) such as tropical endomyocardial fibrosis (EMF), which was first discovered in Uganda (Central Africa) and then comprehensively characterised in the northern regions of Mozambique (East Africa) (Mocumbi et al. 2008), is another poverty-related disease that afflicts predominantly young people in specific rural areas in Africa, Asia and South America. Still of unknown origin, EMF is characterised by the deposition of fibrous tissue in the endomyocardial layers leading to a restrictive physiology/cardiomyopathy (Grimaldi et al. 2016; Beaton and Mocumbi 2017). As the most common form of restrictive cardiomyopathy in the world, EMF most commonly presents as biventricular failure (56% of cases in Mozambique) or as right-sided predominant failure (28% of cases) (Mocumbi et al. 2008). Regardless, as in many of the other conditions described in this chapter, it is commonly insidious and often results in death due to a later presentation. Unfortunately, we don't know what effect climate change will have on this devastating disease until its origins are better understood (Mbanze et al. 2020).

In a world that is going to be overall warmer and more humid, it is hard to ignore the largely ignored threat of more bacterial and fungal infections leading to infective endocarditis, noting that *Staphylococcus aureus* is the most common culprit (Miao et al. 2024). What many people didn't predict in specific terms, was the *COVID-19* pandemic and how it would both cause cardiac complications from both primary infection and some of the vaccinations used to inoculate the world's population (Eftekhar et al. 2024; Razzaque 2024). Over time, COVID-19 will likely transition into a barely commented, endemic infection that mingles and complicates other endemic infectious, respiratory diseases such as *influenza*. What we can't predict is what that means for our collective heart health, although some studies have already found an increased 'signal' of developing atherosclerotic disease in people co-infected with the COVID-19 and influenza viruses (Jalili et al. 2022).

## 7.5  Preparing for the Future

In framing the complexity of threats from altered patterns of infectious causes of heart disease/cardiac complications, it is worth noting that as portrayed in Fig. 7.2, each continent and region of the world is likely to deal with a range of threats that require a 'bespoke' approach to prevention and management. However, as will be described at the individual level, not all countries are able to respond or are able to raise climate change as priority issue and fund potential solutions to the health issues that arise. As succinctly stated in a WHO-funded report on Zimbabwe's preparedness for climate change, 'climatic vulnerability' is not confined to the individual or even community level. Indeed, for poverty-stricken countries such as Zimbabwe (in Southern Africa), vulnerability—*"is a function of the character, magnitude, and rate of climate change and variation to which a system is exposed, its sensitivity, and its adaptive capacity"* (United Nations Environment Programme Zimbabwe MoE 2020). As identified within this high-level report, key issues that will expose this vulnerability will involve the 'geographic redistribution' of malaria to densely populated urban areas, food insecurity, chronic malnutrition and HIV/AIDs. Once again, it is worth noting that this report, whilst identifying increasing climate threats, identified potential positives (from climate change) from many perspective (United Nations Environment Programme Zimbabwe MoE 2020).

Such climate-driven threats (and potential positives), along with national/regional vulnerability are not confined to sub-Saharan Africa. A recent report focussing on the highly populous and diverse region of South-East Asia (World Health Organisation 2008), highlighted similar themes and issues to that outlined above—*"Low income countries and areas where malnutrition is widespread, education is poor, and infrastructures are weak will have most difficulty adapting to climate change and related health hazards. Vulnerability is also determined by geography and is higher in areas with a high endemicity of climate-sensitive diseases, water stress, low food production and isolated populations"* (World Health Organisation 2008). Once again, the importance of where people live (in terms of the interaction between geography, climate and where the indirect impact of climate will be most evident) was highlighted, with those greatest at risk (in South-East Asia) being people living on islands, in the mountains, water stress and coastal regions and mega-cities with densely populations.

In contrast to the above (from the size and density of affected populations to its socio-economic resources), there is evidence that the high-income/relatively small population (~27 million people) of Australia not only has the willingness to become more 'climatically resilient'—but can afford to make that happen. This is evidenced by a recent government report, that included a comprehensive plan to address years of non-funding for health research with a climatic focus and four key objectives—(1) Building a climate-resilient health system, (2) Health system decarbonisation, (3) International collaboration and, (4) Including health (and the issue of climate change) into all government policies (Australian Government Department of Health and Aged Care 2024).

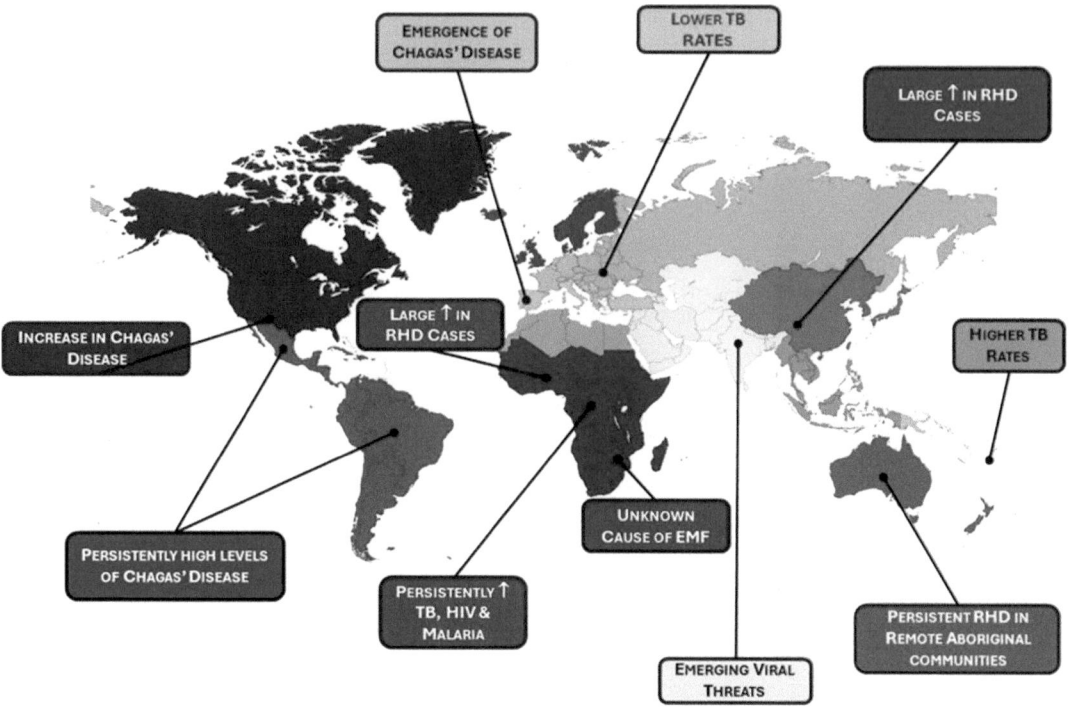

**Fig. 7.2** A global 'threat board' of infectious diseases with cardiac complications. In addition to the specific 'threats' highlighted in text, this graph outlines additional threats such as Dengue Fever (2 billion people) (Coker et al. 2011), the enduring problem of RHD and related valvopathies/cardiomyopathies in Central Australia resulting in significant gaps in survival rates compared to the rest of the Australian population (Brown et al. 2014) and the evolving threat from respiratory viral infections, arboviral infections and bat-borne viral infection on the Indian sub-continent (Mourya et al. 2019)

## 7.6 But Which One (Future) Are We Preparing for?

As this book has already highlighted in earlier chapters, the narrative of heat-related climate provocations has overshadowed that posed by cold-related provocations (and more importantly, dynamic weather changes). This signals a dangerous (and perhaps delusional or just politicised) perspective that gravitates towards simplistic messaging and responses around climate change. This certainly appears to be the case around the Australian policy document highlighted above that pays scant regard to the full spectrum of threats posed by climate change. The multiple and emerging threats to our health (and specifically cardiovascular health) from the confluence of more challenging climatic and environmental conditions, requires a different mind-set from top to bottom. Once again, it is beyond the scope of the author and this book, to delve into geo-politics and global health policies! Nevertheless, a central theme of this book has focussed on a key part of the 'matrix' of responses needed to care and treat people presenting with the antecedents of, or established CVD—that is, the need for a new paradigm of health care delivery in relation to climate change embraced by clinicians and health policy makers. Unfortunately, the evidence-base is extremely thin in this regard. However, the next two chapters will explore what the science and evidence tell us in terms of being able to prevent current and future cardiac events by proactively addressing *climatic vulnerability* and promoting/replacing it with a higher degree of *climatic resilience*. Depending on the aspirational goal, the presented evidence and discussions therein will span multiple levels of intervention—whole population, health systems and individualised management.

# References

Abayomi A, Cowan MN. The HIV/AIDS epidemic in South Africa: convergence with tuberculosis, socio-ecological vulnerability, and climate change patterns. S Afr Med J. 2014;104:583.

Adefuye MA, Manjunatha N, Ganduri V, Rajasekaran K, Duraiyarasan S, Adefuye BO. Tuberculosis and cardiovascular complications: an overview. Cureus. 2022;14:e28268.

American Lung Association. State of the air report: health impact of air pollution; 2024. https://www.lung.org/research/sota/health-risks. Accessed May 2024.

Australian Government Department of Health and Aged Care. National health and climate strategy; 2024.

Beaton A, Mocumbi AO. Diagnosis and management of endomyocardial fibrosis. Cardiol Clin. 2017;35:87–98.

Brown A, Carrington MJ, McGrady M, Lee G, Zeitz C, Krum H, Rowley K, Stewart S. Cardiometabolic risk and disease in indigenous Australians: the heart of the heart study. Int J Cardiol. 2014;171:377–83.

Castro-Rodriguez B, Espinoza-Andrade S, Franco-Sotomayor G, Benitez-Medina JM, Jimenez-Pizarro N, Cardenas-Franco C, Granda JC, et al. A first insight into tuberculosis transmission at the border of Ecuador and Colombia: a retrospective study of the population structure of mycobacterium tuberculosis in Esmeraldas province. Front Public Health. 2024;12:1343350.

Ceccarelli S, Rabinovich JE. Global climate change effects on venezuela's vulnerability to chagas disease is linked to the geographic distribution of five triatomine species. J Med Entomol. 2015;52:1333–43.

Coker RJ, Hunter BM, Rudge JW, Liverani M, Hanvoravongchai P. Emerging infectious diseases in southeast Asia: regional challenges to control. Lancet. 2011;377:599–609.

Conrad K. The era of climate change medicine-challenges to health care systems. Ochsner J. 2023;23:7–8.

Cordero RR, Feron S, Damiani A, Carrasco J, Karas C, Wang C, Kraamwinkel CT, Beaulieu A. Extreme fire weather in Chile driven by climate change and El Nino-Southern Oscillation (ENSO). Sci Rep. 2024;14:1974.

Denholm J. Seasonality, climate change and tuberculosis: new data and old lessons. Int J Tuberc Lung Dis. 2020;24:469.

Dinkele R, Khan PY, Warner DF. Mycobacterium tuberculosis transmission: the importance of precision. Lancet Infect Dis. 2024;65:323–93.

Eftekhar Z, Haybar H, Mohebbi A, Saki N. Cardiac complications and COVID-19: a review of life-threatening co-morbidities. Curr Cardiol Rev. 2024;56:8795.

Flores-Lopez CA, Moo-Llanes DA, Romero-Figueroa G, Guevara-Carrizales A, Lopez-Ordonez T, Casas-Martinez M, Samy AM. Potential distributions of the parasite *Trypanosoma cruzi* and its vector *Dipetalogaster maxima* highlight areas at risk of Chagas disease transmission in Baja California Sur, Mexico, under climate change. Med Vet Entomol. 2022;36:469–79.

World Economic Forum. Quantifying the impact of climate change on human health; 2024. https://www.weforum.org/press/2024/01/wef24-climate-crisis-health/#:~:text=Davos%2DKlosters%2C%20Switzerland%2C%2016,trillion%20in%20economic%20losses%20worldwide. Accessed May 2024.

Garrido R, Bacigalupo A, Pena-Gomez F, Bustamante RO, Cattan PE, Gorla DE, Botto-Mahan C. Potential impact of climate change on the geographical distribution of two wild vectors of Chagas disease in Chile: *Mepraia spinolai* and *Mepraia gajardoi*. Parasit Vectors. 2019;12:478.

Gomez-Ochoa SA, Rojas LZ, Echeverria LE, Muka T, Franco OH. Global, regional, and national trends of chagas disease from 1990 to 2019: comprehensive analysis of the global burden of disease study. Glob Heart. 2022;17:59.

Grimaldi A, Mocumbi AO, Freers J, Lachaud M, Mirabel M, Ferreira B, Narayanan K, Celermajer DS, Sidi D, Jouven X, Marijon E. Tropical endomyocardial fibrosis: natural history, challenges, and perspectives. Circulation. 2016;133:2503–15.

Hasnain MG, Garcia-Esperon C, Tomari YK, Walker R, Saluja T, Rahman MM, Boyle A, Levi CR, Naidu R, Filippelli G, Spratt NJ. Bushfire-smoke trigger hospital admissions with cerebrovascular diseases: evidence from 2019–20 bushfire in Australia. Eur Stroke J. 2024;12:23969873231223308.

Heaney A, Stowell JD, Liu JC, Basu R, Marlier M, Kinney P. Impacts of fine particulate matter from wildfire smoke on respiratory and cardiovascular health in California. Geohealth. 2022;6:e2021GH000578.

Hidron A, Vogenthaler N, Santos-Preciado JI, Rodriguez-Morales AJ, Franco-Paredes C, Rassi A. Cardiac involvement with parasitic infections. Clin Microbiol Rev. 2010;23:324–49.

Hu FB. Obesity in the USA: diet and lifestyle key to prevention. Lancet Diabetes Endocrinol. 2023;11:642–3.

Hu Y, Tong Z, Huang X, Qin JJ, Lin L, Lei F, Wang W, Liu W, Sun T, Cai J, She ZG, Li H. The projections of global and regional rheumatic heart disease burden from 2020 to 2030. Front Cardiovasc Med. 2022;9:941917.

Hu FH, Tang XL, Ge MW, Jia YJ, Zhang WQ, Tang W, Shen LT, Du W, Xia XP, Chen HL. Mortality of children and adolescents co-infected with tuberculosis and HIV: a systematic review and meta-analysis. AIDS. 2024;78:12574.

Humbert M, Kovacs G, Hoeper MM, Badagliacca R, Berger RMF, Brida M, Carlsen J, Coats AJS, et al. 2022 ESC/ERS guidelines for the diagnosis and treatment of pulmonary hypertension. Eur Respir J. 2023;61:11542.

Jalili M, Sayehmiri K, Ansari N, Pourhossein B, Fazeli M, Jalilian AF. Association between influenza and COVID-19 viruses and the risk of atherosclerosis: meta-analysis study and systematic review. Adv Respir Med. 2022;90:338–48.

Jegasothy E, Hanigan IC, Van Buskirk J, Morgan GG, Jalaludin B, Johnston FH, Guo Y, Broome RA. Acute health effects of bushfire smoke on mortality in Sydney, Australia. Environ Int. 2023;171:107684.

Johns Hopkins University of Medicine. Understanding coronavirus vaccinations; 2023.

Karthikeyan G, Watkins D, Bukhman G, Cunningham MW, Haller J, Masterson M, Mensah GA, et al. Research priorities for the secondary prevention and management of acute rheumatic fever and rheumatic heart disease: a national heart, lung, and blood institute workshop report. BMJ Glob Health. 2023;8:98.

Kharwadkar S, Attanayake V, Duncan J, Navaratne N, Benson J. The impact of climate change on the risk factors for tuberculosis: a systematic review. Environ Res. 2022;212:113436.

Khraishah H, Alahmad B, Ostergard RL, AlAshqar A, Albaghdadi M, Vellanki N, Chowdhury MM, Al-Kindi SG, Zanobetti A, Gasparrini A, Rajagopalan S. Climate change and cardiovascular disease: implications for global health. Nat Rev Cardiol. 2022;19:798–812.

Liu D, Huang F, Li Y, Mao L, He W, Wu S, Xia H, He P, Zheng H, Zhou Y, Zhao B, Ou X, Song Y, Song Z, Mei L, Liu L, Zhang G, Wei Q, Zhao Y. Transmission characteristics in Tuberculosis by WGS: nationwide cross-sectional surveillance in China. Emerg Microbes Infect. 2024;13:2348505.

Marijon E, Mocumbi A, Narayanan K, Jouven X, Celermajer DS. Persisting burden and challenges of rheumatic heart disease. Eur Heart J. 2021;42:3338–48.

Masri S, Jin Y, Wu J. Compound risk of air pollution and heat days and the influence of wildfire by SES across California, 2018–2020: implications for environmental justice in the context of climate change. Climate. 2022;10:323.

Mbanze J, Cumbane B, Jive R, Mocumbi A. Challenges in addressing the knowledge gap on endomyocardial fibrosis through community-based studies. Cardiovasc Diagn Ther. 2020;10:279–88.

Medrano BA, Lee M, Gemeinhardt G, Yamba L, Restrepo BI. High all-cause mortality and increasing proportion of older adults with tuberculosis in Texas, 2008–2020. Epidemiol Infect. 2024;32:1–34.

Miao H, Zhang Y, Zhang Y, Zhang J. Update on the epidemiology, diagnosis, and management of infective endocarditis: a review. Trends Cardiovasc Med. 2024;10:215.

Mocumbi AO, Ferreira MB, Sidi D, Yacoub MH. A population study of endomyocardial fibrosis in a rural area of Mozambique. N Engl J Med. 2008;359:43–9.

Mocumbi A, Humbert M, Saxena A, Jing ZC, Sliwa K, Thienemann F, Archer SL, Stewart S. Pulmonary hypertension. Nat Rev Dis Primers. 2024;10:1.

Moncayo A. Chagas disease: current epidemiological trends after the interruption of vectorial and transfusional transmission in the Southern Cone countries. Mem Inst Oswaldo Cruz. 2003;98:577–91.

Mourya DT, Yadav PD, Ullas PT, Bhardwaj SD, Sahay RR, Chadha MS, Shete AM, Jadhav S, Gupta N, Gangakhedkar RR, Khasnobis P, Singh SK. Emerging/re-emerging viral diseases and new viruses on the Indian horizon. Indian J Med Res. 2019;149:447–67.

Muflihah H, Yulianto FA, Rina S, Sampurno E, Ferdiana A, Rahimah SB. Tuberculosis coinfection among COVID-19 patients: clinical presentation and mortality in a tertiary lung hospital in Indonesia. Int J Mycobacteriol. 2024;13:58–64.

Naqvi HR, Mutreja G, Shakeel A, Singh K, Abbas K, Naqvi DF, Chaudhary AA, Siddiqui MA, Gautam AS, Gautam S, Naqvi AR. Wildfire-induced pollution and its short-term impact on COVID-19 cases and mortality in California. Gondwana Res. 2023;114:30–9.

Ordunez P, Martinez R, Soliz P, Giraldo G, Mujica OJ, Nordet P. Rheumatic heart disease burden, trends, and inequalities in the Americas, 1990–2017: a population-based study. Lancet Glob Health. 2019;7:e1388–97.

Ou Z, Yu D, Liang Y, Wu J, He H, Li Y, He W, Gao Y, Wu F, Chen Q. Global burden of rheumatic heart disease: trends from 1990 to 2019. Arthritis Res Ther. 2022;24:138.

Paul JR, Dixon GL. Climate and rheumatic heart disease: a survey among American Indian school children in northern and southern locaties. JAMA. 1937;108:2096–100.

Razzaque MS. Can adverse cardiac events of the COVID-19 vaccine exacerbate preexisting diseases? Expert Rev Anti Infect Ther. 2024;22:131–7.

Sergi C, Serra N, Colomba C, Ayanlade A, Di Carlo P. Tuberculosis evolution and climate change: how much work is ahead? Acta Trop. 2019;190:157–8.

Sheu JJ, Chiou HY, Kang JH, Chen YH, Lin HC. Tuberculosis and the risk of ischemic stroke: a 3-year follow-up study. Stroke. 2010;41:244–9.

Sliwa K, Wilkinson D, Hansen C, Ntyintyane L, Tibazarwa K, Becker A, Stewart S. Spectrum of heart disease and risk factors in a black urban population in South Africa (the Heart of Soweto Study): a cohort study. Lancet. 2008;371:915–22.

Sliwa K, Carrington M, Mayosi BM, Zigiriadis E, Mvungi R, Stewart S. Incidence and characteristics of newly diagnosed rheumatic heart disease in urban African adults: insights from the heart of Soweto study. Eur Heart J. 2010;31:719–27.

Stewart S, Carrington M, Pretorius S, Methusi P, Sliwa K. Standing at the crossroads between new and historically prevalent heart disease: effects of migration and socio-economic factors in the Heart of Soweto cohort study. Eur Heart J. 2011a;32:492–9.

Stewart S, Mocumbi AO, Carrington MJ, Pretorius S, Burton R, Sliwa K. A not-so-rare form of heart failure in urban black Africans: pathways to right heart failure in the heart of soweto study cohort. Eur J Heart Fail. 2011b;13:1070–7.

Taniguchi J, Aso S, Jo T, Matsui H, Fushimi K, Yasunaga H. Factors affecting in-hospital mortality in patients with miliary tuberculosis: a retrospective cohort study. Respir Investig. 2024;62:520–5.

The VP. WHO world malaria report. Lancet Microbe. 2023;2024(5):e214.

Thienemann F, Dzudie A, Mocumbi AO, Blauwet L, Sani MU, Karaye KM, Ogah OS, Mbanze I, Mbakwem A, Udo P, Tibazarwa K, Damasceno A, Keates AK, Stewart S, Sliwa K. The causes, treatment, and outcome of pulmonary hypertension in Africa: insights from the pan African pulmonary hypertension cohort (PAPUCO) registry. Int J Cardiol. 2016;221:205–11.

Torres RM, Correia D, Nunes M, Dutra WO, Talvani A, Sousa AS, Mendes F, et al. Prognosis of chronic Chagas heart disease and other pending clinical challenges. Mem Inst Oswaldo Cruz. 2022;117:e210172.

United Nations Environment Programme Zimbabwe MoE. Climate, tourism and hospitality industry. Zimbabwe climate change vulnerability assessment: an indicator-based report; 2020. https://wedocs.unep.org/handle/20.500.11822/45301. Accessed May 2024.

Villamizar-Monsalve MA, Lopez-Aban J, Vicente B, Pelaez R, Muro A. Current drug strategies for the treatment and control of schistosomiasis. Expert Opin Pharmacother. 2024;25:409–20.

Vyklyuk Y, Semianiv I, Nevinskyi D, Todoriko L, Boyko N. Applying geospatial multi-agent system to model various aspects of tuberculosis transmission. New Microbes New Infect. 2024;59:101417.

Wongtrakul W, Charoenngam N, Ungprasert P. Tuberculosis and risk of coronary heart disease: a systematic review and meta-analysis. Indian J Tuberc. 2020;67:182–8.

World Health Organisation. Regional health office for southeast Asia. Climate change and health; 2008. https://iris.who.int/bitstream/handle/10665/126809/SEA_ACM_Meet.%20?sequence=1. Accessed May 2024.

World Health Organisation WH. Global tuberculosis report 2021; 2022.

Yacoub S, Mocumbi AO, Yacoub MH. Neglected tropical cardiomyopathies: I. Chagas disease: myocardial disease. Heart. 2008;94:244–8.

Zheng X, Guan Q, Lin X. Changing trends of the disease burden of non-rheumatic valvular heart disease in China from 1990 to 2019 and its predictions: findings from global burden of disease study. Front Cardiovasc Med. 2022;9:912661.

Zuhlke L, Engel ME, Karthikeyan G, Rangarajan S, Mackie P, Cupido B, Mauff K, Islam S, Joachim A, Daniels R, Francis V, et al. Characteristics, complications, and gaps in evidence-based interventions in rheumatic heart disease: the global rheumatic heart disease registry (the REMEDY study). Eur Heart J. 2015;36:1115–1122a.

## Abstract

If the contents and topics covered thus far in this book have proved anything, it is that there is still little definitive science to guide us in relation to optimally managing people living with heart disease from a climatic perspective. And yet, there is ample epidemiological evidence to suggest why we need to change the paradigm of clinical management on this basis. Thus, the title of this chapter might have been extended with a 'thus far'. Moreover, it could well have been structured with a list of (Donald Rumsfeld-inspired) "*known-unknowns*" and even "*unknown, unknowns*". Nevertheless, there is sufficient evidence to start an important conversation around what clinicians operating within a tertiary healthcare setting can do to promote resilience in people presenting with various forms of heart disease. This is the major focus of this chapter, whist providing some critical reflections on where the evidence falls short in protecting people from both current climatic conditions and that resulting from future climatic change.

## Keywords

Seasonal influences · Climatic profiling · Multifaceted interventions · Tertiary care · Disease management · Randomised trial

## 8.1 The Cost of Doing the Same Old Thing

Much of the previous chapters have built a case for why it would be insane (by repeating the same old mistakes in quantifying and responding to a growing epidemic of cardiovascular disease [CVD] inclusive of predominant forms of heart disease) to not consider if there are other factors driving high-levels of morbidity and mortality within cardiac populations worldwide. As noted in previous chapters, non-conventional risk factors such as exposure to 'non-optimal' temperature variations (Burkart et al. 2021) and air pollution (American Lung Association 2024; Al-Kindi et al. 2020) have been added to the list of 'conventional' risk factors such as hypertension, dyslipidaemia and metabolic disorders as important antecedents for heart disease. However, if you (the reader) were to reflect on your own clinical practise or the approach to public health policies, would you immediately recognise them being acknowledged, assessed and addressed? Moreover, would you consider it the clinician's role to proactively address these drivers of poor cardiovascular health from a primary or secondary prevention perspective? Or is that the job of governments to set aspirational targets and climate change mitigation measures and policies far beyond individual patient treatment and management? To place the 'do the same old thing' into context, it has been estimated that by 2035, 130 million people (representing 45.1% of the adult population) in the United States will have developed cardiovascular disease (CVD). At today's prices this patient population will generate a burden of ~$US750billion/annum in health care expenditure (Martin et al. 2024).

This chapter provides truly 'embryonic' attempts to drag the 'climate change and provocation to health narrative' from the type of large-scale population reports described in Chap. 5, down to the level of individualised, tailored management of at-risk men and women hospitalised for the common manifestations (at least within Australia's ethnically diverse population and other high-income countries) of multimorbid heart disease. These are based on the holistic framework/model of vulnerability to resilience presented in Chap. 4. This attempt to change the narrative and paradigm of how clinicians perceive the influence of climatic provocations to health are informed by pilot research of the climatic influence on the health outcomes of multiple cohorts participating in a series of disease management trials (Ball et al. 2013; Carrington et al. 2013; Chan et al. 2023; Gao et al. 2021; Stewart et al. 2011, 2012, 2015a, b) and then a prospective, health services trial that was sadly 'cruelled' by the COVID-19 pandemic (Stewart et al. 2024), but still provides important insights into future directions for health services and research in this area. Critically, at the time of writing the final stages of this book, the primary outcomes of the trial had not been published. Thus, only the insights gained from the public presentation of the trial cohort, type of interventions and influence on seasonal transitions will be provided at this point. Notwithstanding all the caveats around trial reports in terms of selection biases and often frustrating inability to transfer learnings/findings straight into real-world clinical practise (Oertelt-Prigione and Turner 2024), it is the author's sincere hope that many of the findings reported here, will resonate with clinicians reading and reviewing this chapter.

## 8.2  Seasonal 'Frequent Flyers'

One of the key questions one must ask in assessing epidemiological data on hospital admissions and deaths due to cardiovascular disease and then linked to climatic conditions, is—would that have happened if someone is optimally managed and treated? IF yes—why bother with any new approach, simply apply gold-standard evidence. IF no—then we probably need to do something different (i.e. beyond what we've already know works normally)!

It was on this premise that we brought together meticulously curated profiling and outcome data from a series of disease management trials that applied the latest standards of clinical care and beyond [given mostly positive, but never negative, outcomes from the new models of care we applied (Ball et al. 2013; Carrington et al. 2013; Chan et al. 2023; Gao et al. 2021; Stewart et al. 2011, 2012, 2015a, b)]. Based on the different phenotypes outlined in Chap. 4 and reported outcomes, our natural focal point were people hospitalised with heart disease and evidence of multimorbidity (Murphy et al. 2004; Bi et al. 2011; Hatvani-Kovacs et al. 2016; Stewart et al. 2016; Forman et al. 2018; Vardeny et al. 2018; Chen et al. 2019; Dewan et al. 2023; Goodwin et al. 2023; Lang et al. 2024). The specific aim of this retrospective study was to determine the prevalence and characteristics of 'seasonality' (climatic-associated variations) in a large, real-world cohort of patients hospitalised with a combination of coronary artery disease (CAD), heart failure (HF) and/or chronic forms of atrial fibrillation (AF) who had been followed-up for at least 12 months and subject to minimum standards of gold-standard care/treatment—the working hypothesis being that there would negligible evidence (<5% variation) of rises and falls in hospital admissions and deaths within this diverse cohort.

As subsequently reported (Loader et al. 2019) and demonstrated in Fig. 8.1, even when adjusting for different exposure times (to different seasons and climates, given studies were conducted

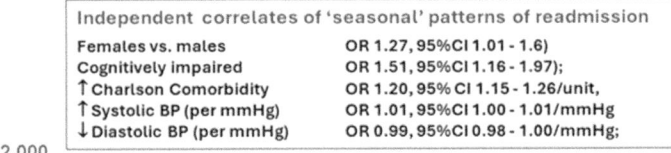

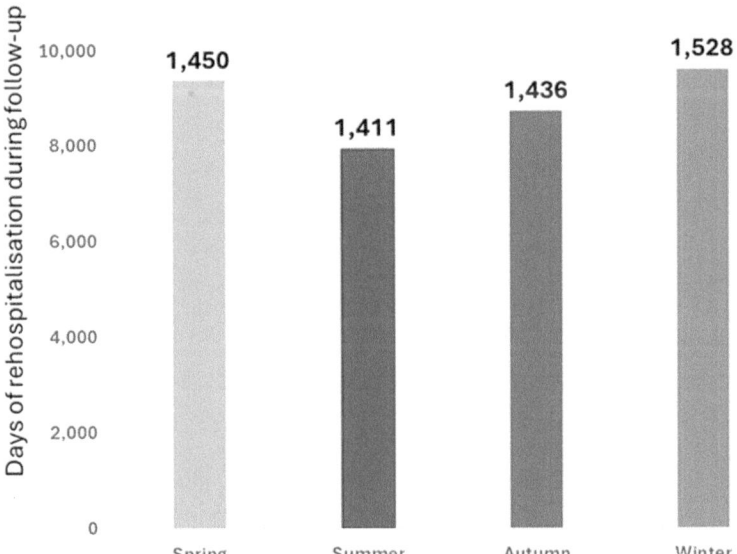

**Fig. 8.1** Pattern of seasonally linked rehospitalisation in people with multimorbid heart disease. This graph shows the total number of readmissions (above each bar) and related hospital stay (in days) according to the season of follow-up in 1598 people discharged from hospital with chronic forms of heart disease and typical patterns of comorbidity. The independent correlates of pre-specified seasonality were derived from multiple logistic regression analyses (expressed as adjusted odds ratios) of a comprehensive range of socio-economic and clinical profiling variables

across 5 different Australian cities >2000 km apart), there was a, now predictable, pattern of seasonal influences on outcomes evident among the 1598 disease-management trial patients (39% female) who were aged $70 \pm 12$ years and with a combination of CAD (58%), HF (54%), AF (50%) and multimorbidity (all) at their index admission. For our analyses, seasonality was prospectively defined as $\geq 4$ hospital admissions for any cause AND >45% of related bed-days occurring in any one season (winter, spring, summer or autumn) during median 988 (IQR 653, 1394) days follow-up, with addition analyses according to the specific month (not presented but confirmatory of the results presented below) (Loader et al. 2019).

According to our prospectively applied criteria, we had confidence in rejecting our preliminary hypothesis (of no climatic influences on outcomes), given that 29.9% of trial participants displayed a clear pattern of seasonality. Further supporting our supposition that these 'vulnerable' individuals are at high-risk for poor outcomes, regardless of their exposure to high-level care/treatment, they were more than twofold more likely to die (adjusted HR 2.16, 95% CI 1.60–2.90). Critically, the highest and lowest number of deaths occurring during spring (31.7%) and summer (19.9%), respectively. It is important to note the realistic proposition that after a winter of cold provocations, vulnerable individuals die shortly thereafter when climatic conditions are more volatile (i.e. in spring) and this phenomenon is consistent with strong evidence of a more cold-related delayed response as opposed to a more immediate heat-related response discussed in earlier chapters. Critically, none of the specific cardiac conditions influenced 'when' a trial participant was readmitted or died. Rather, as also shown in Fig. 8.1, it was the type of personal to physiological factors that appeared to modulate who was more likely to have patterns of ill-health more closely associated with the weather (Loader et al. 2019).

## 8.3    The RESILIENCE Trial

Any retrospective analysis is going to have its flaws. It is for this reason that these pilot data informed a more definitive, single-centre, 'proof-of-concept' trial, that was initially designed to provide much more definitive evidence that it was possible to effectively protect against climatic vulnerability via multifaceted intervention. Unfortunately, despite being completed as planned, much of the trial findings and outcomes described below have to be considered within the context of the COVID-19 pandemic and the fact it was conducted in one of the most locked-down cities in the world—Melbourne in south-eastern Australia (Auton and Sturman 2022; Saul et al. 2022). This meant the trial inevitably fell short in the number of recruited participants overall, had uncontrolled imbalances between groups because of the stop-start nature of the recruitment periods (with more people diagnosed with chronic lung disease in the intervention group) and the intervention (including home visits) was restricted for many patients. Nevertheless, this chapter focusses on the baseline profile of the competitively funded (NHMRC of Australia—GNT1135894) and Medical Research Future Fund of Australia— MRF1175865 **RE**silience to **S**easonal **IL**lness and **I**ncreased **E**mergency admissio**N**s **C**ar**E** (**RESILIENCE**) Trial. Complying with the SPIRIT 2013 Statement (Chan et al. 2013), the trial design/protocol is registered online and will be updated (NCT04614428) when the full

results are published. It also focuses on what type of multifaceted intervention was applied based on the evidence uncovered as part of a major review of the relevant literature and model of climatic vulnerability to resilience published in *Nature Reviews Cardiology* (Stewart et al. 2017) and presented in Chap. 4.

The single-centre RESILIENCE Trial prospectively tested the hypothesis that—*"among patients who present to (*a 671-bed tertiary hospital with specialist Internal Medicine/ Cardiology/ Coronary Care/Intensive Care units/specialists) hospital with chronic heart disease and multimorbidity, an individually tailored health intervention (the RESILIENCE Intervention Program designed to—(1) identify seasonal/environmental vulnerability and therefore risk of recurrent hospitalisation and premature mortality and then (2) coordinate a holistic plan to promote resilience to the environmental challenges of acute weather events/ predictable changes in the weather (seasonal transitions), will increase the person-focussed, composite end-point of days-alive-and out-of-hospital (DAOH) during a minimum 12-months follow-up by ≥10%, when compared to "standard care" (SC)"* (Stewart et al. 2024).

To address the first component of this hypothesis, all study participants randomised into the two arms of the trial, were subject to comprehensive profiling to determine their extent of climatic vulnerability at hospital discharge; noting that those randomised to the RESILIENCE intervention had their home environments and behaviours in the home more comprehensively assessed. The trial cohort comprised 203 hospitalised patients aged $75.7 \pm 10.2$ years and, consistent with an intention of recruiting a representative trial cohort (Oertelt-Prigione and Turner 2024), 51% were women. Consistent with our pilot research and as intended, 94% of the trial cohort were being managed/hospitalised with a combination of CAD, HF and/or AF with concurrent high-levels of multimorbidity [Charlson Index of Comorbidity Score assessed as $6.3 \pm 2.7$ (Prommik et al. 2022)]. Moreover, 8.9% had at least mild frailty according to the Rockwood

Clinical Frailty Scale (Patel et al. 2022; Ambagtsheer et al. 2019, 2020). The likelihood of this patient cohort experiencing climatically provoked events once they returned to their own homes post-discharge appeared high; even without specific profiling to confirm this. This bold statement is based on their clinical profile and especially given the broader context of—(a) past studies demonstrating strong patterns of climatically-influenced cardiovascular event rates in Melbourne (Loughnan et al. 2010, 2014; Price et al. 2023) and (b) the notoriously fickle 'four seasons in one day' weather Melbournians are routinely exposed to. On the latter point, while we did monitor air pollution levels during the trial, the Air Quality Index levels remained relatively favourable (and healthy) when compared to the ones reported and presented in Chap. 1 from other major cities around the world (World Air Quality Index 2024; Real-time 100 Most Polluted Cities in the World 2024). In specific terms, Melbourne (population of ~5 million people) has a relatively benign *Marine West Coast Climate* (mean temperature, 14 °C; range, 9–20 °C, July to January) (The Koppen Climate Classification 2024). However, during a two

and a half year period of follow-up, we counted >100 distinctive weather events (50:50 going from hot to cold or cold to hot) whereby temperatures rapidly rose and fell by 20 °C or more.

Granular profiling, based on that presented in Chap. 4 confirmed, at the very least, that there were multiple points of 'vulnerability' that could be addressed to promote greater climatic resilience, based on the pre-discharge profiling alone (see Fig. 8.2). As such, 19–20% of women and men had a pre-existing pattern of climatically linked hospitalisations (i.e. seasonality) when applying a stricter set of criteria than that applied in the pilot study described in the previous section (Loader et al. 2019). Overall, 48% of women and 40% of men had 3 or more areas of physiological vulnerability (including poor cardio-respiratory fitness, diabetes, depression, impaired cognition, anaemia and vitamin D deficiency to compound their chronic heart disease. Concurrently 15–16% of women and men had 3 or more behavioural traits of concern—including poor thermoregulatory control, poor dietary habits, excessive alcohol, inability/low physical activity levels, tobacco use and low vaccination rates. Overall, 62% of women and 61% of men

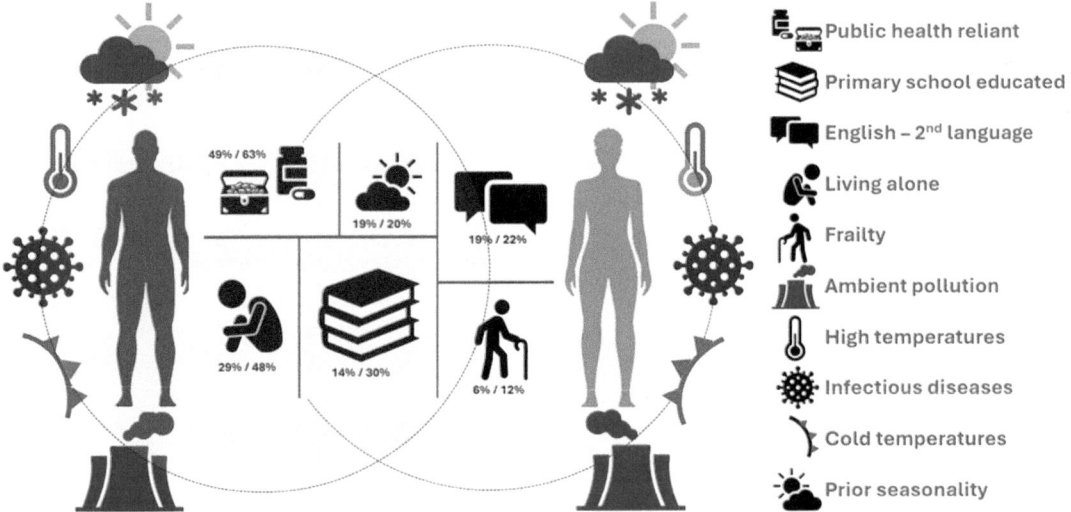

**Fig. 8.2** Vulnerability in people hospitalised with multimorbid heart disease (RESILIENCE trial cohort). Based on the holistic model/framework of climatic vulnerability to resilience presented in Chap. 4, this figure summarises some of the key areas of vulnerability found within the RESILIENCE trial cohort of men (left-sided figures) and women (right-sided figures) with heart disease and multimorbidity, that could (and should) be addressed to prevent further hospitalisations and/or death (Stewart et al. 2024)

were considered high-risk/vulnerable because of multiple factors being evident from a bio-behavioural perspective (Stewart et al. 2024). For ethical reasons, pre-discharge profiling, resulted in up to a third of patients (from both groups) receiving iron studies/replacement therapy, vitamin D supplements and commencing thyroid therapy.

In this context (of finding many points of intervention), as reported in the initial presentation of the trial results, subsequent home visits, where a physical assessment of the housing environment was performed, further revealed that around 5% were living in conditions leaving them physically exposed to weather extremes and up to 25% had a combination of issues relating to poor weather/climatic awareness, a lack of socio-economic resources to implement any recommended changes and/or new areas of concern around their behavioural patterns. Similarly, although the trial intervention will be more fully reported in the primary outcome trial report, it is also worth considering that key components of the RESILIENCE Intervention (overseen by a qualified nurse and Cardiologist) via a dedicated review clinic and follow-up included the following—(a) climatic and weather-based education and goal setting (all), (b) SMS weather alerts, (c) adjustment of pharmacological therapy (~50%), (d) help with securing home heating/cooling subsidies (~50%), organising additional vaccinations (>50%) and vitamin D supplements (~10%).

## 8.4  Which Components of Management Address Vulnerability?

In reviewing the preliminary results of RESILIENCE, the critical questions, at this point, are—(1) Do the additional profiling characteristics we generated for this trial really matter (in terms of influencing subsequent health outcomes), and (2) If could they be routinely detected and addressed, would that result in better outcomes?

Although the full details and outcomes of the RESILIENCE Trial cannot be revealed at this point, Fig. 8.3 shows the results of multiple logistic regression of all baseline profiling variables to determine if there were different correlates of 'when' (in this case during the specific climatic transitions of autumn-to-winter, winter-to-spring, spring-to-summer and summer-to-autumn as part of the cyclical nature of daylight hours/seasonal change experienced by Melbournians) trial participants were readmitted to hospital or died. As can be seen, this was indeed the case thereby providing a definitive answer to the first question posed above and with some interventional effect also evident relative to standard care (providing encouragement beyond/regardless of the headline results of the trial). Some of the more highly influential factors would only have been revealed by our purposeful profiling and this will be a strong recommendation arising from the RESILIENCE Trial—to ensure that a broader range of factors are documented and considered when managing people living with chronic heart disease and multimorbidity.

In the context of this book, beyond the rich amount of profiling data generated by this unique trial (to my/our knowledge, no other multifaceted intervention of this kind has been tested in a prospective trial previously) some important climatic/health 'touchpoints' (many of which are highlighted in Figs. 8.2 and 8.3) are worthy of discussion, relative to the current evidence/literature around preventing climatically provoked health events. In reviewing these points, it's important to note that while there have numerous studies (and even trials) of singular interventions, there remains a dearth of multi-faceted and coordinated strategies that address the complexity of issues encompassed by climatic vulnerability.

- **Supporting People with Limited Socio-economic Resources**: Beyond the social and health inequalities highlighted in Chaps. 1 and 7, there is a wealth of evidence that socio-economic status is a major determinant

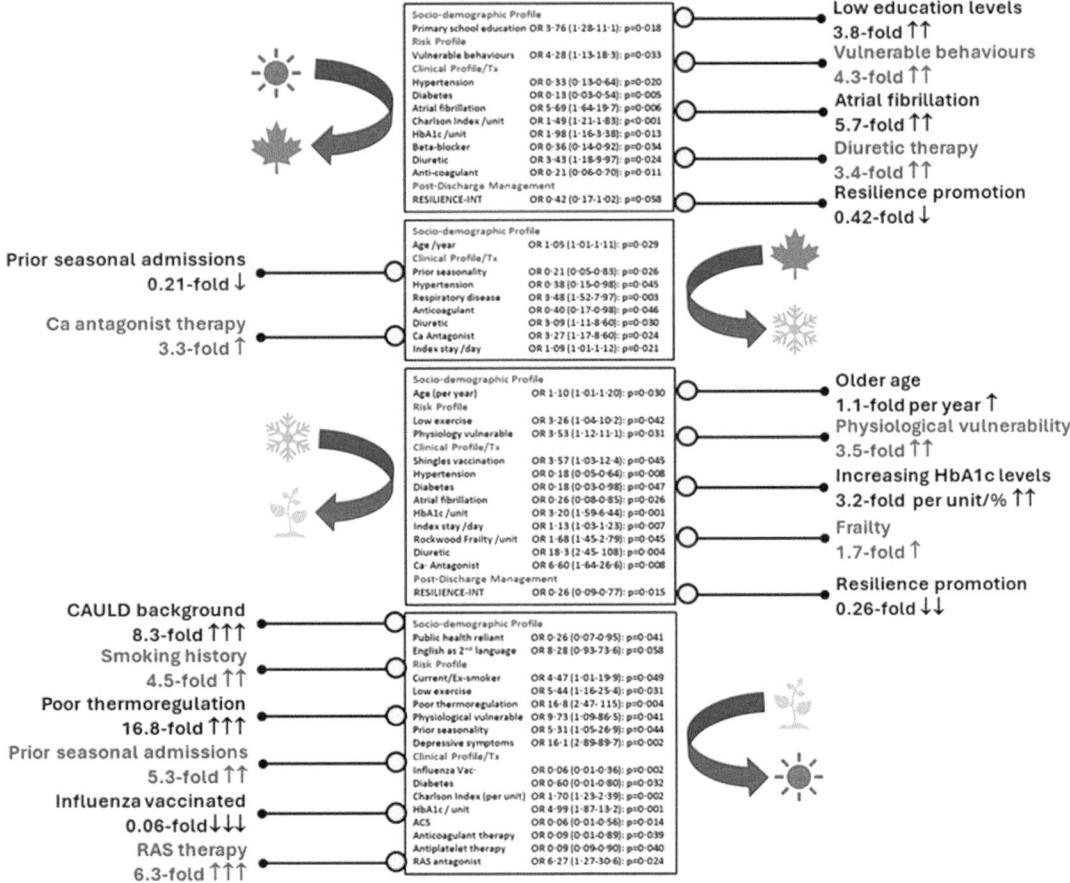

**Fig. 8.3** Correlates of events during different climatic transitions during the RESILIENCE trial follow-up. These data highlight the differential pattern of clinical, socio-demographic and climatic factors measured as part of the RESILIENCE trial, that independently correlated with hospital episodes and deaths spanning contiguous seasonal transitions—from top to bottom, these transitions are summer-to-autumn, autumn-to-winter, winter-to spring and spring-to-summer

of health outcomes (Aimo and Maggioni 2023; Asada et al. 2023; Banwell et al. 2012; Dioun et al. 2023; Ge et al. 2022; Gutman et al. 2019; Hessel 2021; McAlister et al. 2004). Essentially, poorer people have far worse health outcomes than their much fewer and richer counterparts—especially in the context of the development and consequences of chronic heart disease. One of the key issues (especially in relation to the type of intervention we were trialling as part of the RESILIENCE Trial) is the inability of people to make and fund healthy choices. In Europe, it's estimated that around 1 in 10 people are living with 'energy poverty'—as

defined by the European Commission (2016) this is *"the inability to afford basic energy services (heating, cooling, lighting, mobility and power) to guarantee a decent standard of living due to a combination of low income, high energy expenditure and low energy efficiency of homes"* (https://energy-poverty.ec.europa.eu/what-energy-poverty_en). As argued by the WELLBASED Investigators (Stevens et al. 2022), most interventional studies to address the problem have adopted a singular approach (Fenwick et al. 2013). To date, the results of the WELLBASED Intervention study, focussing on the implementation and evaluation of "*a comprehensive urban programme, based on the social-ecological model, to reduce energy poverty and its effects on the citizens' health and wellbeing in six European urban study sites*" is yet to be reported. However, the investigators have adopted to apply a multifaceted approach that is consistent (in a much narrower scope) with the RESILIENCE Intervention (Stewart et al. 2024). Overall, the potential benefits of this logical approach to supporting people to attain and maintain thermoregulatory control are yet to be realised (for a multitude of reasons), whilst remaining a critical barrier to achieving better health outcomes (Fefferman et al. 2021). In countries like Australia (subsidised health care with a public health to private health cover divide) and the USA (where many people rely on a job to have adequate health cover), the capacity to make health choices in terms of preventative care (this extends to dental health) is highly dependent on a person's wealth. This 'divide' was evident with the RESILIENCE Trial cohort.

- *Focussing on Health Literacy*: Although not routinely considered important in clinical practise (a quick PubMed search of heart disease and health literacy reveals only 27 relevant papers) or even dependent on a person's education level (although we used this as a surrogate in RESILIENCE), 'health literacy' is an important determinant of health outcomes (Dinges et al. 2022; Bonner et al. 2022; Peltzer et al. 2020; Magnani et al. 2018; Rowlands et al. 2013; Ibrahim et al. 2008). As defined by the American Heart Association (who noted the historical lack of integrated efforts to address health literacy to improve cardiovascular health outcomes), it is the degree to which someone is able to access (and then process) health information and services—thereby enabling them to participate in health-related decisions (Magnani et al. 2018). In a recent systematic review the following conclusions were made— "*patients with lower HL* (health literacy) *feel less capable to perform lifestyle changes,* *exhibit fewer proactive coping behaviors, are more likely to deny CHD, are generally older, are less often employed, have lower educational levels and lower socioeconomic status, experience faster physical decline, and use the healthcare system less, compared to patients with higher HL*" (Peltzer et al. 2020). Although we made a concerted effort to overcome the barriers of health literacy in RESILIENCE, it was often difficult to do and we had many participants with the complicating factor of coming from culturally and diverse language backgrounds that required additional attention.

- *Addressing Social Isolation*: Social isolation comes in many forms, and in the many studies of disease management I've led (typically focussing on older people with chronic heart disease), living alone is a common trait. Unsurprisingly, given the reported literature (Lee and Singh 2021; Smith et al. 2021; Gronewold et al. 2021; Williams 2016) it is typically associated with poor outcomes, despite out best efforts to intervene and provide more social connections and psychological support. Of course, not everyone who lives alone is socially isolated—like everything discussed in this book, it's more complicated than that! Nevertheless, indicative of its adverse impact on people of all ages, in a large North American study of close to 400,000 people—"*the age-adjusted all-cause mortality risk was 45% higher (hazard ratio [HR] 1.45; 95% confidence interval [CI] 1.40–1.50) and the heart disease mortality risk was 83% higher (HR 1.83; 95% CI 1.67–2.00) among adults aged 18–64 years living alone at the baseline, compared with adults living with others*" (Lee and Singh 2021). It would come as no surprise that during the RESILIENCE Trial (with Melbourne in long periods of lock-down) we struggled to connect with our participants and many people suffered long bouts of social isolation that was undoubtedly detrimental to their health and eroded the potential benefits of higher levels of (often face-to-face) support.

- ***Recognising Frailty***: Of increasing focus (Shen et al. 2024; Tajik et al. 2023a, b; Dovjak 2022; McKechnie et al. 2021), frailty (which can be defined as an age-associated clinical syndrome characterised by decrease physiological and psychological reserves, particularly in times of stress [e.g. operations, infections and/or acute illness]) is a condition/state of increasing interest to researchers and clinicians alike. In the context of defining and framing *climatic vulnerability* it is of high interest (Dovjak 2022). The reported prevalence of frailty ranges from 9–17% among community-dwelling people aged 65 years and over living in high-income countries (Morley et al. 2013). Unsurprisingly, there are many parallels between the detection (and consequences) of frailty and vulnerability to climatic provocations to health (Dent et al. 2020).

- ***Ensuring People are Vaccinated***: As highlighted in this chapter (particularly in the context of low-to-middle income countries) the nexus between climatic conditions, infections affecting the respiratory/cardiopulmonary system and the heart directly is a broad one with many intersecting points. Vaccinations are a critical preventative strategy that are both proven and tragically rejected by many people—the COVID-19 pandemic polarising the public in this respect (Saul et al. 2022; Jassat et al. 2021; Nogareda et al. 2023; Yeow et al. 2024). In a major multinational, blinded, randomised trial, Loeb and colleagues reported that among—"*5129 participants and randomly assigned (1:1) 2560 (50.0%) to influenza vaccine and 2569 (50.0%) to placebo*" no difference in the primary outcomes were observed, nor was there a group difference in mortality. However, for the "*secondary outcomes of all-cause hospitalisations (HR 0.84 [95% CI 0.74–0.97]; p=0.013) and pneumonia (HR 0.58 [0.42–0.80]; p=0.0006) were significantly reduced in the vaccine group compared with the placebo group*" (Loeb et al. 2022). Once again, this (vaccination with the influenza vaccine and other important evidence-based vaccines) was a

major focus/component of the RESILIENCE Intervention.

- ***Counselling People to Wear Masks, Wash Hands and Socially Isolate***: In one of the great ironies (given how adversely affected it was by the pandemic), while the RESILIENCE Intervention was designed before COVID-19 was discovered, the study intervention had a major focus (along with vaccinations) on infection control measures that are now standard practise due to the pandemic (Auton and Sturman 2022; Yang et al. 2024). We will never know, what impact these measure might have had on the pattern of pre-pandemic respiratory infections and their nexus with CVD. However, with increasing public complacency to these infection control measures, this combination of strategies may become very important again in helping people deemed to be climatically vulnerable become more resilient.

- ***Addressing High-Risk Behaviours***: In earlier chapters the case of (mainly men) suffering cardiac events performing the simple act of shovelling snow was described (Nichols et al. 2012). Not surprisingly, this was not a problem for an Australian cohort (snow being at a premium in a limited number of high places during limited times of the year). This highlights the highly contextual nature of finding common behaviours that trigger cardiac events in certain populations under certain climatic conditions. Nevertheless, in Australia and our cohort, venturing out during times of even relatively high temperatures was a major issue. Thus, a major component of education and intervention (relatively cost-free) was to ensure vulnerable individuals reorganised any social outings, trips or outdoor activities to cooler times of the day when required (i.e. when temperature and humidity levels were rising).

- ***Providing Weather Alerts***: Very much related to the above, was our provision of weather alerts, although the nature and relationship of these alerts (based on the findings of the RESILIENCE Trial) will need to be heavily modified. Nevertheless, heat alert systems for Australia (Nicholls et al. 2008) have been developed and routinely applied—noting the observation that this becomes complicated because "*some days with very high maximum temperatures will be affected by the passage of cool changes and cold fronts in the afternoon, leading to a rapid drop in temperature*" (Nicholls et al. 2008). Largely ignored, this very definition of 'volatile' weather conditions is likely to drive many more cardiac events (as revealed by impending reports from RESILIENCE and the population data presented in Chap. 5). In other countries such as the USA, studies of the National Weather Service alerts for extreme heat were not found to have been associated with a reduction in all-cause mortality (Weinberger et al. 2018). In the UK, the "*UK Health Security Agency (UKHSA) Weather-Health Alert system is aimed at health and social care professionals and anyone with a role in reducing health impacts caused by extended periods of hot or cold weather*" based on a traffic-light system (UK Health Service Agency 2024). Once again, the issue of dynamic and/or volatile changes in the weather are unlikely to attract an alert. However, these will become a major feature of climate change.

When considering all of the above and the different ways they can be applied to a single person, one of the major challenges in unpacking the findings from the RESILIENCE Trial (and any other multifaceted intervention trials) will be to determine what works in terms of managing people in whom there is strong evidence of climatically driven events. Further complicating the picture, even a careful analysis of cardiovascular related versus non-cardiovascular events may struggle to delineate between cause and effect of any intervention. Ultimately, we will need to decide what should be strengthened/added or even culled to improve health outcomes as a matter of the 'art and science' of clinical management of climatic vulnerability, as opposed to basing it on a strict 'dose–response' basis. It is here where robust health economic analyses will become extremely important.

## 8.5  Pharmacotherapy for Cardiovascular Resilience—But Which One?

One of the more controversial aspects around the management of people displaying vulnerability to climatic conditions is the role of pharmacotherapy. The natural assumption, and indeed a core component of 'resilience' described in Chap. 4, is the optimisation of pharmacotherapy, with the assumption that device-based therapy (such as pacemakers, cardiac synchronisation and automated defibrillators, along with implanted arrhythmia monitoring) will also be applied from an evidence-based perspective (Konemann et al. 2023; McDonagh et al. 2021). However, unpublished data (unfortunately due to commercial sensitivities no data can be presented a this time) shows that the typical peaks and troughs in HF events shown in Chap. 5 persist, even in the most well-managed clinical trial populations on a world-wide basis. These results mirror those of our own in terms of optimal disease management, including the optimisation of pharmacotherapy (Loader et al. 2019).

With the RESILIENCE Trial, there was a clear intention to optimise pharmacotherapy to match the seasonal transitions in weather conditions and this was performed by a Cardiologist applying the new 'paradigm' of management advocated in this book. However, while the application of treatments such as anti-platelet and anti-coagulation therapy appeared to 'protective' from a seasonal/climatic perspective, diuretic and calcium antagonist therapy did not appear. Such findings need to be interpreted with caution, as this may well reflect the underlying conditions being treated. However, it makes sense that anyone requiring diuresis (due to cardiac overload/high filling pressures) and/or a vasodilator will have a blunted response to thermoregulation. As such, the concept of undertaking trials to both optimise treatment and then adjust to climatic provocations (that might run as interference to the therapeutic goals of treatment) is a logical one. Unfortunately, until major clinical trials (as funded by big pharmaceutical companies) start counting 'when' people have events not just how many, this is unlikely to happen anytime soon.

The bed-rock of managing anyone with hypertension to left ventricular systolic dysfunction has been agents that modulate the renin-angiotensin aldosterone system (RAAS) (Jhund et al. 2014a; Lee et al. 2023; McMurray et al. 2005; McMurray 2001). However, this system is crucial to the maintenance of homeostasis in the context of both cold- and heat-stress responses (Hiramatsu et al. 1984; Kosunen et al. 1976). The recent, positive results of the multicentre, multinational Safety, Tolerability and Efficacy of Rapid Optimization, Helped by N-Terminal Pro-Brain Natriuretic Peptide Testing of Heart Failure Therapies (STRONG-HF) Trial (Celutkiene et al. 2024; Pagnesi et al. 2024; Cotter et al. 2024; Mebazaa et al. 2022; Kimmoun et al. 2019) may well help to allay fears (at least among people hospitalised with acute HF) that RAAS therapy is anything but positive in the context of climatic challenges. As reported by Mebazza and colleagues—*"the high-intensity care group had been up-titrated to full doses of prescribed drugs (renin-angiotensin blockers 278 [55%] of 505 versus 11 [2%] of 497; beta blockers 249 [49%] versus 20 [4%]; and mineralocorticoid receptor antagonists 423 [84%] versus 231 [46%])"* and this was associated with—*"Heart failure readmission or all-cause death up to day 180 occurred in 74 (15.2% down-weighted adjusted Kaplan–Meier estimate) of 506 patients in the high-intensity care group and 109 (23.3%) of 502 patients in the usual care group (adjusted risk difference 8.1% [95% CI 2.9–13.2]; p = 0.0021; risk ratio 0.66 [95% CI 0.50–0.86])"* (Mebazaa et al. 2022). Of course, whether this would be similar for those not presenting with acute HF is not known; noting higher levels of adverse events with more aggressive therapy. Nor do we know if this approach modulated any climatic-driven variations in outcomes—once again, something that remains under-reported (i.e. when events occur) and of little interest to those leading major clinical trials.

We do at least know that the landscape of those presenting with HF associated with 'preserved' systolic function [including many women with an EF >50% but significant diastolic dysfunction (Playford and Strange 2021; Playford et al. 2021; Stewart et al. 2021)] has been radically altered with the introduction of the angiotensin-neprilysin-inhibition across the spectrum of systolic-to-diastolic dysfunction (Mc Causland et al. 2020, 2022; Docherty et al. 2021a, b; Solomon et al. 2019; Damman et al. 2018; Jhund et al. 2014b). As indicated previously, this new therapy has potentially transformed the management of many women presenting with the syndrome (Lam and Myhre 2023; Ravera et al. 2022; Zaman et al. 2021; Jin et al. 2020; McMurray et al. 2020; Beale et al. 2018). However, whether this translates to greater 'resilience' in the face of climate change among both men and women with the syndrome and various forms of left ventricular dysfunction (noting the discussion around pulmonary hypertension in the previous chapter) remains unknown.

Finally, in the context of previously described, high-levels of diabetes and associated cardio-metabolic disease along with rising obesity rates worldwide along (noting paradoxical rises in malnutrition at the other end of the spectrum) (Belancic et al. 2020; Chong et al. 2023; Tan et al. 2023; Gona et al. 2024; McKenzie et al. 2024; Tumas and Lopez 2024), it would be remiss not to consider the evidence (and hype) surrounding sodium-glucose transporter 2 (SGLT-2) inhibitors and newer agents within that class (Peng et al. 2022; Vaduganathan et al. 2022). Originally developed as a specific 'metabolic' treatment for people with diabetes (Talha et al. 2023), this class of agents has rapidly been applied to a broad spectrum of cardiovascular conditions (Mavrakanas et al. 2023; Rivera et al. 2023) and, of course, as a 'weight-loss' drug. The latter (weight loss) as one of (presumably) a multi-system effect of SGLT-2 has stimulated more intense (remembering that 'weight loss in a pill' has always been the holy grail for pharma companies!) in other agents such as the Glucagon-Like Peptide 1 (GLP-1) inhibitors. In a recent head-to-head study reported by Wehrman and colleagues, among people with Type 2 diabetes, those treated with GLP-1 inhibition—*"averaged a 0.65% reduction in A1c compared to a 1.05% reduction in the SGLT-2 group (P = 0.1397)"* and the resulting weight reduction was—*"4.1 kg in the GLP-1 group compared to 3.6 kg in the SGLT-2 group (P = 0.6993)"* (Wehrman et al. 2024). Of course, issues such as the 'obesity paradox' [i.e. the potential for protective effects against premature mortality in older people who have already developed CVD but not younger people in whom this is yet to occur (Chen et al. 2024a, b; Reinhardt et al. 2024; Alebna et al. 2024)] and specific issues around the mechanisms of action (Lopez-Candales et al. 2023; La Grotta et al. 2022; Georgianos et al. 2021; Li et al. 2020) of this new class of agent (particularly around rebound weight gain/sustainable weight loss/fat versus muscle mass loss), are yet to be resolved (Cheong et al. 2022). Nevertheless, it is be hoped that the net effect of more people with greater cardio-metabolic health as a result of reduced weight and a parallel ability to improve their cardio-respiratory fitness (Kokkinos et al. 2023; Williamson et al. 2021; Safdar and Mangi 2020; Ortega et al. 2010; Fulton et al. 2009), will mean a more 'resilient' population in the face of climate change.

## 8.6 Still Much More to Unlearn and Re-learn!

If the (preliminary) results of the RESILIENCE Trial tell us anything, it is that changing the course of those who have already become vulnerable to provocative climatic conditions will be very difficult. Regardless, the attempt needs to be made and further research and trials are needed in this space. In parallel to the efforts to refine and improve the multifaceted intervention applied in the RESILIENCE Trial, there needs to be greater efforts to preserve levels of *climatic resilience* in people who are yet to suffer a major cardiac event. This will undoubtedly require a change in mind-set for may clinicians (essentially discarding redundant thinking about the 'static' state of heart disease and embracing the concept of dynamic clinical fluctuations in response to prevailing climatic conditions). The next chapter will focus on how that might be achieved, with a major focus on the role of primary care in, perhaps, delivering what tertiary care cannot.

## References

Aimo A, Maggioni AP. The burden of socio-economic inequalities on etiology, management and outcomes of patients with heart failure: the G-CHF registry. G Ital Cardiol. 2023;24:767–8.

Alebna PL, Mehta A, Yehya A, et al. Update on obesity, the obesity paradox, and obesity management in heart failure. Prog Cardiovasc Dis. 2024;82:34–42.

Al-Kindi SG, Brook RD, Biswal S, Rajagopalan S. Environmental determinants of cardiovascular disease: lessons learned from air pollution. Nat Rev Cardiol. 2020;17:656–72.

Ambagtsheer RC, Beilby JJ, Visvanathan R, Dent E, Yu S, Braunack-Mayer AJ. Should we screen for frailty in primary care settings? A fresh perspective on the frailty evidence base: a narrative review. Prev Med. 2019;119:63–9.

Ambagtsheer RC, Beilby J, Seiboth C, Dent E. Prevalence and associations of frailty in residents of Australian aged care facilities: findings from a retrospective cohort study. Aging Clin Exp Res. 2020;32:1849–56.

American Lung Association. State of the air report: health impact of air pollution; 2024. https://www.lung.org/research/sota/health-risks. Accessed May 2024.

Asada Y, Grignon M, Hurley J, Stewart SA, Smith NK, Kirkland S, McMillan J, Griffith LE, Wolfson C, et al. Trajectories of the socioeconomic gradient of mental health: results from the CLSA COVID-19 questionnaire study. Health Policy. 2023;131:104758.

Auton JC, Sturman D. Individual differences and compliance intentions with COVID-19 restrictions: insights from a lockdown in Melbourne (Australia). Health Promot Int. 2022;37:987.

Ball J, Carrington MJ, Wood KA, Stewart S, Investigators S. Women versus men with chronic atrial fibrillation: insights from the standard versus atrial fibrillation specific management study (SAFETY). PLoS ONE. 2013;8:e65795.

Banwell C, Dixon J, Bambrick H, Edwards F, Kjellstrom T. Socio-cultural reflections on heat in Australia with implications for health and climate change adaptation. Glob Health Action. 2012;5:687.

Beale AL, Meyer P, Marwick TH, Lam CSP, Kaye DM. Sex differences in cardiovascular pathophysiology: why women are overrepresented in heart failure with preserved ejection fraction. Circulation. 2018;138:198–205.

Belancic A, Klobucar Majanovic S, Stimac D. The escalating global burden of obesity following the COVID-19 times: are we ready? Clin Obes. 2020;10:e12410.

Bi P, Williams S, Loughnan M, Lloyd G, Hansen A, Kjellstrom T, Dear K, Saniotis A. The effects of extreme heat on human mortality and morbidity in Australia: implications for public health. Asia Pac J Public Health. 2011;23:27S – 36.

Bonner C, Batcup C, Ayre J, Cvejic E, Trevena L, McCaffery K, Doust J. The impact of health literacy-sensitive design and heart age in a cardiovascular disease prevention decision aid: randomized controlled trial and end-user testing. JMIR Cardio. 2022;6:e34142.

Burkart KG, Brauer M, Aravkin AY, Godwin WW, Hay SI, He J, Iannucci VC, Larson SL, Lim SS, Liu J, Murray CJL, Zheng P, Zhou M, Stanaway JD. Estimating the cause-specific relative risks of nonoptimal temperature on daily mortality: a two-part modelling approach applied to the global burden of disease study. Lancet. 2021;398:685–97.

Carrington MJ, Ball J, Horowitz JD, Marwick TH, Mahadevan G, Wong C, Abhayaratna WP, Haluska B, Thompson DR, Scuffham PA, Stewart S. Navigating the fine line between benefit and risk in chronic atrial fibrillation: rationale and design of the standard versus atrial fibrillation specific management study (SAFETY). Int J Cardiol. 2013;166:359–65.

Celutkiene J, Cerlinskaite-Bajore K, Cotter G, Edwards C, Adamo M, Arrigo M, Barros M, Biegus J, Chioncel O, et al. Impact of rapid up-titration of guideline-directed medical therapies on quality of life: insights from the STRONG-HF trial. Circ Heart Fail. 2024;17:e011221.

Chan AW, Tetzlaff JM, Gotzsche PC, Altman DG, Mann H, Berlin JA, Dickersin K, Hrobjartsson A, Schulz KF, Parulekar WR, Krleza-Jeric K, Laupacis A, Moher D. SPIRIT 2013 explanation and elaboration: guidance for protocols of clinical trials. BMJ. 2013;346:e7586.

Chan YK, Stickland N, Stewart S. An inevitable or modifiable trajectory towards heart failure in high-risk individuals: insights from the nurse-led intervention for less chronic heart failure (NIL-CHF) study. Eur J Cardiovasc Nurs. 2023;22:33–42.

Chen L, Chan YK, Busija L, Norekval TM, Riegel B, Stewart S. Malignant and benign phenotypes of multimorbidity in heart failure: implications for clinical practice. J Cardiovasc Nurs. 2019;34:258–66.

Chen Y, Koirala B, Ji M, Commodore-Mensah Y, et al. Obesity paradox of cardiovascular mortality in older adults in the United States: a cohort study using 1997–2018 national health interview survey data linked with the national death index. Int J Nurs Stud. 2024a;155:104766.

Chen QF, Ni C, Katsouras CS, Liu C, Yao H, Lian L, Shen TW, Shi J, Zheng J, Shi R, Yujing W, Lin WH, Zhou XD. Obesity paradox in patients with acute coronary syndrome: is malnutrition the answer? J Nutr. 2024b;116:45871.

Cheong AJY, Teo YN, Teo YH, Syn NL, Ong HT, Ting AZH, Chia AZQ, Chong EY, et al. SGLT inhibitors on weight and body mass: a meta-analysis of 116 randomized-controlled trials. Obesity. 2022;30:117–28.

Chong B, Jayabaskaran J, Kong G, Chan YH, Chin YH, Goh R, et al. Trends and predictions of malnutrition and obesity in 204 countries and territories: an analysis of the global burden of disease study 2019. EClinicalMedicine. 2023;57:101850.

Cotter G, Deniau B, Davison B, Edwards C, Adamo M, Arrigo M, Barros M, Biegus J, Celutkiene J, et al. Optimization of evidence-based heart failure medications after an acute heart failure admission: a secondary analysis of the STRONG-HF randomized clinical trial. JAMA Cardiol. 2024;9:114–24.

Damman K, Gori M, Claggett B, Jhund PS, Senni M, Lefkowitz MP, et al. Renal effects and associated outcomes during angiotensin-neprilysin inhibition in heart failure. JACC Heart Fail. 2018;6:489–98.

Dent E, Ambagtsheer RC, Beilby J, Stewart S. Editorial: frailty and seasonality. J Nutr Health Aging. 2020;24:547–9.

Dewan P, Ferreira JP, Butt JH, Petrie MC, Abraham WT, Desai AS, Dickstein K, Kober L, et al. Impact of multimorbidity on mortality in heart failure with reduced ejection fraction: which comorbidities matter most? An analysis of PARADIGM-HF and ATMOSPHERE. Eur J Heart Fail. 2023;25:687–97.

Dinges SM, Krotz J, Gass F, Treitschke J, Fegers-Wustrow I, Geisberger M, Esefeld K, et al. Cardiovascular risk factors, exercise capacity and health literacy in patients with chronic ischaemic heart disease and type 2 diabetes mellitus in Germany: baseline characteristics of the lifestyle intervention in chronic ischaemic heart disease and type 2 diabetes study. Diab Vasc Dis Res. 2022;19:14791641221113780.

Dioun S, Chen L, Hillyer GC, Tatonetti NP, May BL, Melamed A, Wright JD. Association between neighborhood socioeconomic status, built environment and SARS-CoV-2 infection among cancer patients treated at a Tertiary Cancer Center in New York City. Cancer Rep. 2023;6:e1714.

Docherty KF, Campbell RT, Brooksbank KJM, Dreisbach JG, Forsyth P, Godeseth RL, et al. Effect of neprilysin inhibition on left ventricular remodeling in patients with asymptomatic left ventricular systolic dysfunction late after myocardial infarction. Circulation. 2021a;144:199–209.

Docherty KF, Campbell RT, Brooksbank KJM, Godeseth RL, Forsyth P, McConnachie A, Roditi G, et al. Rationale and methods of a randomized trial evaluating the effect of neprilysin inhibition on left ventricular remodelling. ESC Heart Fail. 2021b;8:129–38.

Dovjak P. Frailty in older adults with heart disease. Z Gerontol Geriatr. 2022;55:465–70.

European Commission E. What is Energy Poverty? https://energy-poverty.ec.europa.eu/energy-poverty-observatory/what-energy-poverty_en. Accessed May 2024.

Fefferman N, Chen CF, Bonilla G, Nelson H, Kuo CP. How limitations in energy access, poverty, and socioeconomic disparities compromise health interventions for outbreaks in urban settings. iScience. 2021;24:103389.

Fenwick E, Macdonald C, Thomson H. Economic analysis of the health impacts of housing improvement studies: a systematic review. J Epidemiol Commun Health. 2013;67:835–45.

Forman DE, Maurer MS, Boyd C, Brindis R, Salive ME, Horne FM, Bell SP, Fulmer T, Reuben DB, Zieman S, Rich MW. Multimorbidity in older adults with cardiovascular disease. J Am Coll Cardiol. 2018;71:2149–61.

Fulton JE, Simons-Morton DG, Galuska DA. Physical activity: an investment that pays multiple health dividends: comment on "combined effects of cardiorespiratory fitness, not smoking, and normal waist girth on morbidity and mortality in men," "physical activity and survival in male colorectal cancer survival," "effects of a television viewing reduction on energy intake and expenditure in overweight and obese adults," and "physical activity and rapid decline in kidney function among older adults." Arch Intern Med. 2009;169:2124–7.

Gao L, Scuffham P, Ball J, Stewart S, Byrnes J. Long-term cost-effectiveness of a disease management program for patients with atrial fibrillation compared to standard care: a multi-state survival model based on a randomized controlled trial. J Med Econ. 2021;24:87–95.

Ge Y, Zhang L, Gao Y, Wang B, Zheng X, China PCG. Socio-economic status and 1 year mortality among patients hospitalized for heart failure in China. ESC Heart Fail. 2022;9:1027–37.

Georgianos PI, Vaios V, Dounousi E, Salmas M, Eleftheriadis T, Liakopoulos V. Mechanisms for cardiorenal protection of SGLT-2 inhibitors. Curr Pharm Des. 2021;27:1043–50.

Gona P, Gona C, Ballout S, Mapoma C, Rao S, Mokdad A. Trends in the burden of most common obesity-related cancers in 16 Southern Africa development community countries, 1990–2019. Findings from the global burden of disease study. Obes Sci Pract. 2024;10:e715.

Goodwin NP, Clare RM, Harrington JL, Badjatiya A, Wojdyla DM, Udell JA, Butler J, Januzzi JL, et al. Morbidity and mortality associated with heart failure in acute coronary syndrome: a pooled analysis of 4 clinical trials. J Card Fail. 2023;10:548.

Gronewold J, Engels M, van de Velde S, Cudjoe TKM, Duman EE, Jokisch M, Kleinschnitz C, Lauterbach K, Erbel R, Jockel KH, Hermann DM. Effects of life events and social isolation on stroke and coronary heart disease. Stroke. 2021;52:735–47.

Gutman SJ, Costello BT, Papapostolou S, Iles L, Ja J, Hare JL, Ellims A, Marwick TH, Taylor AJ. Impact of sex, socio-economic status, and remoteness on therapy and survival in heart failure. ESC Heart Fail. 2019;6:944–52.

Hatvani-Kovacs G, Belusko M, Pockett J, Boland J. Can the excess heat factor indicate heatwave-related morbidity? A case study in Adelaide, South Australia. Ecohealth. 2016;13:100–10.

Hessel FP. Overview of the socio-economic consequences of heart failure. Cardiovasc Diagn Ther. 2021;11:254–62.

Hiramatsu K, Yamada T, Katakura M. Acute effects of cold on blood pressure, renin-angiotensin-aldosterone system, catecholamines and adrenal steroids in man. Clin Exp Pharmacol Physiol. 1984;11:171–9.

Ibrahim SY, Reid F, Shaw A, Rowlands G, Gomez GB, Chesnokov M, Ussher M. Validation of a health literacy screening tool (REALM) in a UK population with coronary heart disease. J Public Health. 2008;30:449–55.

Jassat W, Mudara C, Ozougwu L, Tempia S, Blumberg L, et al. Difference in mortality among individuals admitted to hospital with COVID-19 during the first and second waves in South Africa: a cohort study. Lancet Glob Health. 2021;9:e1216–25.

Jhund PS, Claggett BL, Voors AA, Zile MR, Packer M, Pieske BM, Kraigher-Krainer E, Shah AM, et al. Elevation in high-sensitivity troponin T in heart failure and preserved ejection fraction and influence of treatment with the angiotensin receptor neprilysin inhibitor LCZ696. Circ Heart Fail. 2014a;7:953–9.

Jhund PS, Claggett B, Packer M, Zile MR, Voors AA, Pieske B, Lefkowitz M, Shi V, Bransford T, McMurray JJ, Solomon SD. Independence of the blood pressure lowering effect and efficacy of the angiotensin receptor neprilysin inhibitor, LCZ696, in patients with heart failure with preserved ejection fraction: an analysis of the PARAMOUNT trial. Eur J Heart Fail. 2014b;16:671–7.

Jin X, Chandramouli C, Allocco B, Gong E, Lam CSP, Yan LL. Women's participation in cardiovascular clinical trials from 2010 to 2017. Circulation. 2020;141:540–8.

Kimmoun A, Cotter G, Davison B, Takagi K, Addad F, Celutkiene J, Chioncel O, et al. Safety, tolerability and efficacy of rapid optimization, helped by NT-proBNP and GDF-15, of heart failure therapies (STRONG-HF): rationale and design for a multicentre, randomized, parallel-group study. Eur J Heart Fail. 2019;21:1459–67.

Kokkinos P, Faselis C, Samuel IBH, Lavie CJ, Zhang J, Vargas JD, Pittaras A, et al. Changes in cardiorespiratory fitness and survival in patients with or without cardiovascular disease. J Am Coll Cardiol. 2023;81:1137–47.

Konemann H, Ellermann C, Zeppenfeld K, Eckardt L. Management of ventricular arrhythmias worldwide: comparison of the latest ESC, AHA/ACC/HRS, and

CCS/CHRS guidelines. JACC Clin Electrophysiol. 2023;9:715–28.

Kosunen KJ, Pakarinen AJ, Kuoppasalmi K, Adlercreutz H. Plasma renin activity, angiotensin II, and aldosterone during intense heat stress. J Appl Physiol. 1976;41:323–7.

La-Grotta R, Frige C, Matacchione G, Olivieri F, de Candia P, Ceriello A, Prattichizzo F. Repurposing SGLT-2 inhibitors to target aging: available evidence and molecular mechanisms. Int J Mol Sci. 2022;23:895.

Lam CSP, Myhre PL. Left ventricular ejection fraction in women: when normal isn't normal. Heart. 2023;109:1584–5.

Lang JJ, Prince SA, Merucci K, Cadenas-Sanchez C, Chaput JP, Fraser BJ, Manyanga T, McGrath R, Ortega FB, Singh B, Tomkinson GR. Cardiorespiratory fitness is a strong and consistent predictor of morbidity and mortality among adults: an overview of meta-analyses representing over 20.9 million observations from 199 unique cohort studies. Br J Sports Med. 2024;58:556–66.

Lee H, Singh GK. Social isolation and all-cause and heart disease mortality among working-age adults in the United States: the 1998–2014 NHIS-NDI record linkage study. Health Equity. 2021;5:750–61.

Lee SH, Rhee TM, Shin D, Hong D, Choi KH, Kim HK, Park TK, Yang JH, Song YB, Hahn JY, et al. Prognosis after discontinuing renin angiotensin aldosterone system inhibitor for heart failure with restored ejection fraction after acute myocardial infarction. Sci Rep. 2023;13:3539.

Li W, Yu K, Sun S. Novel oral hypoglycemic agents SGLT-2 inhibitors: cardiovascular benefits and potential mechanisms. Pharmazie. 2020;75:224–9.

Loader J, Chan YK, Hawley JA, Moholdt T, McDonald CF, Jhund P, Petrie MC, McMurray JJ, Scuffham PA, Ramchand J, Burrell LM, Stewart S. Prevalence and profile of "seasonal frequent flyers" with chronic heart disease: analysis of 1598 patients and 4588 patient-years follow-up. Int J Cardiol. 2019;279:126–32.

Loeb M, Roy A, Dokainish H, Dans A, Palileo-Villanueva LM, Karaye K, Zhu J, et al. Influenza vaccine to reduce adverse vascular events in patients with heart failure: a multinational randomised, double-blind, placebo-controlled trial. Lancet Glob Health. 2022;10:e1835–44.

Lopez-Candales A, Sawalha K, Drees BM, Norgard NB. In search of mechanisms to explain the unquestionable benefit derived from sodium-glucose cotransporter-2 (SGLT-2) inhibitors use in heart failure patients. Postgrad Med. 2023;135:323–6.

Loughnan ME, Nicholls N, Tapper NJ. The effects of summer temperature, age and socioeconomic circumstance on acute myocardial infarction admissions in Melbourne, Australia. Int J Health Geogr. 2010;9:41.

Loughnan M, Tapper N, Loughnan T. The impact of "unseasonably" warm spring temperatures on acute myocardial infarction hospital admissions in Melbourne, Australia: a city with a temperate climate. J Environ Public Health. 2014;2014:483785.

Magnani JW, Mujahid MS, Aronow HD, Cene CW, Dickson VV, et al. Health literacy and cardiovascular disease: fundamental relevance to primary and secondary prevention: a scientific statement from the American heart association. Circulation. 2018;138:e48–74.

Martin SS, Aday AW, Almarzooq ZI, Anderson CAM, Arora P, Avery CL, Baker-Smith CM, et al. 2024 heart disease and stroke statistics: a report of US and global data from the American heart association. Circulation. 2024;149:e347–913.

Mavrakanas TA, Tsoukas MA, Brophy JM, Sharma A, Gariani K. SGLT-2 inhibitors improve cardiovascular and renal outcomes in patients with CKD: a systematic review and meta-analysis. Sci Rep. 2023;13:15922.

Mc Causland FR, Lefkowitz MP, Claggett B, Anavekar NS, Senni M, Gori M, Jhund PS, McGrath MM, et al. Angiotensin-neprilysin inhibition and renal outcomes in heart failure with preserved ejection fraction. Circulation. 2020;142:1236–45.

Mc Causland FR, Lefkowitz MP, Claggett B, Packer M, Senni M, Gori M, Jhund PS, et al. Angiotensin-neprilysin inhibition and renal outcomes across the spectrum of ejection fraction in heart failure. Eur J Heart Fail. 2022;24:1591–8.

McAlister FA, Murphy NF, Simpson CR, Stewart S, MacIntyre K, Kirkpatrick M, Chalmers J, Redpath A, Capewell S, McMurray JJ. Influence of socioeconomic deprivation on the primary care burden and treatment of patients with a diagnosis of heart failure in general practice in Scotland: population based study. BMJ. 2004;328:1110.

McDonagh TA, Metra M, Adamo M, Gardner RS, Baumbach A, Bohm M, Burri H, Butler J, Celutkiene J, Chioncel O, Cleland JGF, et al. 2021 ESC guidelines for the diagnosis and treatment of acute and chronic heart failure. Eur Heart J. 2021;42:3599–726.

McKechnie DGJ, Papacosta AO, Lennon LT, Ellins EA, Halcox JPJ, Ramsay SE, Whincup PH, Wannamethee SG. Subclinical cardiovascular disease and risk of incident frailty: the British regional heart study. Exp Gerontol. 2021;154:111522.

McKenzie BL, Pinho-Gomes AC, Woodward M. Addressing the global obesity burden: a gender-responsive approach to changing food environments is needed. Proc Nutr Soc. 2024;32:1–9.

McMurray JJ. Angiotensin II receptor antagonists for the treatment of heart failure: what is their place after ELITE-II and Val-HeFT? J Renin Angiotensin Aldosterone Syst. 2001;2:89–92.

McMurray J, Cohen-Solal A, Dietz R, Eichhorn E, Erhardt L, Hobbs FD, Krum H, Maggioni A, et al. Practical recommendations for the use of ACE inhibitors, beta-blockers, aldosterone antagonists and angiotensin receptor blockers in heart failure:

putting guidelines into practice. Eur J Heart Fail. 2005;7:710–21.

McMurray JJV, Jackson AM, Lam CSP, Redfield MM, Anand IS, Ge J, et al. Effects of sacubitril-valsartan versus valsartan in women compared with men with heart failure and preserved ejection fraction: insights from PARAGON-HF. Circulation. 2020;141:338–51.

Mebazaa A, Davison B, Chioncel O, Cohen-Solal A, Diaz R, Filippatos G, Metra M, Ponikowski P, Sliwa K, et al. Safety, tolerability and efficacy of up-titration of guideline-directed medical therapies for acute heart failure (STRONG-HF): a multinational, open-label, randomised, trial. Lancet. 2022;400:1938–52.

Morley JE, Vellas B, van Kan GA, Anker SD, Bauer JM, et al. Frailty consensus: a call to action. J Am Med Dir Assoc. 2013;14:392–7.

Murphy NF, Stewart S, MacIntyre K, Capewell S, McMurray JJ. Seasonal variation in morbidity and mortality related to atrial fibrillation. Int J Cardiol. 2004;97:283–8.

Nicholls N, Skinner C, Loughnan M, Tapper N. A simple heat alert system for Melbourne, Australia. Int J Biometeorol. 2008;52:375–84.

Nichols RB, McIntyre WF, Chan S, Scogstad-Stubbs D, Hopman WM, Baranchuk A. Snow-shoveling and the risk of acute coronary syndromes. Clin Res Cardiol. 2012;101:11–5.

Nogareda F, Regan AK, Couto P, Fowlkes AL, Gharpure R, Loayza S, et al. Effectiveness of COVID-19 vaccines against hospitalisation in Latin America during three pandemic waves, 2021–2022: a test-negative case-control design. Lancet Reg Health Am. 2023;27:100626.

Oertelt-Prigione S, Turner B. Tackling biases in clinical trials to ensure diverse representation and effective outcomes. Nat Commun. 2024;15:1407.

Ortega FB, Lee DC, Sui X, Kubzansky LD, Ruiz JR, Baruth M, Castillo MJ, Blair SN. Psychological well-being, cardiorespiratory fitness, and long-term survival. Am J Prev Med. 2010;39:440–8.

Pagnesi M, Vilamajo OAG, Meirino A, Dumont CA, Mebazaa A, Davison B, Adamo M, et al. Blood pressure and intensive treatment up-titration after acute heart failure hospitalization: insights from the STRONG-HF trial. Eur J Heart Fail. 2024;26:638–51.

Patel K, Shrier WEJ, Sengupta N, Hunt DCE, Hodgson LE. Frailty, assessed by the rockwood clinical frailty scale and 1-year outcomes following ischaemic stroke in a non-specialist UK stroke centre. J Stroke Cerebrovasc Dis. 2022;31:106451.

Peltzer S, Hellstern M, Genske A, Junger S, Woopen C, Albus C. Health literacy in persons at risk of and patients with coronary heart disease: a systematic review. Soc Sci Med. 2020;245:112711.

Peng Y, Qin D, Wang Y, Xue L, Qin Y, Xu X. The effect of SGLT-2 inhibitors on cardiorespiratory fitness capacity: a systematic review and meta-analysis. Front Physiol. 2022;13:1081920.

Playford D, Strange G, Celermajer DS, Evans G, Scalia GM, Stewart S, et al. Diastolic dysfunction and mortality in 436,360 men and women: the National Echo Database Australia (NEDA). Eur Heart J Cardiovasc Imaging. 2021;22:505–15.

Playford D, Strange G, et al. Preserved ejection fraction and structural heart disease in 446,848 patients investigated with echocardiography. ESC Heart Fail. 2021;8:1687–90.

Price D, Hughes KM, Dona DW, Taylor PE, Morton DAV, Stevanovic S, Thien F, Choi J, Torre P, Suphioglu C. The perfect storm: temporal analysis of air during the world's most deadly epidemic thunderstorm asthma (ETSA) event in Melbourne. Ther Adv Respir Dis. 2023;17:17534666231186726.

Prommik P, Tootsi K, Saluse T, Strauss E, Kolk H, Martson A. Simple excel and ICD-10 based dataset calculator for the Charlson and Elixhauser comorbidity indices. BMC Med Res Methodol. 2022;22:4.

Ravera A, Santema BT, de Boer RA, Anker SD, Samani NJ, Lang CC, Ng L, Cleland JGF, et al. Distinct pathophysiological pathways in women and men with heart failure. Eur J Heart Fail. 2022;24:1532–44.

Real-time 100 Most Polluted Cities in the World. https://www.aqi.in/au/real-time-most-polluted-city-ranking. Accessed May 2024.

Reinhardt M, Schupp T, Abumayyaleh M, Lau F, Schmitt A, Abel N, Akin M, Rusnak J, Akin I, Behnes M. Obesity paradox in heart failure with mildly reduced ejection fraction. Pragmat Obs Res. 2024;15:31–43.

Rivera FB, Tang VAS, De Luna DV, Lerma EV, Vijayaraghavan K, Kazory A, Shah NS, Volgman AS. Sex differences in cardiovascular outcomes of SGLT-2 inhibitors in heart failure randomized controlled trials: a systematic review and meta-analysis. Am Heart J plus. 2023;26:11542.

Rowlands GP, Mehay A, Hampshire S, Phillips R, Williams P, Mann A, Steptoe A, Walters P, Tylee AT. Characteristics of people with low health literacy on coronary heart disease GP registers in South London: a cross-sectional study. BMJ Open. 2013;3:10098.

Safdar B, Mangi AA. Survival of the fittest: impact of cardiorespiratory fitness on outcomes in men and women with cardiovascular disease. Clin Ther. 2020;42:385–92.

Saul A, Scott N, Spelman T, Crabb BS, Hellard M. The impact of three progressively introduced interventions on second wave daily COVID-19 case numbers in Melbourne, Australia. BMC Infect Dis. 2022;22:514.

Shen Z, Zhang Y, Zhou D, Lv J, Huang C, Chen Y, Zhang Y, Lin Y. Prevalence, factors and early outcomes of frailty among hospitalized older patients with valvular heart disease: a prospective observational cohort study. Nurs Open. 2024;11:e2122.

Smith RW, Barnes I, Green J, Reeves GK, Beral V, Floud S. Social isolation and risk of heart disease and stroke: analysis of two large UK prospective studies. Lancet Public Health. 2021;6:e232–9.

Solomon SD, McMurray JJV, Anand IS, Ge J, Lam CSP, Maggioni AP, Martinez F, Packer M, Pfeffer MA, et al. Angiotensin-neprilysin inhibition in heart failure with preserved ejection fraction. N Engl J Med. 2019;381:1609–20.

Stevens M, Raat H, Ferrando M, Vallina B, Lucas R, Middlemiss L, Redon J, Rocher E, van Grieken A. A comprehensive urban programme to reduce energy poverty and its effects on health and wellbeing of citizens in six European countries: study protocol of a controlled trial. BMC Public Health. 2022;22:1578.

Stewart S, Carrington MJ, Marwick TH, Davidson PM, Macdonald P, Horowitz JD, Krum H, Newton PJ, Reid C, Chan YK, Scuffham PA. Impact of home versus clinic-based management of chronic heart failure: the WHICH? (which heart failure intervention is most cost-effective and consumer friendly in reducing hospital care) multicenter, randomized trial. J Am Coll Cardiol. 2012;60:1239–48.

Stewart S, Ball J, Horowitz JD, Marwick TH, Mahadevan G, Wong C, Abhayaratna WP, Chan YK, Esterman A, Thompson DR, Scuffham PA, Carrington MJ. Standard versus atrial fibrillation-specific management strategy (SAFETY) to reduce recurrent admission and prolong survival: pragmatic, multicentre, randomised controlled trial. Lancet. 2015a;385:775–84.

Stewart S, Chan YK, Wong C, Jennings G, Scuffham P, Esterman A, et al. Impact of a nurse-led home and clinic-based secondary prevention programme to prevent progressive cardiac dysfunction in high-risk individuals: the Nurse-led Intervention for less chronic heart failure (NIL-CHF) randomized controlled study. Eur J Heart Fail. 2015b;17:620–30.

Stewart S, Riegel B, Boyd C, Ahamed Y, Thompson DR, Burrell LM, et al. Establishing a pragmatic framework to optimise health outcomes in heart failure and multimorbidity (ARISE-HF): a multidisciplinary position statement. Int J Cardiol. 2016;212:1–10.

Stewart S, Keates AK, Redfern A, McMurray JJV. Seasonal variations in cardiovascular disease. Nat Rev Cardiol. 2017;14:654–64.

Stewart S, Playford D, Scalia GM, Currie P, Celermajer DS, Prior D, Codde J, et al. Ejection fraction and mortality: a nationwide register-based cohort study of 499,153 women and men. Eur J Heart Fail. 2021;23:406–16.

Stewart S, Patel SK, Lancefield TF, Rodrigues TS, Doumtsis N, Jess A, Vaughan-Fowler ER, Chan YK, Ramchand J, Yates PA, Kwong JC, McDonald CF, Burrell LM. Vulnerability to environmental and climatic health provocations among women and men hospitalized with chronic heart disease: insights from the RESILIENCE TRIAL cohort. Eur J Cardiovasc Nurs. 2024;23:278–86.

Stewart S, Carrington MJ, Marwick T, Davidson PM, Macdonald P, Horowitz J, Krum H, Newton PJ, Reid C, Scuffham PA. Which heart failure intervention is most C-e and consumer friendly in reducing Hospital care trial I. The WHICH? Trial: rationale and design of a pragmatic randomized, multicentre comparison of home- versus clinic-based management of chronic heart failure patients. *Eur J Heart Fail.* 2011;13:909–16.

Tajik B, Voutilainen A, Lyytinen A, Kauhanen J, Lip GYH, Tuomainen TP, Isanejad M. Frailty predicts incident atrial fibrillation in women but not in men: the kuopio ischaemic heart disease risk factor study. Cardiology. 2023a;148:574–80.

Tajik B, Voutilainen A, Sankaranarayanan R, Lyytinen A, Kauhanen J, Lip GYH, Tuomainen TP, Isanejad M. Frailty alone and interactively with obesity predicts heart failure: Kuopio ischaemic heart disease risk factor study. ESC Heart Fail. 2023b;10:2354–61.

Talha KM, Anker SD, Butler J. SGLT-2 inhibitors in heart failure: a review of current evidence. Int J Heart Fail. 2023;5:82–90.

Tan W, Lu X, Xiao T. The influence of neighborhood built environment on school-age children's outdoor leisure activities and obesity: a case study of Shanghai central city in China. Front Public Health. 2023;11:1168077.

The Köppen Climate Classification. https://www.mindat.org/climate.php. Accessed April 2024.

Tumas N, Lopez SR. Double burden of underweight and obesity: insights from new global evidence. Lancet. 2024;403:998–9.

UK Health Service Agency. Cold weather alerts; 2024. https://www.metoffice.gov.uk/weather/warnings-and-advice/seasonal-advice/cold-weather-alerts. Accessed May 2024.

Vaduganathan M, Docherty KF, Claggett BL, Jhund PS, et al. SGLT-2 inhibitors in patients with heart failure: a comprehensive meta-analysis of five randomised controlled trials. Lancet. 2022;400:757–67.

Vardeny O, Claggett B, Kachadourian J, Pearson SM, Desai AS, Packer M, Rouleau J, Zile MR, Swedberg K, Lefkowitz M, Shi V, McMurray JJV, Solomon SD. Incidence, predictors, and outcomes associated with hypotensive episodes among heart failure patients receiving Sacubitril/Valsartan or Enalapril: the PARADIGM-HF trial (prospective comparison of angiotensin receptor neprilysin inhibitor with angiotensin-converting enzyme inhibitor to determine impact on global mortality and morbidity in heart failure). Circ Heart Fail. 2018;11:e004745.

Wehrman G, Halton M, Riveland B, Potter E, Gaddy M. Comparison of A1c reduction, weight loss, and changes in insulin requirements with addition of GLP-1 agonists versus SGLT-2 inhibitors in patients using multiple daily insulin injections. J Pharm Pract. 2024;37:311–7.

Weinberger KR, Zanobetti A, Schwartz J, Wellenius GA. Effectiveness of national weather service heat alerts in preventing mortality in 20 US cities. Environ Int. 2018;116:30–8.

Williams RB. Loneliness and social isolation and increased risk of coronary heart disease and stroke: clinical implications. Heart. 2016;102:2016.

Williamson T, Moran C, Chirico D, Arena R, Ozemek C, Aggarwal S, Campbell T, Laddu D. Cancer and cardiovascular disease: the impact of cardiac rehabilitation and cardiorespiratory fitness on survival. Int J Cardiol. 2021;343:139–45.

World Air Quality Index. https://waqi.info/. Accessed May 2024.

Yang B, Lin Y, Xiong W, Liu C, Gao H, Ho F, Zhou J, Zhang R, Wong JY, Cheung JK, Lau EHY, et al. Comparison of control and transmission of COVID-19 across epidemic waves in Hong Kong: an observational study. Lancet Reg Health West Pac. 2024;43:100969.

Yeow RY, O'Leary MP, Reddy AR, Kamdar NS, Hayek SS, de Lemos JA, Sutton NR. Survival characteristics of older patients hospitalized with COVID-19: insights from the American heart association COVID-19 cardiovascular disease registry. J Am Med Dir Assoc. 2024;25:348–50.

Zaman S, Chow C, Lam CSP, Saw J, Nicholls SJ, Figtree GA. Heart disease in women: where are we now and what is the future? Heart Lung Circ. 2021;30:1–2.

# Future Proofing Our Hearts to Climate Change

**9**

## Abstract

In the last chapter, the embryonic evidence (essentially derived from one, relatively positive, COVID-19 cruelled randomised trial) in respect to the possibility of restoring/promoting resilience in people hospitalised with multimorbid heart disease was presented. However, prevention is always better than cure! Thus, just as the 'interconnectedness' of the different spheres that comprise our planet was highlighted in terms of the impact of pollution earlier in this book and considering the tenant of the UNSDGs, it is critical for us to proactively address climate vulnerability from multiple perspectives. Specifically, at multiple levels, there is urgent need for the world/us to—(1) Recognise the pre-existing to evolving threat (due to climate change) of climatic provocations to heart health, (2) Develop a more systematic approach to recognising which regions and communities are most 'vulnerable' to climatic challenges (from multiple perspectives), (3) Implement a range of public health measures from raising public awareness to implementing broad public health strategies to promote and enhance climatic resilience at the population level, and, beyond the need to reframe the clinical management of people hospitalised for heart disease, to reduce their risk of readmissions and premature mortality due to pre-existing vulnerability to climate provocations—(4) Strengthen primary care health care teams/services to proactively detect and then manage at risk/vulnerable individuals before they are hospitalised and/or die prematurely due to climatic provocations to their cardiovascular health. Such primary care capacity would also ensure the optimal, post-discharge management of those hospitalised with heart disease from this novel perspective. Bringing all the current evidence in this regard together, this chapter provides a critical review of the progress made thus far around these four key points.

## Keywords

High-risk · Winter peaks · Global health priorities · Primary care · Citizen science · Health care expenditure

## 9.1 Investing in Climatic Resilience

In introducing this topic, it goes without saying that any of the regions of the world highlighted in Chaps. 1 and 8, are at highly vulnerable to the adverse effects of climatic change—with a reduced capacity to respond and improve

S. Stewart, *Heart Disease and Climate Change*, Sustainable Development Goals Series, https://doi.org/10.1007/978-3-031-73106-8_9

their individual to population-based resilience. But what of those countries that can? What should they be investing in? As a health services researcher with broad interests in the origins to population impact of heart disease, it should come as no surprise that I strongly believe that global health research organisations (from the UK-based Wellcome Trust to the US-based Gates Foundation and National Institutes for Health) would be wise to recognise the importance of this topic to low-to-middle income countries and invest a good portion of their funds to establish evidence-based programmes that promote climatic resilience and can be funded through the World Bank/International Development Association and International Monetary Fund etc. Perhaps I am biased but having had many excellent projects focusing on the growing threat of non-communicable forms of cardio-pulmonary disease with a strong environmental and climatic focus in some of the poorest regions of sub-Saharan Africa rejected, I'm not too optimistic in this regard. But what about high-income countries who, in theory at least, can afford to address the problem of climate change and it's looming impact on the heart health of their predominantly ageing populations in whom antecedent cardiovascular risk is at historical highs? (Roth et al. 2020). The following sections outline a range of public health to clinical management strategies that I believe are important steps to at least prepare us for the prospect of more climatically-induced provocations to our heart health. At the end of this chapter/book, I will also outline a comprehensive list of 'unknowns' and key issues/questions derived from the broad range of climatic-related topics covered and nominate where funding needs to be directed for more research. A recent report on Australia's funding of climate and health research [along with what has been identified for priority funding (Beggs et al. 2024)] provides a fascinating background to these priority recommendations for action.

## 9.2   Identifying High-Risk Regions and Communities

As a quasi-geographer (and some will guess a highly amateur meteorologist without the math skills to be one!), my research has always been predicated on trying to reveal not only if a problem exists and its size, but specifically 'who and why' is most affected by that problem (Stewart 2023). My seasonal to climatic focussed research, of course, has extended to the 'when'. Nevertheless, the first component of these ('who is most affected and therefore a priority target') is informed by a healthy dose of pragmatism. In a past life as a clinician working in an Intensive Care/Coronary Care environment, I quickly learned to apply a triage approach when confronted by multiple cardiac arrests and/or clinical crises. Then, as someone trying to establish the evidence for heart failure management programmes (Stewart and Blue 2004; Stewart 2017; Stewart et al. 1999, 2011) and applying new health services based on the evidence that we [as the original 'pioneers' of this new approach (Sochalski et al. 2009)] had provided, I quickly realised that we couldn't help everyone with the resources and personnel we had available. Thus, there had to be a priority list informed by a system to sort the wheat-from-chaff for attention—particularly in using resources wisely. If someone needed help, they were provided with that help. If they chose not follow advice or accept help, we moved on to the next person who did. Of course, in a circular argument of 'interconnectedness', moving on to the next one who heeds advice or accepts help doesn't work at all well in the face of global climate change. While humans build artificial boundaries, they can't control the atmosphere or hydrosphere—thus climate change for one person/country is 'climate change for everyone' (Global Climate Change 2021).

It was on the original premise (determining who has the greatest needs) that the CARDIAC-ARIA project to map-out where Australians might access cardiac services (Clark et al. 2007, 2012), along with granular burden of disease reports specifically identifying 'where' people with varying forms of risk factors (Carrington et al. 2010, 2012) and established forms of heart disease live [with a focus on those living in rural-remote communities (Carrington et al. 2012)] were conceived.

In this context, I believe there is a strong rationale for every government and jurisdiction with public health institutions and ready access to high-level partnerships with academic institutions, to invest in projects that map-out the 'hot-spots' for climatic vulnerability. These would mimic similar projects in respect to identifying socio-economic inequalities (Bragg and Lacey 2024; Department of Health, Federal Government of Australia 2023) to high rates of cardiac morbidity and mortality (Chan et al. 2016), but within the much broader perspective of climatic conditions, evolving climate change and their capacity to provoke otherwise preventable cardiovascular events. The following section describes a project that, using the best available information, mapped-out where in Australia people are most at risk of being hospitalised and then dying from heart failure (HF) and, from all forms of cardiovascular disease (CVD) from a climatic perspective. Although rudimentary in many respects, it does demonstrate the type of broader thinking and perspective needed to address climate provocations to cardiovascular health (and health more broadly) from a public health to clinical perspective.

### 9.2.1   Determining the 'Heart' of Peak Winter in Australia

This section describes the main features and findings from an unpublished, commercially-funded research report (Burrell et al. 2019) completed in 2019, that described the burden of heart failure (HF) and, more generally, CVD in Australia in a completely different way than previously reported. The report presented the findings of an ongoing programme of research

tracking the evolving characteristics of HF and other forms of CVD in Australia—the latest iteration of which has focussed on the unique insights provided by the National Echo Database of Australia (NEDA) (Stewart et al. 2021, 2022a, b; Strange et al. 2019a, b, 2021). It had three main purposes:

1. Highlight and explain the issue of climatically provoked cardiac events (in this case from a 'seasonality' perspective) from a population to individual level perspective—with the express purpose of dispelling any misconceptions (much of perpetuated by the media and academia alike) that on such a warm continent, we might not be exposed to the risk of cold weather events.

2. Translate the one-dimensional figures typically used to describe the burden of any disease (e.g. number of related admissions per year) into a more dynamic picture of seasonal ebbs and flows in hospital admissions across the country; with a particular focus on its impact on the pattern of hospital admissions among vulnerable individuals affected by the potentially deadly and disabling syndrome HF.

3. Outline the ongoing costs and consequences of not adequately responding to a phenomenon that will be exacerbated by climate change; noting (at the time of the report) a likely increase in those susceptible to climatic provocations due to the progressing ageing of the population in whom the risk of developing HF remained high.

In generating the key facts and figures contained within this report, it was explicitly acknowledged there were many caveats to be considered. Many of these (caveats) inform the research issues and questions highlighted and posed at the end of this book. However, it remained deliberately provocative to stimulate an interest among the public and health professionals alike, in a phenomenon that, paradoxically, is just as likely to occur in the warm and temperate climes of Australia, as it is in the wintry climes of Scandinavia.

To complete the report, each (Australian) region of interest was firstly categorised from a climatic/weather perspective using the Köppen Climate Classification system (The Koppen Climate Classification 2024). We then specifically focussed (in order of priority) on—(1) Describing the burden and seasonal pattern of HF-related admissions in Australia overall, (2) Further describing that burden/pattern within the six major States of Australia with large populations exposed to distinctive climatic transitions linked to seasonal change and then, (3) Providing examples of distinctive populations within each of these states (plus the Australian Capital Territory) with varying levels of risk of climatic vulnerability from multiple perspectives (see below). For population profiling, we relied on the latest Australian Bureau of Statistics (ABS) data on the estimated demographic structure of the overall Australian population and for each State/Territory according to sex and in 10-year age groups above the age of 25 years for the year 2018 (Australian Bureau of Statistics 2024). These same demographic data were then examined with increasing granularity (smaller areas) using data derived from the 2016 Australian Census. The ABS also publish data on the projected population of Australia for the foreseeable future (Australian Bureau of Statistics 2024). To conservatively estimate the burden of HF within 5–10 years, we used the same population data for the years 2025 and 2030 (the now and near future from the perspective of this book). For baseline event rates, the Australian Institute of Health and Welfare had published data on the age and sex-specific rate of CVD and HF in 2015–2016 (Australian Institute of Health and Welfare 2024). For the analyses, these rates were combined with ABS population data for Australia and the smaller jurisdictional/geographic areas of interest for the year of the report (2019) in addition to the years 2025 and 2030 for future projections of HF-related admissions. All methods conformed to the GATHER standards for the reporting of burden of disease data.

## 9.2.2  A 5-Point System to Determine Regional and Climatic Vulnerability

As described in Chap. 4 of this book and then specifically demonstrated within the RESILIENCE Trial cohort (Stewart et al. 2024) (see Chap. 8), several largely non-modifiable (but not exclusively so) factors will likely contribute to an increased signal of climatic vulnerability in those at risk of, or already affected by CVD. This 'signal' is magnified in respect to HF. Many of these same factors can be assessed at the population level to assess the vulnerability of whole regions to higher levels of climatic vulnerability, with the prospect of high health service demands and resulting dysfunction/stress (including ambulance ramping, bed-block, and cancellation of elective procedures/admissions) when provocative climatic conditions occur. It is on this basis that a 5-Point rating system to evaluate the potential for higher-levels of climatic provocations to heart health within each geographic region of interest was generated—see Fig. 9.1. Although we were unable (at the time of the report and even at the time of the writing of this book), to determine the relative influence of each of these five factors on cardiac events (this requires further research) we logically assumed that the more areas of vulnerability, the more urgent the action needed.

## 9.2.3  A Shared Lithosphere and Atmosphere with Varying Climatic Provocations

Based on the population profile of Australia in 2019 [noting that post-pandemic Australia's population has rapidly expanded to nearly 27 million people in 2024 (Australian Bureau of Statistics 2024)], our research suggested that the Australian health care system was contending with close to 600,000 hospital admissions linked to CVD; with a predominance of male admissions. The projected difference between peak

**High proportion of men and women within the high-risk age bracket for developing CVD and HF**

**Higher levels of poverty and fewer socio-economic resources to achieve optimal thermoregulation**

**Region identified by the NHFA as a CVD "hot-spot" based on higher levels of CVD-related admissions**

**High proportion of individuals living in large urban centres with associated exposure to air pollution**

**Geographic area that is exposed to a broad range of climatic conditions regardless of mean temperatures**

**Fig. 9.1** A regional, 5-point rating system of climatic vulnerability. This is the 5-point system of 'climatic vulnerability' we developed to assess different regions (big to small) of Australia (see figures below). To determine if one of these five symbols be applied, we compared: (1) ABS population data from a regional to whole Australian perspective (Australian Bureau of Statistics 2024); (2) Contemporary data on the indices of socio-economic wealth across jurisdictions (Australian Institute of Health and Welfare 2022); (3) The National Heart Foundation of Australia's (NHFA) "hot-spot" report on CVD admissions (National Heart Foundation of Australia 2024); (4) The pattern of urban and rural dwelling communities on a regional basis (Australian Bureau of Statistics 2024); and (5) The predominant range of climatic conditions affecting each geographic area (The Koppen Climate Classification 2024)

winter versus summer lows/troughs in hospital activity was likely to be *>40,000 admissions nationally*; a 32% absolute difference in hospital admissions between these two climatically different seasons. Based on *sensitivity analyses*, this variance ranged from a possible low-to-high of *36,000 (28% difference)* to *47,500 (36% difference)*. When comparing the overall pattern of CVD admissions in the summer months (markedly lower than every other season) we estimated that in 2019, climatic transitions and different weather conditions would *contribute to an additional 71,000 more admissions due to CVD in the months of autumn, winter, and spring, combined* (Burrell et al. 2019).

Although the likely number of hospital admissions due to HF is much higher (150,000 or more/annum where HF is listed as a primary or secondary cause), in this report we estimated there were *~70,000 hospital admissions* where HF was the listed primary cause in Australia during 2019. Overall, the projected difference between 'peak' winter versus summer 'lows/troughs' in hospital activity linked to a primary diagnosis of HF, was likely to be *>5000 admissions* nationally (a 32% increase); the low-to-high range being *4400 (27%)* to *5900 (36%)*. Once again, compared to the summer months, it was estimated that seasonal fluctuations in the remaining months would *contribute to an* *additional 8800 more admissions due to HF.* The impact of such climatically-driven variations in cardiovascular-related (including HF-specific events) obviously depends on the size of the population at risk and the capacity of the local health care system—Fig. 9.2 shows the overall number of CVD admission in Australia and the overall pattern of HF-specific admissions in the most populous States (New South Wales [including Sydney], Victoria [Melbourne] and Queensland [Brisbane]) exposed to a wide-range of different climatic conditions. Accordingly, simply providing these 'raw' numbers doesn't provide the full picture. Thus, in profiling where people live in Australia and applying our 5-point vulnerability index, we identified two smaller States (South Australia and Tasmania) with vastly different climates that would be at most at risk from—(a) current climatic conditions and (b) future climate change (Fig. 9.3 and 9.4) (Burrell et al. 2019).

One of the main driving factors in any political decision-making process is money! What does something cost (beyond human lives and the health of a population). It was for this reason that a health economic analysis of the potential savings from eliminating climatic variability in cardiovascular events (and once again, specifically those linked to HF) in Australia was performed. As shown on an age- and sex-specific basis in Fig. 9.5 (CVD-related hospitalisation) and Fig. 9.6 (HF-specific hospitalisation), when assuming a conservative cost of $5000 per admission (equivalent to 2–3 days stay) CVD-related admission expenditure among men and women (in 2019) was estimated to be *$AUD 1.73 billion and $AUD1.23 billion*, respectively In considering the total expenditure of *$3 billion* in hospital costs attributable to CVD each year, we estimated that if it were indeed possible to eliminate/attenuate greater rates of hospitalisation in winter, autumn and spring linked to

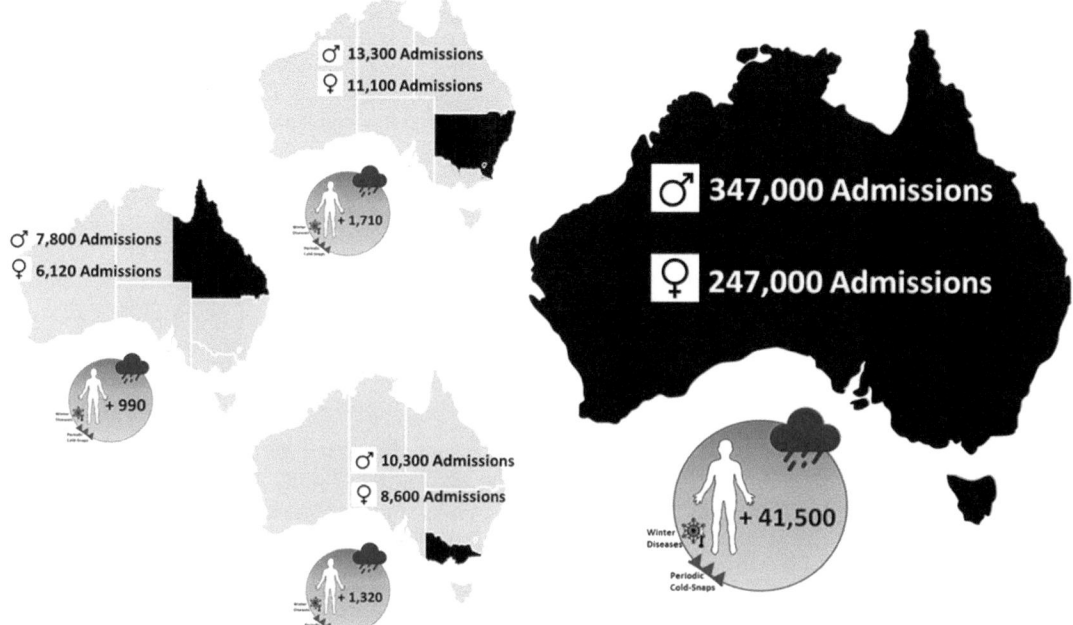

**Fig. 9.2** Seasonal fluctuation in CVD and HF admissions in Australia. This figure shows the estimated total of sex-specific CVD-related admissions in Australia in 2019 (left), plus those admissions related to HF within the 3 largest States in Australia (Victoria, Queensland and New South Wales). The lower symbols indicate the degree of variability (men and women combined) between the warmer summer months (low/trough levels) and cooler winter months (peak levels) in hospital activity

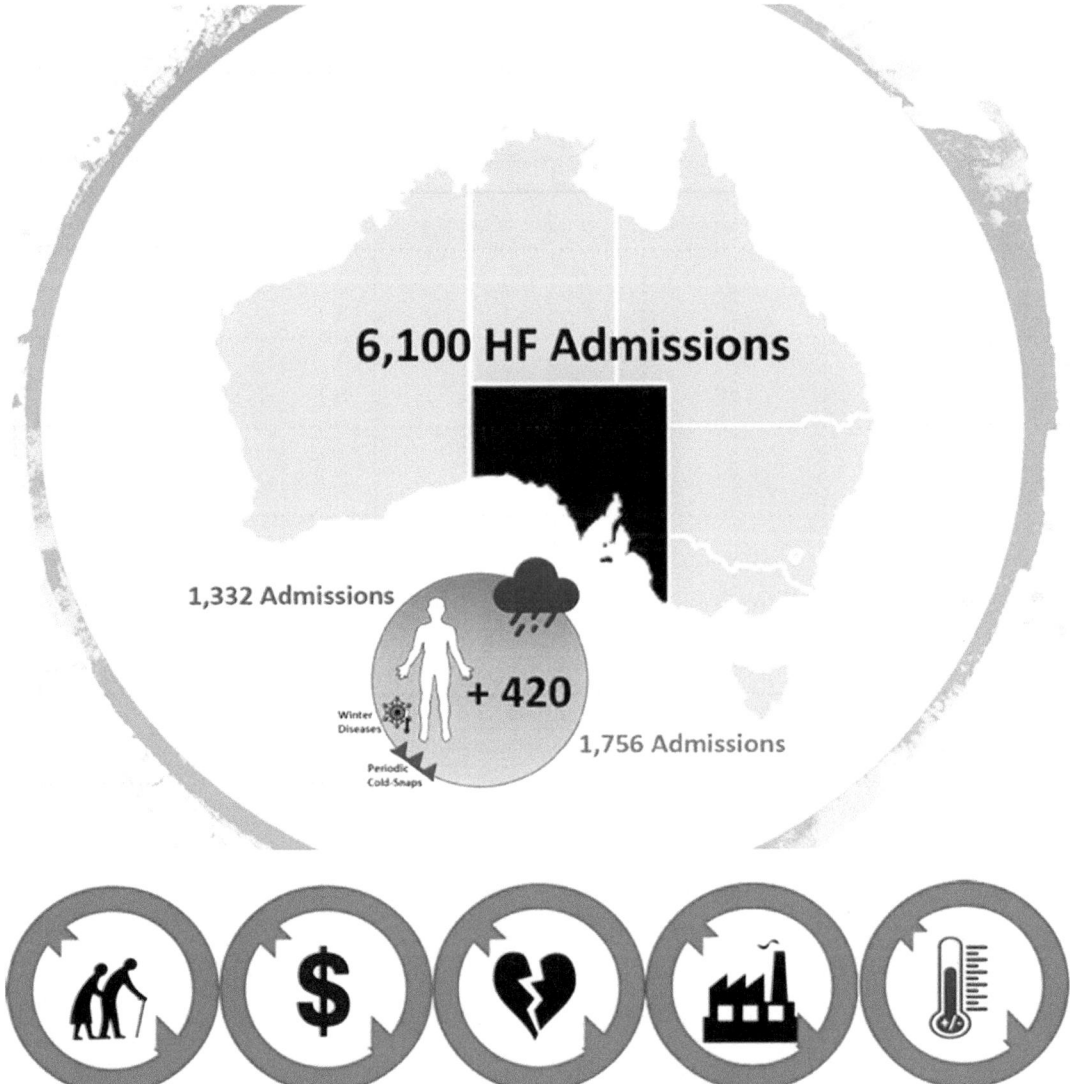

**Fig. 9.3** Climatic vulnerability—the case of South Australia. The climate in Adelaide (where the majority of South Australian's live) is mild, and generally warm and temperate; classified as Csa by Köppen-Geiger. The average annual temperature is ~16 °C and the annual average rainfall is ~550 mm. The driest month is February (with 15 mm of rain) compared to July (average rainfall of 76 mm). January is the warmest month (average temperature ~22 °C) and July is the coldest month (average temperature ~11 °C). Overall, South Australia has a relatively older population compared to the national average, has higher levels of poverty, has been identified as a key hot-spot for higher CVD-related hospitalisation (>55 admissions per 10,000) compared to the national average and has a relatively concentrated population more likely to be exposed to urban pollution. Moreover, despite its warmer climate, the people of South Australian contend with a marked variation in wintry versus summer climatic conditions (monthly 61 mm and 11.3 °C rain and temperature differentials, respectively) that is likely to provoke *Seasonality*; particularly among vulnerable groups (The Koppen Climate Classification 2024; Australian Bureau of Statistics 2024; Australian Institute of Health and Welfare 2022; National Heart Foundation of Australia 2024)

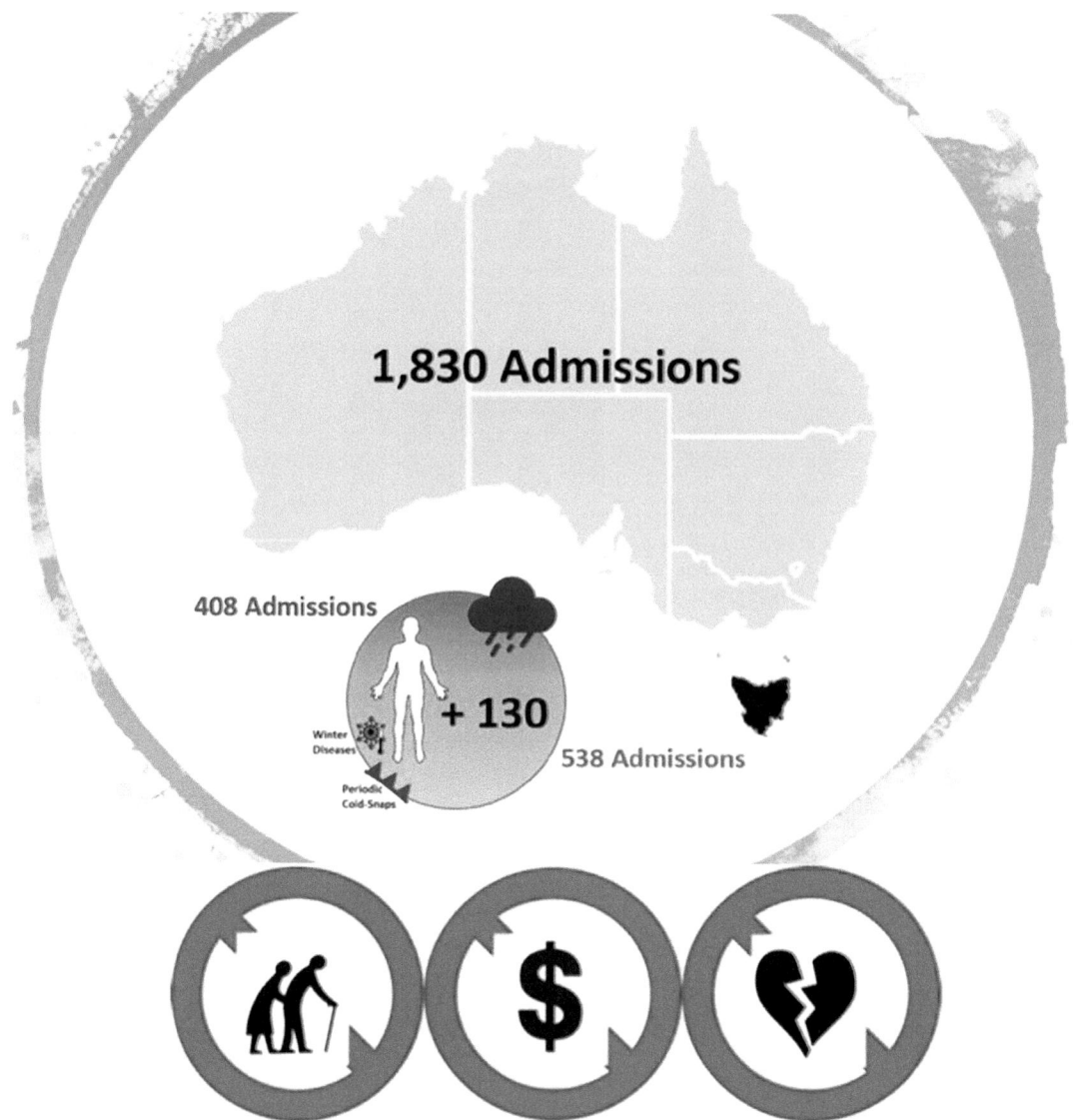

**Fig. 9.4** Climatic vulnerability—the case of Tasmania. The climate in the smaller but geographically dispersed State of Tasmania is still considered warm and temperate; classified as Cfb by Köppen-Geiger. The average annual temperature in the main urban centre of Hobart is ~12 °C and the annual average rainfall is ~600 mm. The driest month is January (with 40 mm of rain) compared to August (average rainfall of 57 mm). January is the warmest month (average temperature ~17 °C) and July is the coldest month (average temperature ~8 °C). Like South Australia, Tasmania has a markedly older demographic profile when compared to the national average. It also has the highest reported levels of poverty and, in 2015/2016, had a Gross State Product per Capita that was ~$16,000 below that of New South Wales. The rate of CVD-related admissions in Tasmania is >50 per 10,000. Alternatively, given its less urbanised population distribution and wilderness areas, the people of Tasmania are less likely to be exposed to urban pollution compared to those living on the mainland. Moreover, despite impressions of a colder and harsher climate compared to continental Australia, the weather patterns in Tasmania have less variance in respect to precipitation and mean temperatures. The local population is probably more at risk of heat waves than cold snaps given the greater predictability in overall weather conditions (The Koppen Climate Classification 2024; Australian Bureau of Statistics 2024; Australian Institute of Health and Welfare 2022; National Heart Foundation of Australia 2024)

**Fig. 9.5** Cost of climatic variations in CVD-related admissions in Australia. These figures show the increasing cost of seasonal variations in CVD-related hospital admissions on an age-specific basis for men (top panel **a**) and women (bottom panel **b**) in Australia

climatic provocations to heart health, there were a total of *$AUD 360 million in cost-savings per annum* to work with (in terms of funding new programmes to promote greater resilience within the highest risk communities).

In the context of climate change (noting the recurring argument that we still don't know if it will have a 'net' positive, neutral of negative effect!) projections showed that by 2030, the total number of CVD-related admissions in Australia will have risen to 645,000. Moreover, climatic-induced variability would contribute to an increased differential of *45,000 hospitalisations* between the coldest and warmest times of the year. From a HF perspective, local variations in the population profile and climatic conditions (with some academics claiming that "*lethal heat will make parts of Australia un-inhabitable*") will either exacerbate or attenuate the problem. However, given that a HF-related hospitalisation

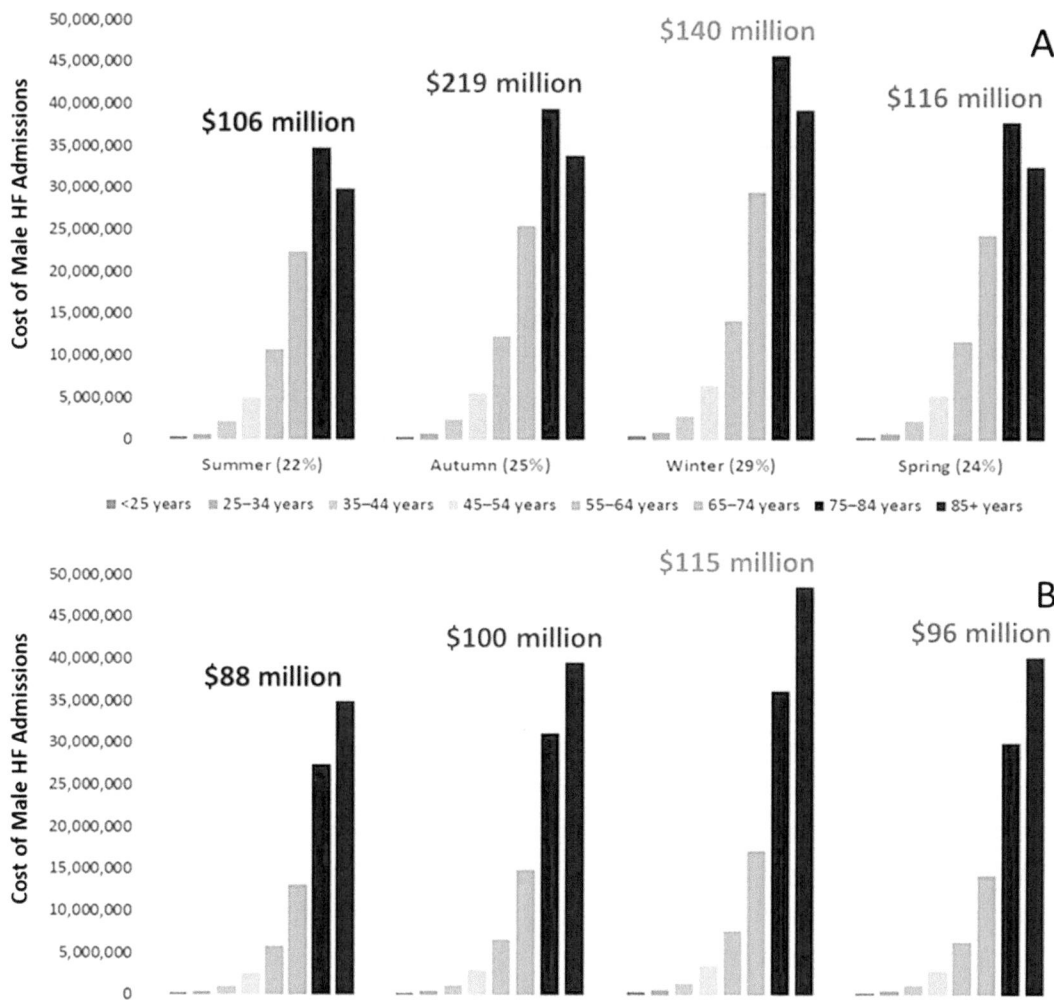

**Fig. 9.6** Cost of climatic variations in HF-related admissions in Australia. These figures show the increasing cost of seasonal variations in HF-related hospital admissions on an age-specific basis for men (top panel **a**) and women (bottom panel **b**) in Australia

is associated with an increased risk of mortality [it remains 'more malignant' (Stewart et al. 2001, 2010) than many common forms of cancer in this respect (Nakao et al. 2023; Sabanoglu et al. 2023; Banerjee et al. 2022; Tribouilloy et al. 2008)] the cost in life-years lost/QALY from this climatic-related phenomenon is enormous and requires a whole of system response.

## 9.3    National to Global Health Priorities

If we are to be more proactive, what does this mean? In Chap. 8, some of the evidence around the impact of broader public health [e.g. vaccinations (Avramidis et al. 2024; Bell et al. 2022; Global Burden of Disease Investigators 2021)] and specific policy decisions/government support [subsidised funding to overcome fuel poverty (Champagne et al. 2023; Lawler et al. 2023)] were discussed and then examined on an individual basis—to address climatic vulnerability at the individual level. The cost-dynamics of an individually targeted versus a national to regional policy-driven approach to addressing points of vulnerability is something that needs to be more fully explored.

Nevertheless, founded on a key figure included in a previously mentioned 'State-of-the-Art' review of CVD and air pollution (Rajagopalan et al. 2018), Fig. 9.7 provides a list

**Global Action**

| ACTION | DESIRED OUTCOME |
|---|---|
| Climate change recognised as global threat | • Every government/jurisdiction acknowledges the problem as a shared responsibility to address<br>• Funding is specifically apportioned to meet this threat |
| Realistic targets established with strict timeframes | • The world works towards a common goal to limit climatic variations of human origin<br>• Every person on the planet has the capacity to maintain homeostasis and comfort |
| Global framework to reduce the human footprint on the planet with agreed action plan | • Commercial development of low-polluting, renewable energy sources (e.g. to close coal-fired power plants & open solar, wind and geothermal plants) and transport options.<br>• Large-scale "carbon capture" and pollution traps developed and implemented.<br>• Deforestation and other ecological damage halted and reversed through rehabilitation.<br>• Specific funds for low-to-middle income countries to transform their growing economies using 'environmentally-friendly' and cost-efficient technologies |

**Societal/Government Action**

| ACTION | DESIRED OUTCOME |
|---|---|
| Fiscal and structural policies to promote climatic resilience at a regional to national level | • Invest in the healthy built environment - removed from polluting sources (e.g. roads), addition of green spaces and, critically, build environmentally-friendly, sustainable and efficient housing<br>• Evaluation, reclamation and reproposing of land-use and waterways.<br>• Fiscal rewards for low-emissions and carbon-capture<br>• Funding of regional and community maps of climatic resilience to vulnerability<br>• Eradication of 'fuel poverty' to insulate, heat or cool households<br>• Establish open-access/community 'thermal comfort zones'<br>• Provision of 'safe' working environments to reduce exposure to provocative conditions |
| Investing in public health policies and research | • Renewed efforts to address rising rates of obesity, sedentary behaviours and poor food choices exacerbated by the dynamics of food-supply and cost of processed food/sweetened drinks<br>• Renewed efforts to address high rates of hypertension and other major risk factors (including salt-reduction measures and earlier detection in high-risk populations/cohorts)<br>• Purposed funding of the mechanisms and impact of climate change (including modelling of local climate changes/trends and pollution) on the health of national to local populations<br>• Support public awareness and health campaigns<br>• Implement and embed routine health alerts linked to climatic conditions (e.g. heat and cold alerts that are not just about the 'now' but proactive in planning/preparation) |
| Repurposing health care systems to recognise and address climatic provocations to health | • All health professional training curriculum includes a focus on climatic provocations and climate change and how it can be best detected and managed within a holistic framework<br>• Renewed focus on proactive screening and evidence-based clinical management of the common antecedents of cardio-pulmonary disease.<br>• Evidence-based, integrated, primary care programs that readily recognise and manage those vulnerable to climatic provocations (including anyone with high cardiovascular risk or heart disease) to prevent (re-)hospitalisation and premature mortality in the future<br>• Evidence-based, tailored tertiary care programs that readily recognise and manage those admitted to hospital with a previous or likely pattern of climatically provoked clinical crises and/or premature mortality in the immediate future. |

**Individual Action**

| ACTION | DESIRED OUTCOME |
|---|---|
| Provide education to help everyone to understand the importance of climatic provocations to their overall health/heart health and modulate their behaviours accordingly | • Every person and their family/significant others is aware of their interactions with the environment, broader climatic conditions and those acute episodes (e.g. heatwave and/or cold spells) that may provoke a clinical crisis and/or adversely affect their health in the longer-term<br>• Every person and their family/significant others is willing to shape their behaviours to minimise climatic provocations to their health and to maintain their comfort levels/homeostasis |
| Provide resources to respond to current and future climatic provocations to their health | • Every person/household has the capacity (regardless of their socio-economic circumstances) to insulate and climate-proof their home as well as where clothing that is appropriate to climatic conditions<br>• Every person has the capacity to maintain homeostasis and comfort without, socio-economic, physical or psychological distress |
| Ensure those with diminished capacity to understand and/or act are provided with the care they need | • Highly vulnerable and frail individuals can readily access support programs (from dependent to fully dependent) and dedicated facilities (including climatic safe-havens) when needed.<br>• The most vulnerable of the world's population and within discrete nations, regions and communities are identified and supported to be more climatically resilient. |

**Fig. 9.7** Applying different levels of action to combat climate change

of 'ideal' actions and outcomes—from a global health to individual perspective, based on many of the issues, topics and potential solutions surrounding climate change and heart health covered in this book. It's important to recognise that this list is not exhaustive but indicative of the type of mind-set/paradigm shift in thinking needed (at all levels) to understand that humans are intimately linked/adapted to our environment, and we are steadily eroding our hard-earned (through physiological and cultural adaptations to technological advances) resilience and ability to maintain homeostasis by damaging both the planet and, indeed ourselves through personal and collective inaction. Some of the actions required are well beyond most people's 'pay-grade'. However, from a health professional/clinician's perspective, we can do much more to provide *climatically aware health care*. The last chapter focussed on the individual admitted to hospital. The next section outlines (as an example of what is needed) a new programme of research designed to explore the role of primary care in preventing climatically provoked cardiovascular events.

## 9.4  Addressing a Missing Link in the Chain of Protection— Primary Care!

Despite some research on seasonal/climatic-induced variations in primary care consultations (Blashki et al. 2007; Wolk and Porter 2024; Xie et al. 2021), like that of tertiary-care focussed programmes, there appears to be is insufficient evidence to confirm if the type of variances described in major events (i.e. emergency hospitalisation/premature mortality described in Chap. 5) linked to climatic challenges in people with a combination of antecedent (hypertension and diabetes) and established forms of heart disease extends to primary care—although this is likely the case. We also don't know (yet) how primary care services can sustainably reduce the need for hospital care and prolong life in this growing patient cohort. This is particularly true for the growing number of people (worldwide) who are surviving to older ages but living with debilitating, multimorbid heart disease (Stewart 2023). As highlighted above, a whole of system/population approach to managing the critical issues discussed in this book is needed. As part of this systematic approach, it is unfortunate that the role of primary care (in the proactive detection and management of vulnerable individuals with heart disease) remains poorly defined overall. Consequently, primary care represents a missing link in the chain of protection needed for many people vulnerable to the growing impact of climate change.

### 9.4.1  The AMBIENCE Project

It was in the above context, that we recently initiated (noting the research programme remains in its infancy) the Multicentre, primary care **B**ased **I**nitiative to **E**nhance resilie**N**ce to **C**limatic **E**vents—**The AMBIENCE Project.** The primary objective of this project is to understand and explain how we can best equip primary care teams to *proactively recognise* those most vulnerable to provocative climatic conditions (those living with diabetes, hypertension, and heart disease at concurrent risk from respiratory illnesses) to develop practical, cost-effective health programmes to improve their health, both now and in the future, as the impact of climate change grows. The specific aims and related hypotheses of this pilot project are to:

1.  **Aim**: Document the pattern of primary care contacts across 3 diverse Australian GP/primary care clinics to reveal expected variations in the primary health care needs of people at high risk of (due to hypertension/diabetes) or with heart disease (and co-existent lung disease) in urban and rural/regional communities according to the climatic variations they experience (seasonal change/heat waves/cold-spells/bushfires) via a clinical record/surveillance study of >13,000 people over 5 years.

**Hypothesis**: There is a strong correlation (including at each location) between climatic conditions and acute weather events with the frequency and specific causes (including non-cardiac) of primary care presentations among a representative cohort of adults being treated for the common antecedents and/or established forms of heart disease living in urban and rural-regional communities of Australia.

2. **Aim**: At the individual/household level, document the spectrum of *vulnerability-resilience* to climatic provocations to health based on our holistic model (see Chap. 4) via *a prospective, cohort study of 150 randomly selected individuals/households being treated for heart disease.*

**Hypothesis**: Among a representative cohort of 150 men and women (1:1) aged ≥50 years living in a diverse range of communities and being actively managed for heart disease and common comorbidities (e.g. diabetes and respiratory disease) by a GP, at least one in three (33%) will demonstrate a high-level of vulnerability to climatic provocations to their health in two opposing seasons (baseline/6 months).

3. **Aim**: Develop a primary-focussed intervention specifically designed to promote resilience to climatic provocations to health among a diverse cohort of people being actively treated for the common forms of heart disease via a prospective pilot trial of 150 people to inform a more definitive/appropriately powered randomised trial with hard endpoints.

**Hypothesis**: Among a representative cohort of 150 men and women (1:1) aged ≥50 years living in a diverse range of

for common forms of heart disease and comorbidity by a GP, when compared to standard care (SC), a *Practice Nurse Coordinated Management Program* (PN-CMP) will result in greater (objectively measured) climatic resilience as well as fewer GP visits and hospital episodes during both 12-month (grant duration) and longer-term follow-up.

Consistent with the fundamental purpose of this book (in stimulating a new approach to how people living with heart disease are managed within the context of climate change, the AMBIENCE Project is designed to answer the following (over-arching) research question—*What adaptations to primary health care services (including profiling and management from a climatic health perspective) will promote resilience and better health outcomes in vulnerable Australians affected by heart disease?*

Once again, it is beyond the scope of this book to outline the finer details of this innovative project (our interdisciplinary team of investigators being unaware of similar projects worldwide). However, it is important to note that in planning the project we carefully considered 'where' and 'when' health care can best mitigate the harmful effects of climate change and who would benefit most from any intervention. The project also considered *key learnings* from the RESILIENCE Trial outlined in Chap. 8 (Stewart et al. 2024)—the findings of which re-emphasised the need to find vulnerable people and act before they are hospitalised and die prematurely due to climatic provocations. Some of the key decisions and aspects of the project have relevance to future research (and the provision of any dedicated health services) in this area.

**Identifying a Diversity of Vulnerable Communities**: As indicated by Figs. 9.8, 9.9 and 9.10, we selected a diverse range of communities, in whom the issues highlighted at a regional level in the section above, can be investigated and addressed in far more detail than the

**Fig. 9.8** Profile of a diverse and poorer, urban community in Australia. Urban Australia—Salisbury, South Australia is a (relatively) densely populated suburb, located in the northern parts of Adelaide. Close to 150,000 people live in the area, with an average household of 2.6 people. Just less than half of the population were born overseas, and a similar proportion are non-English speaking as a primary language. A diversity of cultures from Europe, the Middle East, Central Asia, South Asia, South-East Asia and Africa are represented within the community. The latest iteration of the *Australian Social Health Atlas* identifies key socio-economic and subsequent health disparities in Salisbury and neighbouring (urban/peri-urban) suburbs. Adelaide is exposed to a *Mediterranean climate (Köppen climate classification Csa)*, with mild to cool winters (mean max. 15.5 °C) with moderate rainfall and hot/dry summers (mean max. 29 °C) with hotter temperatures recorded in the flat Adelaide Plains (where Salisbury is located). In 2019, the mercury rose to 46.6 °C and in 2008 stayed above a max of 35 °C for 15 days (Australia city records) (The Koppen Climate Classification 2024; Australian Bureau of Statistics 2024; Australian Institute of Health and Welfare 2022; National Heart Foundation of Australia 2024)

broader HF-focussed report we produced for the whole of Australia. As such, it represents a natural progression in this research space and will, hopefully, inform some of the 'desirable' actions and outcomes outlined in Fig. 9.7.

**Developing and Testing New Ways of Managing Vulnerable People**: Those assigned to the PN-CMP will receive a primary care adapted version of the RESILIENCE Intervention described in Chap. 8. This will include titrated follow-up as needed and a focus on both physical and mental health [e.g. following extreme weather events (Hasnain et al. 2024; Usher et al. 2024; Walter et al. 2020; Mellish et al. 2024)] of study participants:

- *Individualised case-management*—in consultation with their GP and taking into consideration the additional findings on climatic vulnerability and that person's stated health priorities.

**Fig. 9.9** Profile of a town and linked communities in regional Australia. Regional Australia—Kyneton (and Daylesford) Victoria is a small but concentrated regional town located in the Macedon Ranges 87 km north of Melbourne. Just over 5000 people live in the town's footprint, while there are many more farming and community hamlets located in an increasingly popular "tree change" region with many tourists regularly visiting local thermal springs. Most people living in Kyneton, and surrounding communities have an Australian, Anglo-Saxon and Northern European background. The latest iteration of the *Australian Social Health Atlas* identifies key socio-economic and subsequent health disparities in Keyneton and neighbouring regional communities. At 616 m above sea level, Keyneton is exposed to a *relative cool and wet temperate climate (Köppen climate classification Cfb)*. Summer temperatures range markedly from 10 to 37 °C and winter temperatures are typically 1–2 to 9 °C. Annual rainfall fluctuates from 445 to 1350 mm with snowfall not uncommon—thereby providing a marked counterpoint to Adelaide and Horsham (The Koppen Climate Classification 2024; Australian Bureau of Statistics 2024; Australian Institute of Health and Welfare 2022; National Heart Foundation of Australia 2024)

- *Promoting physiological homeostasis*—this will include individualised provision of— vitamin D/iron replacement therapy, flexible diuretic/fluid plans, optimal asthma/COPD treatment/action plans, strengthening infection prevention measures (e.g. hand hygiene to prevent seasonal viral infections and proactive vaccination optimization) and applying the principles of cardiac (and as appropriate, pulmonary) rehabilitation (Piepoli et al. 2022).

- *Behavioural adaptions to seasonal/environmental challenges*: Education plus counselling sessions on how to avoid environmental provocations, routine 'cold spell', 'heat wave' and 'asthma' alerts access to the PN for day-to-day support and (if appropriate) sourcing warm clothing and blankets.
- *Optimising the home environment*: Facilitated access to government subsidies for home insulation/power bills social-support services, along with subsidised access to more efficient

**Fig. 9.10** Profile of remote, farming town/hub in rural Australia. Remote/Rural Australia—Horsham, Victoria is a relatively small but concentrated regional/rural town located 300 km north of Melbourne. Just over 20,000 people live in the town footprint, while there are many more farming and community hamlets located in a large but sparsely populated region of Grampian Ranges and tourists who regularly visit them. Many people living in Horsham and surrounding regions have an Australian, Anglo-Saxon and Northern European background. The latest iteration of the *Australian Social Health Atlas* identifies key socio-economic and subsequent health disparities in Horsham and neighbouring rural communities. Horsham is exposed to a *cold-semi arid climate (Köppen climate classification BSk)*. With a mean annual temperature of 14.7 °C and annual rainfall of 380 mm, due to its topography and confluence of adjacent Oceanic/Mediterranean climates, it is one of the coolest climates on the continent. However, in 2019 the mercury rose to 47.9 °C and in Feb 2024 a major bushfire burnt close to 5000 ha (The Koppen Climate Classification 2024; Australian Bureau of Statistics 2024; Australian Institute of Health and Welfare 2022; National Heart Foundation of Australia 2024)

sources of heating/cooling (with a dedicated budget quarantined for this purpose).

This programme will be evaluated and improved with a nested health economic and consumer-based consultations (with a strong emphasis on the co-design of future services/trials as needed). As detailed below, it also involves an important component of public/consumer engagement designed to promote greater community awareness of the issues being addressed—stressing once again, the solution to climate change and health won't be found in one simple package of intervention.

### 9.4.2  A Citizen Science Project to Explore Climatic Experiences and Perceptions

A critical component of the AMBIENCE project will be the involvement of participants as *'citizen scientists'* during the study. Citizen science refers to the process of involving non-scientists (volunteers or community members) in the scientific process, commonly for the purpose of data collection. Citizen science [as well as citizen juries (Wells et al. 2021; Ambagtsheer et al. 2024)] has become increasingly valued within the scientific community as it holds the potential to extend the bounds of data collection and encourages meaningful interaction between researchers, key stakeholders such as policy-makers, and the general public (Garcia-Rojas et al. 2022; Miller-Rushing et al. 2019). During AMBIENCE, participants will be engaged as citizen scientists to collect and monitor environmental conditions within their houses via the use of household weather monitoring stations, along with keeping a weekly diary relating to their perceptions of environmental conditions. All participants will receive training relating to the proper use of the equipment, honouring the principles of research, and enhancing consumer engagement capacity. Their (and their families'/significant others) perceptions of weather/climatic conditions relative to official statistics and their home conditions will provide important information for future planning, considering two important and interrelated factors to consider—(1) As described earlier in the book, as people get older and live with chronic disease, their perceptions of, and responses to temperature variations are effectively 'blunted' at both ends of the cold-to-heat spectrum and, (2) As also described, there is increasing focus on the reporting of measures of "comfort" such as the *Universal Thermal Climate Index* (which was developed to quantify thermal comfort in relation to the ability of the body to maintain homeostasis through self-thermoregulation) (Brode et al. 2013) Critically, households are typically filled with people at different life-stages, varying levels of health and perceptions of ambient conditions and their comfort levels. This important component of the AMBIENCE Project will compare these within households and across households experiencing similar weather conditions and acute events.

## 9.5    Looking to the Future

Covering a broad range of topics, but with the express purpose of pointing the way to better management of people at risk of, or living with heart conditions/disease worldwide, this book has probably raised more questions than answers in relation to climate change. Unfortunately this reflects two truisms—(1) Modern medicine/health care has large ignored the pervasive impact of climatic conditions on human health and, most specifically, the cardiovascular system and is therefore ill-equipped to generate specific therapeutic programmes to prevent and attenuate the negative effects of this phenomenon (many are positive ones!) and, (2) Climate change is altering the dynamics of these human-climate interactions and will continue to do so until we can halt and reverse it. Thus, we are working with a paucity of information to make substantive gains in health, both now and in the future. The last chapter reviews what we do know and what our priorities might be for the future—especially around growing the evidence-base for cost-effective therapies and programmes to protect people from climate change.

## References

Ambagtsheer RC, Hurley CJ, Lawless M, Braunack-Mayer A, Visvanathan R, Beilby J, Stewart S, Cornell V, Leach MJ, Taylor D, et al. IMPAACT: improving the participation of older people in policy decision-making on common health conditions: a study protocol. BMJ Open. 2024;14:e075501.

Australian Bureau of Statistics. Statistics/census; 2024. https://www.abs.gov.au/. Accessed April 2024.

Australian Institute of Health and Welfare. https://www.aihw.gov.au/reports/heart-stroke-vascular-diseases/hsvd-facts/contents/about. Accessed April 2024.

Australian Institute of Health and Welfare. Health across socio-economic groups; 2022. https://www.aihw.gov.au/reports/australias-health/health-across-socioeconomic-groups. Accessed May 2024.

Avramidis I, Pagkozidis I, Domeyer PJ, Papazisis G, Tirodimos I, Dardavesis T, Tsimtsiou Z. Exploring perceptions and practices regarding adult vaccination against seasonal influenza, tetanus, pneumococcal disease, herpes zoster and COVID-19: a mixed-methods study in Greece. Vaccines. 2024;12:695.

Banerjee A, Pasea L, Chung SC, Direk K, Asselbergs FW, Grobbee DE, Kotecha D, Anker SD, Dyszynski T, Tyl B, Denaxas S, Lumbers RT, Hemingway H. A population-based study of 92 clinically recognized risk factors for heart failure: co-occurrence, prognosis and preventive potential. Eur J Heart Fail. 2022;24:466–80.

Beggs PJ, Trueck S, Linnenluecke MK, Bambrick H, Capon AG, Hanigan IC, Arriagada NB, Cross TJ, et al. The 2023 report of the MJA-lancet countdown on health and climate change: sustainability needed in Australia's health care sector. Med J Aust. 2024;220:282–303.

Bell S, Campbell J, Lambourg E, Watters C, O'Neil M, Almond A, Buck K, Carr EJ, Clark L, Cousland Z, Findlay M, Joss N, Metcalfe W, Petrie M, Spalding E, Traynor JP, Sanu V, Thomson P, Methven S, Mark PB. The impact of vaccination on incidence and outcomes of SARS-CoV-2 infection in patients with kidney failure in Scotland. J Am Soc Nephrol. 2022;33:677–86.

Blashki G, McMichael T, Karoly DJ. Climate change and primary health care. Aust Fam Physician. 2007;36:986–9.

Bragg F, Lacey B. Social and spatial inequalities in premature mortality across Europe. Lancet Public Health. 2024;9:e148–9.

Brode P, Blazejczyk K, Fiala D, Havenith G, Holmer I, Jendritzky G, Kuklane K, Kampmann B. The universal thermal climate index UTCI compared to ergonomics standards for assessing the thermal environment. Ind Health. 2013;51:16–24.

Burrell LM, Beilby J, Chan YK, Stewart S. A report on the seasonal impact of heart failure in Australia; 2019.

Carrington MJ, Jennings GL, Stewart S. Pattern of blood pressure in Australian adults: results from a national blood pressure screening day of 13,825 adults. Int J Cardiol. 2010;145:461–7.

Carrington MJ, Jennings GL, Clark RA, Stewart S. Assessing cardiovascular risk in regional areas: the healthy hearts beyond city limits program. BMC Health Serv Res. 2012;12:296.

Champagne SN, Phimister E, Macdiarmid JI, Guntupalli AM. Assessing the impact of energy and fuel poverty on health: a European scoping review. Eur J Public Health. 2023;33:764–70.

Chan YK, Tuttle C, Ball J, Teng TK, Ahamed Y, Carrington MJ, Stewart S. Current and projected burden of heart failure in the Australian adult population: a substantive but still ill-defined major health issue. BMC Health Serv Res. 2016;16:501.

Clark RA, Eckert KA, Stewart S, Phillips SM, Yallop JJ, Tonkin AM, Krum H. Rural and urban differentials in primary care management of chronic heart failure: new data from the CASE study. Med J Aust. 2007;186:441–5.

Clark RA, Coffee N, Turner D, Eckert KA, van Gaans D, Wilkinson D, Stewart S, et al. Application of geographic modeling techniques to quantify spatial access to health services before and after an acute cardiac event: the cardiac accessibility and remoteness index for Australia (ARIA) project. Circulation. 2012;125:2006–14.

Department of Health, Federal Government of Australia. Social health Atlas of Australia; 2023. https://data.gov.au/dataset/ds-sa-9fb5195b-e696-45bb-a277-866e2efa62d5/distribution/dist-sa-7da720f1-bc9e-4487-a89c-3fd695bb8faf/details?q=. Accessed May 2024.

Garcia-Rojas MI, Keatley MR, Roslan N. Citizen science and expert opinion working together to understand the impacts of climate change. PLoS ONE. 2022;17:e0273822.

Global Burden of Disease Investigators. Measuring routine childhood vaccination coverage in 204 countries and territories, 1980–2019: a systematic analysis for the global burden of disease study 2020, release 1. Lancet. 2021;398:503–21.

Global Climate Change. Population displacement and public health. New York: Springer; 2021.

Hasnain MG, Garcia-Esperon C, Tomari YK, Walker R, Saluja T, Rahman MM, Boyle A, Levi CR, Naidu R, Filippelli G, Spratt NJ. Bushfire-smoke trigger hospital admissions with cerebrovascular diseases: evidence from 2019–20 bushfire in Australia. Eur Stroke J. 2024;12:23969873231223308.

Lawler C, Sherriff G, Brown P, Butler D, Gibbons A, Martin P, Probin M. Homes and health in the outer hebrides: a social prescribing framework for addressing fuel poverty and the social determinants of health. Health Place. 2023;79:102926.

Mellish S, Ryan JC, Litchfield CA. Short-term psychological outcomes of Australia's 2019/20 bushfire season. Psychol Trauma. 2024;16:292–302.

Miller-Rushing AJ, Gallinat AS, Primack RB. Creative citizen science illuminates complex ecological responses to climate change. Proc Natl Acad Sci USA. 2019;116:720–2.

Nakao YM, Nakao K, Nadarajah R, Banerjee A, Fonarow GC, Petrie MC, Rahimi K, Wu J, Gale CP. Prognosis, characteristics, and provision of care for patients with the unspecified heart failure electronic health record phenotype: a population-based linked cohort study of 95,262 individuals. EClinicalMedicine. 2023;63:102164.

National Heart Foundation of Australia. Australian heart maps; 2024. https://www.heartfoundation.org.au/for-professionals/australian-heart-maps. Accessed May 2024.

Piepoli MF, Adamo M, Barison A, Bestetti RB, Biegus J, Bohm M, Butler J, Carapetis J, Ceconi C, Chioncel O, Coats A, Crespo-Leiro MG, et al. Preventing heart failure: a position paper of the heart failure association in collaboration with the European association of preventive cardiology. Eur J Heart Fail. 2022;24:143–68.

Rajagopalan S, Al-Kindi SG, Brook RD. Air pollution and cardiovascular disease: JACC state-of-the-art review. J Am Coll Cardiol. 2018;72:2054–70.

Roth GA, Mensah GA, Johnson CO, Addolorato G, Ammirati E, Baddour LM, Barengo NC, Beaton AZ, Benjamin EJ, Benziger CP, Bonny A, Brauer M, et al. Global burden of cardiovascular diseases and risk factors, 1990–2019: update from the GBD 2019 study. J Am Coll Cardiol. 2020;76:2982–3021.

Sabanoglu C, Sinan UY, Akboga MK, Coner A, Gok G, Kocabas U, Bekar L, Gazi E, Cengiz M, Kilic S, Inanc IH, Cakmak HA, Zoghi M. Long-term prognosis of patients with heart failure: follow-up results of journey HF-TR study population. Anatol J Cardiol. 2023;27:26–33.

Sochalski J, Jaarsma T, Krumholz HM, Laramee A, McMurray JJ, Naylor MD, Rich MW, Riegel B, Stewart S. What works in chronic care management: the case of heart failure. Health Aff. 2009;28:179–89.

Stewart S. Home is where the heart is when it comes to transitional care in heart failure, but is it the only way to improve health outcomes? Eur J Heart Fail. 2017;19:1444–6.

Stewart S, Blue L. Improving outcomes in chronic heart failure: Specialist nurse intervention from research to practice. 2nd ed. London: BMJ Books; 2004.

Stewart S, Marley JE, Horowitz JD. Effects of a multi-disciplinary, home-based intervention on unplanned readmissions and survival among patients with chronic congestive heart failure: a randomised controlled study. Lancet. 1999;354:1077–83.

Stewart S, MacIntyre K, Hole DJ, Capewell S, McMurray JJ. More 'malignant' than cancer? Five-year survival following a first admission for heart failure. Eur J Heart Fail. 2001;3:315–22.

Stewart S, Ekman I, Ekman T, Oden A, Rosengren A. Population impact of heart failure and the most common forms of cancer: a study of 1,162,309 hospital cases in Sweden (1988 to 2004). Circ Cardiovasc Qual Outcomes. 2010;3:573–80.

Stewart S, Carrington MJ, Marwick T, Davidson PM, Macdonald P, Horowitz J, Krum H, Newton PJ, Reid C, Scuffham PA. Which heart failure intervention is most C-e and consumer friendly in reducing Hospital care trial I. The WHICH? Trial: rationale and design of a pragmatic randomized, multicenter comparison of home- versus clinic-based management of chronic heart failure patients. Eur J Heart Fail. 2011;13:909–16.

Stewart S, Playford D, Scalia GM, Currie P, Celermajer DS, Prior D, Codde J, et al. Ejection fraction and mortality: a nationwide register-based cohort study of 499,153 women and men. Eur J Heart Fail. 2021;23:406–16.

Stewart S, Chan YK, Playford D, et al. Mild pulmonary hypertension and premature mortality among 154,956 men and women undergoing routine echocardiography. Eur Respir J. 2022a;59:5879.

Stewart S, Chan YK, Playford D, et al. Incident aortic stenosis in 49,449 men and 42,229 women investigated with routine echocardiography. Heart. 2022b;108:875–81.

Stewart S, Patel SK, Lancefield TF, Rodrigues TS, Doumtsis N, Jess A, Vaughan-Fowler ER, Chan YK, Ramchand J, Yates PA, Kwong JC, McDonald CF, Burrell LM. Vulnerability to environmental and climatic health provocations among women and men hospitalized with chronic heart disease: insights from the RESILIENCE TRIAL cohort. Eur J Cardiovasc Nurs. 2024;23:278–86.

Stewart S. Charting and navigating the "scylla and charybdis" conundrum of our ageing hearts—heart failure and atrial fibrillation ageing hearts—heart failure and atrial fibrillation; 2023.

Strange G, Stewart S, Celermajer D, Prior D, Scalia GM, Marwick T, Ilton M, Joseph M, Codde J, et al. Poor long-term survival in patients with moderate aortic stenosis. J Am Coll Cardiol. 2019a;74:1851–63.

Strange G, Stewart S, Celermajer DS, Prior D, Scalia GM, Marwick TH, Gabbay E, Ilton M, Joseph M, Codde J, et al. Threshold of pulmonary hypertension associated with increased mortality. J Am Coll Cardiol. 2019b;73:2660–72.

Strange G, Playford D, Scalia GM, Celermajer DS, Prior D, Codde J, Chan YK, Bulsara MK, et al. Change in ejection fraction and long-term mortality in adults referred for echocardiography. Eur J Heart Fail. 2021;23:555–63.

The Köppen Climate Classification. https://www.mindat.org/climate.php. Accessed April 2024.

Tribouilloy C, Rusinaru D, Mahjoub H, Souliere V, Levy F, Peltier M, Slama M, Massy Z. Prognosis of heart failure with preserved ejection fraction: a 5 year prospective population-based study. Eur Heart J. 2008;29:339–47.

Usher K, Rice K, Williams J. Editorial for IJMHN: an application of the 'one health' approach for extreme weather events and mental health—can the adoption of a 'one health' approach better prepare us for the predicted drought in parts of rural Australia? Int J Ment Health Nurs. 2024;33:220–3.

Walter CM, Schneider-Futschik EK, Knibbs LD, Irving LB. Health impacts of bushfire smoke exposure in Australia. Respirology. 2020;25:495–501.

Wells R, Howarth C, Brand-Correa LI. Are citizen juries and assemblies on climate change driving democratic climate policymaking? An exploration of two case studies in the UK. Clim Change. 2021;168:5.

Wolk D, Porter R. Climate change and policy reforms: a view from the primary care clinic. J Am Board Fam Med. 2024;37:19–20.

Xie E, Howard C, Buchman S, Miller FA. Acting on climate change for a healthier future: critical role for primary care in Canada. Can Fam Physician. 2021;67:725–30.

# Preventing an Anti-climatic Response to Climate Change

10

**Abstract**

This book has provided a wide-ranging review (from global health issues to individuals traits of climatic vulnerability) of the key issues surrounding climate change and heart disease. In the process it has argued for a '**paradigm change**' in how clinicians and the broader health system consider 'where' people with antecedent risk factors and established forms of heart disease live and work and also 'when' climatic conditions in that location, might provoke a clinical crisis and even death. This requires a new mindset around clinical management (including specific sections/reference to climate factors in expert guidelines) and extends to how disease statistics are reported and presented. However, it would be disingenuous to present all the facts, figures and opinions proffered in this book as the unbridled truth or beyond contest. There is still so much to learn and understand about heart disease and other forms of cardiovascular disease— especially beyond high-income populations with well-developed health systems. This truism is even more stark when considering our knowledge (and potential response) to the impact of climate change and how it will affect the global burden of heart disease and other common forms of cardiovascular disease sensitive to external factors. Thus, this final chapter will critically reflect on the topics covered in each chapter and then identify what is needed in terms of action (including new research and resources) to ensure that climate-provoked cardiac events don't overwhelm already capacity constrained health systems worldwide.

**Keywords**

Cardiopulmonary health · Climate-focussed healthcare · Climate extremes · Climate awareness · Public health policy · Population prevention

## 10.1 Investing in Those Poor in Money but Rich in Ingenuity and Purpose

For whatever reason, there is a broad expectation that any solutions or therapies to the world's problems will be solved by the richest and most well-resourced countries in the world. Yet, as described in Chap. 2, have inflicted global problems that they continue to contribute to (as well as self-inflict upon themselves) high income countries are 'tone deaf' to the plight of poor countries who suffer the worst consequences of climate change. Moreover, the collectively

S. Stewart, *Heart Disease and Climate Change*, Sustainable Development Goals Series,
https://doi.org/10.1007/978-3-031-73106-8_10

ignore the fact that any solutions developed with ingenuity and minimal resources can be readily adapted to richer, more well-resourced countries. Thus, there is an expectation that any solution to heart disease and climate change (two of our biggest current and future health threats) will filter down to the poorer nations of the world, rather than vice-versa. Clearly, this is not working for those individuals, communities and countries for whom the UNSDGs were formulated!

As reflected in Fig. 1.3 (Chap. 1), action on aspirational health and climate goals are unlikely to be achieved without addressing 'inequality'. This is specifically true for any aspirational goals to mitigate a rising tide of heart disease in the poorer regions of the world because of increasingly provocative climatic conditions and persistently high levels of climatic vulnerability from an individual to population level. Whilst acknowledging some amazing funding schemes and levels of organisational support, one of the persistent areas of inequality in this space is the lack of investment and support of research teams in low-income countries to better understand and respond to cardiopulmonary disease in their part of the world. From personal experience, available funding represents a 'drop-in-the-ocean' for what is needed, and the assessment of grant applications is often patronising ("*we already know this*" and "*why are you doing it this way?*"). Consequently, some of the most talented and dedicated researchers in the world (most of whom ignore the bright lights of money and prestige offered by the world's leading institutions), remain in their communities and do their best with what they have. It is these people and their teams who—(a) will experience the brunt of climate change and its impact on the cardiovascular system and heart and, (b) have the latent capacity to develop programmes and therapeutics that would benefit a world yet to experience what they are (in effect acting like a 'time-machine' before climate change progresses). In essence, they can offer high-income countries with future solutions to a growing problem.

As just one example, consider a currently, non-funded initiative to bring together a range of countries, people and institutions in Southern Africa and beyond to address an evolving epidemic of cardio-pulmonary disease in the region—see Fig. 10.1. This 'South-North' initiative had its genesis in the 'Heart of Soweto' and then 'MOZART' (Mocumbi et al. 2019) studies described earlier in the book. Both studies (particularly the former with a Lancet publication and many more high-level reports) have helped reshape thinking about cardiopulmonary disease in the region—and yet neither were peer-review funded. Instead, they relied on philanthropic and commercial funding. Nevertheless, these studies led to the connections and capacity building to enable (with further funding) this 'coalition of the willing' to undertake a meaningful programme of capacity building and research that will benefit the entire regions—see Fig. 10.2. With climate change an increasing threat to the whole world, it would make sense to invest in those who sit at the 'coalface' of its impact and allow them to develop cost-effective strategies (given their low-resource environment) such as micro-financing of cleaner fuels, that can be scaled-up and adapted worldwide.

Some of the key research questions a global network of research centres might address would be:

- What is the local pattern of cardiopulmonary disease (including antecedents and natural history) and how does it compare to the rest of the region/world?
- What role do current climatic conditions and weather events play on the pattern/burden of disease?
- What impact will predicted/evolving climate change have on the pattern/burden of disease.
- What are the characteristics of 'climatic vulnerability' versus 'climatic resilience' in the region and how can the latter be developed?
- What preventative measures to management strategies will be most effective in reducing the burden of climatic-provoked cardio-pulmonary events/ill health and, therefore, the overall burden of disease?

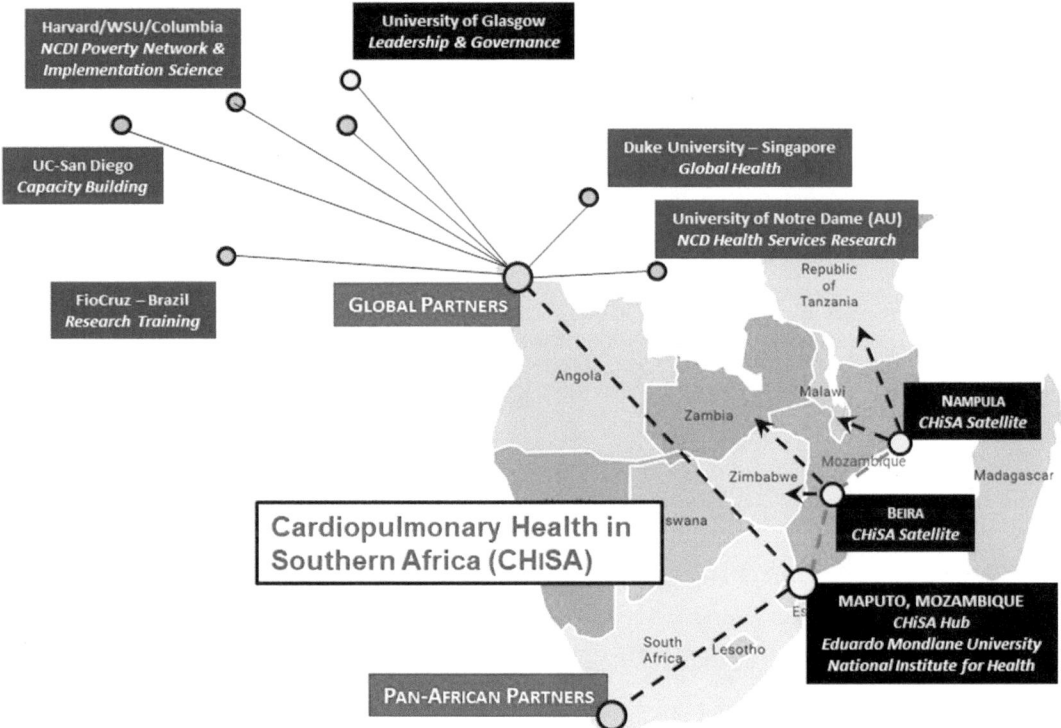

**Fig. 10.1** A major initiative to address high rates of cardiopulmonary disease in Southern Africa. This figure shows the major regional and international partners seeking to improve the cardiopulmonary health of the diverse peoples living in Southern Africa with a major focus on addressing climate change

## 10.2 Heralding a New Era of Climate-Focussed Medicine

In sometimes provocative fashion, Chap. 2 highlighted some of the biggest problems the world is facing, pointing out that the weather and climate change 'have no borders' and whatever 'we' do, eventually affects everyone else on the planet. Sometimes a problem is too big to address, unless of course it is broken-down into smaller, more achievable pieces. For the average clinician/healthcare worker, there is very little leverage to encourage whole governments and populations to change their activities and habits. However, as already described in this book the intersection between heart disease, climatic conditions and climate change is real. Moreover, there is way clinicians and health professionals, worldwide can do their best to mitigate the adverse effects of challenging climatic conditions and climate change on the heart health of the people they care for and treat.

Firstly, on reading this book and with some reflection, the reader might already recognise that these connections have already likely influenced 'who', 'where' and, indeed, 'when' someone presented to their healthcare facility/practise with a cardiac event. Recognising a problem is the first step to any solution. Once this phenomenon has been recognised (or more fully realised) it makes sense to adjust our collective thinking around the assessment of people's wellbeing and their health management and usher in *a new paradigm of climate focused healthcare*. As succinctly put by advocates for a more specific (but highly congruent) "*Era of Climate Change Medicine*" (Conrad 2023). In the context of climate change—"*New diseases are being identified, existing ones are being exacerbated, and traditional health care delivery is being challenged*" (Conrad 2023; Davies and Bhutta 2022).

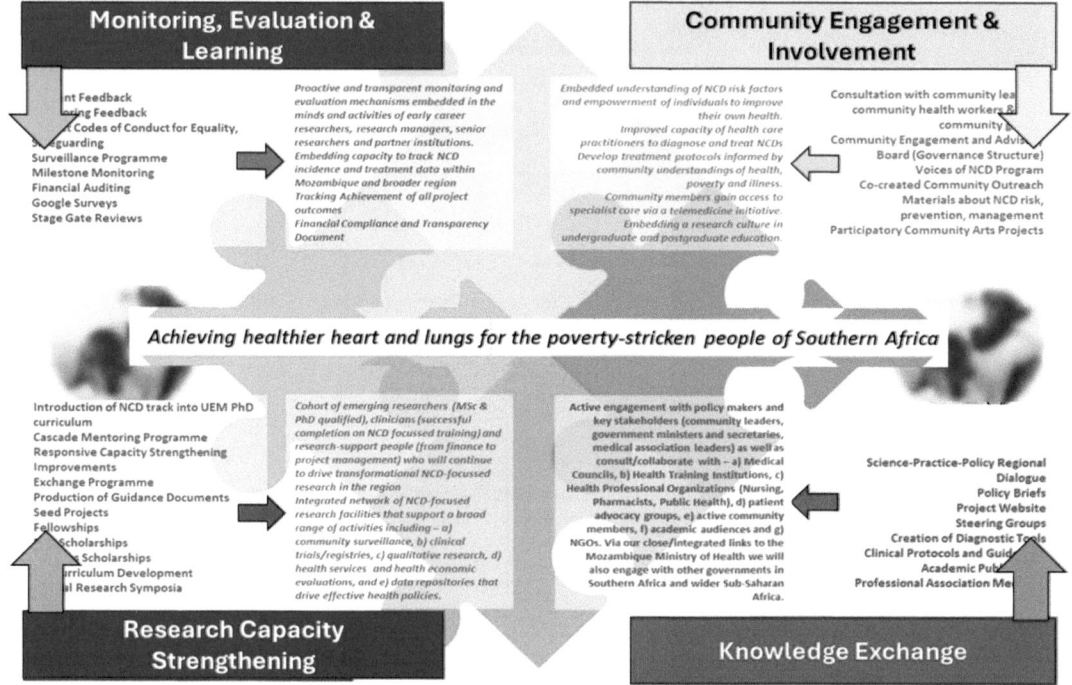

**Fig. 10.2** A major programme of capacity building and research in Southern Africa. This figure shows the major components of a carefully designed framework (highlighting its main components and a myriad of enabling goals and activities) to enhance the kind of capacity building, engagement and research needed to improve the cardiopulmonary health of people living in Southern Africa

Beyond the specific research questions and evidence needed to support new therapies/strategies (see below), it seems clear we need new 'champions' across all health care professions to advocate for the changes needed to address an otherwise neglected driver of ill-health worldwide—climate provocations to health (with the cardiovascular system a core consideration). This includes challenging everyone already working as a health professional/clinician to reconsider what they understand about this broad topic and, ensuring the next generation of health professionals have a comprehensive grasp of the issues covered within this book.

## 10.3   How Far Can We Adapt to Climate Extremes?

Chapter 3 outlined some of the biological to psychological imperatives that drive our physiological responses and impulses to maintain thermoregulation and homeostasis. There is a myriad of issues that arise from the fact that climate change is challenging multi-generational adaptations to ambient weather conditions and climatic/seasonal transitions within a single generation. These intersect with other major phenomena such as rising obesity levels and increasingly more sophisticated technology to maintain thermoregulatory control/homeostasis Some important issues/research questions arising from this are as follows:

- Is it true that otherwise inhabitable parts of the planet will become uninhabitable with climate change, or will our ability to physiologically and technologically adapt overcome climatic extremes?
- What (more) can we learn from traditional cultures (e.g. the Sami to Aboriginal Australians) about climate adaptations?
- Will children who become obese and develop early forms of CVD become highly vulnerable to climate change?

- What activities should we be promoting versus discouraging to better adapt to our changing climate?

## 10.4 How Do We Recognize Climatic Vulnerability?

Chapter 4 introduced the important concept(s) of *climatic vulnerability* and *climatic resilience*. It is hoped that the contents of the is book (taken a whole), has convinced the reader that profiling at risk/vulnerable individuals using the framework/model presented in Fig. 4.2, is something that should be routinely performed in nearly all health care settings. Nevertheless, beyond the question of *"what to do with someone who appears to be on the high-end vulnerable spectrum?"* (see below!), there are still many issues/research questions to be addressed in this regard. These include:

- Is there really a spectrum of vulnerability to resilience, or is it more a dichotomy? If so, what is the threshold at which someone passes from resilient to vulnerable?
- What are the key components to determining vulnerability and resilience from a cardiovascular perspective—what additional components need to be considered or proffered components/indicators disregarded?
- Is it possible to develop a simple screening tool?
- What does the spectrum of vulnerability and resilience look like in different patient populations and people living in different regions of the world under varied socio-economic and cultural circumstances?

In considering the variety of geographic challenges (from a climate perspective) we face as a global community, I was reminded that in many established cultures around the world, there are many 'notorious winds' that are specifically identified, named and then associated (from folklore to contemporary times) with provocative and often harmful conditions. This includes the Föhn (essentially a change from wet and cold conditions on one side of a mountain, to warmer and drier conditions on the other that can extend for hundreds of kilometres either side) that is known as the 'Chinook' or "snow eater" of the North American Rocky Mountains, the 'Zonda' of the South American Andes and the 'Helm' wind of the English Pennines. These winds contrast with those winds classically affecting coastal regions. In Southern Spain (near the narrow passage/transition of the Atlantic Ocean into the Mediterranean Sea—the Straits of Gibraltar) for example, it's common for the locals to talk about the infamous 'Levante' and its counterpart the 'Poniente'. As climate change progresses, it will be interesting to see if new 'winds' occur and if we create new names and associations (both good and bad) with their occurrence.

## 10.5 Confronting an Inconvenient Truth About Cold Versus Heat!

As mentioned earlier in this book, I was inspired to research the impact of seasons and climate on heart health, when first hearing (first-hand) the pioneering research of people like the EuroWinter and MONICA Investigators—I'll be honest here and state that I don't know what prompted them to challenge conventional thinking, but I'm glad they did! Thus, I remain frustrated with those (unfortunately a loud majority of voices telling governments and funding bodies alike the narrative they expect to hear and fund, respectively) that 'heat' is our single biggest enemy. This 'polarising' narrative (much like modern-day politics) ignores the full-spectrum and complexity of climatic challenges to human health.

As described in Chap. 5, heat is indeed our (growing) enemy, but it's not the only or worse (yet) problem from a health perspective. This is due to three irrefutable facts—(1) There are more deaths from cold climatic conditions than hot conditions, (2) Paradoxically, cold kills just as many (or even more people) in milder climates because of the way we adapt to the most prevailing weather conditions (just think of trying to find an air-conditioned hotel room in London or a hotel room with a heater in Singapore!) and (3) Climate change is all about volatility (noting the fatal incident of air-turbulence affecting a Singapore Airlines flight just shortly before I completed this book!) and it will result in more weather extremes (thereby challenging our ability to maintain homeostasis)—including more relatively cold weather.

If I had one simple recommendation arising from the research that has informed the contents of this book, therefore, it would be for scientists and health professionals alike, to adopt a more balanced appreciation of the role of heat and cold in provoking past, current and future cardiac events. As noted above, you can deal with a problem if you don't know if exists! Consider a very recent report from the UK, aiming to examine the pattern of deaths due to extreme heat and cold events during the COVID-19 pandemic (Lo et al. 2024). Overall, during the study period (2020–2022) the researchers documented 8481 (95% CI 6387–10,493) excess deaths due to heat extremes versus 128,533 (95% CI 107,430–153,642) excess deaths due to cold extremes—with the pre-Christmas/festive date (not the coldest time of winter), the 15th December, being the deadliest day. As intended, the authors 'debunked' the notion that climate change is deadlier than COVID-19 (noting that we have a vaccine for one and not the other!). However, in doing so (i.e. highlighting the headline results), the fact that cold resulted in 15-fold more deaths were completely lost in translation from a scientific to public health perspective.

In this context, it is imperative that more community-to-population surveillance studies (with accurate and proportional reporting of the effects of cold versus heat effects, as per some of the studies highlighted earlier in the book) of the impact of climatic conditions, climate change and cardiopulmonary health be conducted—with a priority of conducting studies in populous, low-income countries. We also desperately need to quantify the 'climatic' effect worldwide (including the cost to health care expenditure, quality of life and premature

mortality) of excess cardiac and broader cardiovascular events triggered by climatic conditions. Such data will 'incentivise' more research and, eventually, evidence-based health programmes to both prevent and then mitigate the problem.

## 10.6  Peering Through the Fog of Pollution

While the world tends to think of climate change and pollution as a case of cause and effect, as explained in Chap. 6, this couldn't be further from the truth. It's an unfortunate fact that there is an adverse synergistic relationship between the two, with serious cardiovascular consequences. While health professionals and the public alike tend to focus on external air pollution (a case of recognise and address what you can see and breathe), the avenues to which we are poisoning ourselves is growing, while poorly addressed issues such as indoor pollution remain so. Apart from recommending when and how to avoid pollution (remembering that for many people, including the wealthy, this is unavoidable), health professionals are largely powerless to address the broader issues around pollution.

Despite, the inability to change things (beyond advocacy), health professionals can at least improve their knowledge of 'when' pollution is likely be at its worst (according to climatic conditions) and provide the requisite advice to the people they care for/treat. Concurrently, understanding the pathology of pollution-induced cardiovascular damage will be increasingly important. Unfortunately, there are still much we don't know about the role of pollution (in all its forms) in triggering and exacerbating cardiac events and more research in this area in needed—including those living in the most polluted urban areas of the planet. Simultaneously, there are very few therapeutics that have been developed to counter the impact of pollution on the cardiopulmonary system and this should be an ongoing area of research with commercial backing given the scale of the problem and the investments needed to bring such therapeutics to market.

## 10.7 Achieving Equity in Knowledge and Understanding

As a simple extension of the conclusions made in Sects. 10.1 and 10.5 above, Chap. 7 highlighted the global burden of heart complications generated by a range of endemic infectious diseases that will affect many more (noting some diseases may well decline) people worldwide as climatic conditions rapidly evolve—including those living in high-income countries. A simple example of this would be the spread of Chagas' Disease into Spain and the southern parts of the US. Like the debate about the threat of heat versus cold, there appears to be a real disconnect in how the threats of traditional (i.e. those framed by the Framingham Study in a mainly white semi-rural community in North America!) versus non-traditional pathways to heart disease are perceived and reported. This includes health professionals (at least those living in high-income countries) and the public alike.

Once again, therefore, it is imperative that more community-to-population surveillance studies of the impact of climatic conditions, climate change and cardiopulmonary health be conducted in the most vulnerable peoples in the world—those living in low-to-middle income countries in whom the aspirational goals of the UNSDG's are most relevant (Pineiro et al. 2023; World Heart Federation 2023).

## 10.8 Managing the Cardiac Population with Existing Climatic Vulnerabilities

In Chap. 8, a fledgling research programme (at least in terms of exploring the effects of a multifaceted intervention directed towards those hospitalised with multi-morbid heart disease) was presented. As discussed, for many reasons, the RESILIENCE Trial failed to provide definitive answers, whilst posing more questions around how best to manage vulnerable individuals who are clearly experiencing cardiac events under certain climatic conditions. Beyond the research priorities and issues discussed in the sections above, it is perhaps this area of research that is most pressing—for the simple fact that we have—(a) unequivocable evidence that 'excess' cardiac events are linked to both 'ordinary' and extreme climatic conditions (that perhaps contribute to as much as 20% of cardiac events worldwide) and, (b) there are certain traits (in this book, they are labelled modifiable and non-modifiable vulnerabilities) that render those admitted to hospital with chronic forms of heart disease, most vulnerable to this costly, debilitating (in terms of disease progression) and deadly phenomenon.

Thus, there are some fundamental research questions to be addressed:

- Is it possible, relative to gold-standard management/treatment, to further improve health outcomes among the growing patient population of older people with chronic heart disease, by implementing singular-to-multifactorial strategies designed to address climatic vulnerability?
- If so, what is the magnitude of effect and are such programmes cost-effective?
- In which people do they work best and in which people do they not work—predicating a different approach?
- Is it feasible to integrate climatic-focussed programmes into 'ordinary' disease management programmes?
- Will such programmes work in lower-resource health settings and for younger individuals.

In simple terms, we are at the infancy of developing programmes to reduce high rates of morbidity and mortality in those with chronic heart disease, but this remains a priority given the combination of a more complex cardiac patient population now being increasingly challenged by climate change.

## 10.9 Targeted Prevention at the Population Level

Along with the imperative to address established vulnerabilities in people already hospitalised with heart disease, there is a clear mandate to assess what has been implemented previously in respect to protecting people from climate/weather-induced discomfort to ill-health in the context of climate chance and the increased likelihood of more provocations. The following strategies, have their relative strengths and weaknesses and need to be considered against the capacity of the society/community in question to afford their implementation. Once again, this means that many of the strategies listed below (and any new ones developed thereafter) will remain largely unobtainable to the poorest nations of the world. Nevertheless, this remains a useful list of 'aspirational' goals:

- Subsidised clothing to high-risk individuals living in colder climates and those where cold-snaps are ill-prepared for—*this would be a costly and difficult strategy to implement and would probably need formal assessment of climatic vulnerability and socio-economic circumstances.*
- Subsidised heating and cooling—*this is a commonly applied strategy in many jurisdictions of the world (mostly high-income) but, in many cases, it requires people to apply for support—predicating knowledge of the reasons it is implemented beyond 'personal comfort'.*
- Monitoring diurnal skin and ambient temperatures—*when applied to high-risk people with an alert-based system, this strategy could potentially streamline both societal and individually-based strategies to avoid exposure to environmental extremes (particularly in the home).*
- Monitoring and supplementing vitamin D levels (either pharmacologically or with light-based therapy)—*this is an increasingly common therapeutic strategy, but the target population(s) and benefits are not yet clear-cut.*
- Detecting and treating seasonal affective disorder—*like the strategy above, this may well be best applied to those living in higher latitude countries.*
- Annual influenza vaccinations, single-dose pneumococcal vaccination, structured COVID-19 vaccinations based on the latest evidence around sub-types and infections risks AND priority distribution of established and newly developed vaccinations for endemic diseases associated with cardiac complications—*preventative and targeted (in this case older individuals with high antecedent risk for heart disease or established forms of cardiovascular disease and those living in high-risk infectious areas) vaccination remains central to any strategy to limit the impact of climate change.*
- Public health alert systems for cold weather, heatwaves, dynamic weather conditions, and high levels of air pollution—*this requires a whole of population approach, educating the public on why alerts are issued and what needs to be done to avoid exposure to provocative conditions. Such alerts have the potential to provoke anxiety and/or be largely ignored if too many are generated, or health literacy levels are sub-optimal. Thus, a more tailored approach might work better.*
- Providing public shelter against weather extremes—*it is noticeable that during the pandemic there were more deaths from temperature extremes. One possible influence (beyond social isolation) was the inability of people to travel to public places with climatic control, i.e. where a comfortable temperature was maintained. This may well need to be formalised in the future, with invitations to cost-effectively 'protect' people against climate extremes.*
- Rescheduling social, educational and work schedules to avoid temperature extremes—*already in practice by many occupations routinely exposed to outdoor environments, there*

*may well come a time when people travel, socialise and work at different times (e.g. early morning/late evenings) as a mimic of many cultural practices and traditions among those already living in locations with cold-to-hot extremes.*

Like the multifaceted intervention applied in the RESILIENCE Trial, we simply don't know what 'cocktail' of interventions will work best. However, it is notable that many of the strategies listed above, would be best (and cost-effectively) applied via primary health care teams with the ready capacity to determine who is displaying the early features/characteristics of *climatic vulnerability* (remembering the discussion around 'spectrum' versus 'threshold') and would therefore benefit from a combination of these strategies. This is why the programme of primary care-based research described in Chap. 9 is so important moving forward. In this respect, Fig. 10.3 shows how the AMBIENCE Project described in that chapter focuses on the needs of urban-, regional- and rural-dwelling communities in a thoughtful manner.

## 10.10 Final Remarks—Can We Handle the Truth About Climate Change?

There is a great the irony that in researching and thinking about the contents of this book, I have travelled far and wide around the world in planes that make a substantial contribution (an estimated 2.5% of $CO_2$ emissions) (https://ourworldindata.org/global-aviation-emissions#:~:text=Aviation%20accounts%20for%202.5%25%20of,to%20global%20warming%20to%20date.&text=Flying%20is%20one%20of%20the,of%20the%20world's%20carbon%20emissions). In doing so, I form a new generation that takes travelling the world for granted, selecting a list of 'bucket list' places I'd like to visit for both academic and personal reasons. Fundamentally, I don't want to give that up. However, having seen the grinding congestion of tourism in a post-pandemic world, it's easy to see why there are calls for travel quotas worldwide—especially in the most beautiful parts of the Mediterranean. Unfortunately, that means regressing to a less

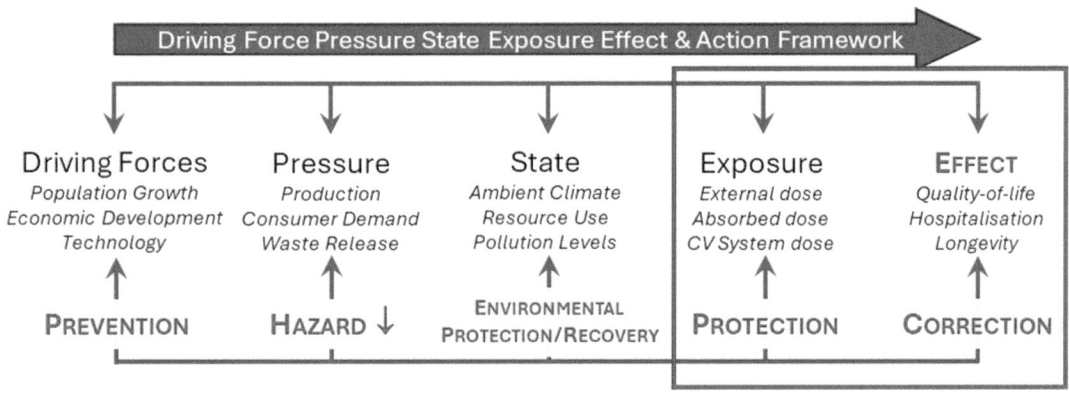

**Fig. 10.3** A theoretical framework for re-orientating primary care to address climate vulnerability. This figure frames the AMBIENCE project within the driving force-pressure-state-exposure-effect-action (DPSEEA) framework. This is used by the environmental protection agency (and other international health agencies) to identify the key points of intervention in respect to climate change (Boylan et al. 2018). Consistent with the discussion around a key role for primary care, it shows where a more proactive approach to protecting people living with elevated cardiac risk factors or established heart disease from 'exposure' to climate provocations to their health, might (and should) have a positive effect

mobile world—joining many past generations who lived and travelled within proximity of where they were born. It seems truly unrealistic to put the 'genie back in the bottle' when it comes to all the technology that has broadened our horizons. Instead, it appears that the unpalatable truth is that we (the human race) are over-populating and polluting the planet and we have to adapt what we do to reduce the harm that produces—knowing that climate change is here to stay. The question is this—in a world of increasing polarities (just think of Democrats versus Republicans and Left versus Right) and a diminishing sense of nuanced debate, can we handle the truth about what causes climate change and how we really don't know where it will lead? It seems that scientists have now been forced into a corner—we either know something or we don't, otherwise the public is not interested. This erosion of public interest and trust in scientific integrity leads many scientists to generate increasingly strident claims about climate change. As discussed in this book, much of this appears to pander to popular opinion, rather than in stated facts.

From a health perspective, we are far behind other disciplines in acknowledging the importance of climatic provocations to health, and particularly the cardiovascular system. Perhaps this is a good thing, if we can generate new knowledge that is based on a 'null hypothesis', rather than assuming we only need to consider heatwaves or negative consequences. As I specifically sought inspiration for the closing remarks on the book, I was struck by the shouts of a little boy as he peered down into the depths of the Bosphorus River in Istanbul Turkey (a location heralded as 'East meets West' but shares a contiguous climate and topography, no matter how much we like to romanticise and compartmentalise our world). At first glance the river looked pristine and a vibrant aquamarine, and indeed, the little boy was shouting excitedly about the jellyfish he'd just spied in the river. Alas, he was mistaken. Instead, he'd spied a white floating plastic bag that was destined to join the other flotsam of rubbish discarded in the river. Of course, no one was prepared to correct him and point-out a problem that will become increasingly obvious to him and is his generation as he grows into adulthood.

So, climate change represents a generational challenge to us (of all ages) and many more generations thereafter. Unlike those yet to discover the truth of what is happening and likely to come (good or bad), anyone invested in promoting better health outcomes needs to be realistic and prepare to mitigate the worst of its effects. Thus, all health professionals, particularly those caring for/treating people with high antecedent risk for, or established forms of heart disease should be cognisant of the pre-existing impact of climatic conditions on the heart and cardiovascular system. As an extension of this, they should also be aware of some of the therapeutic strategies they might apply to reduce the probability of a future cardiac event. This does require a change in mindset (a true paradigm change in health care delivery) for many clinicians. Hopefully, this book has at least prompted some reflection on this and the challenges that need to be met in the future.

# References

Boylan S, Beyer K, Schlosberg D, et al. A conceptual framework for climate change, health and wellbeing in NSW, Australia. Public Health Res Pract. 2018;28:e2841826.

Conrad K. The era of climate change medicine-challenges to health care systems. Ochsner J. 2023;23:7–8.

Davies B, Bhutta MF. Geriatric medicine in the era of climate change. Age Ageing. 2022;51:3595.

Lo YTE, Mitchell DM, Gasparrini A. Compound mortality impacts from extreme temperatures and the COVID-19 pandemic. Nat Commun. 2024;15:4289.

Mocumbi AO, Cebola B, Muloliwa A, Sebastiao F, Sitefane SJ, Manafe N, Dobe I, Lumbandali N, Keates A, Stickland N, Chan YK, Stewart S. Differential patterns of disease and injury in Mozambique: new perspectives from a pragmatic, multicenter, surveillance study of 7809 emergency presentations. PLoS ONE. 2019;14:e0219273.

Pineiro DJ, Codato E, Mwangi J, Eisele JL, Narula J. Accelerated reduction in global cardiovascular disease is essential to achieve the sustainable development goals. Nat Rev Cardiol. 2023;20:577–8.

What share of global $CO_2$ emissions come from aviation? https://ourworldindata.org/

global-aviation-emissions#:~:text=Aviation%20 accounts%20for%202.5%25%20of,to%20global%20 warming%20to%20date.&text=Flying%20is%20 one%20of%20the,of%20the%20world's%20 carbon%20emissions. Accessed June 2024.

World Heart Federation. World heart report 2023: full report; 2023. https://world-heart-federation.org/ resource/world-heart-report-2023/. Accessed June 2024.

# Index